Laser Cladding Coatings: Microstructure, Properties, and Applications

Laser Cladding Coatings: Microstructure, Properties, and Applications

Guest Editors

Kaiming Wang
Dingding Xiang
Jiang Ju
Zhenlin Zhang

MDPI

Basel • Beijing • Wuhan • Barcelona • Belgrade • Novi Sad • Cluj • Manchester

Guest Editors

Kaiming Wang
College of Mechanical and Vehicle Engineering
Changsha University of Science and Technology
Changsha
China

Dingding Xiang
School of Mechanical Engineering and Automation
Northeastern University
Shenyang
China

Jiang Ju
Department of Materials Science and Engineering
City University of Hong Kong
Hong Kong
China

Zhenlin Zhang
School of Materials Science and Engineering
Southwest Jiaotong University
Chengdu
China

Editorial Office
MDPI AG
Grosspeteranlage 5
4052 Basel, Switzerland

This is a reprint of the Special Issue, published open access by the journal *Coatings* (ISSN 2079-6412), freely accessible at: https://www.mdpi.com/journal/coatings/special_issues/laser_clad_coat.

For citation purposes, cite each article independently as indicated on the article page online and as indicated below:

Lastname, A.A.; Lastname, B.B. Article Title. *Journal Name* **Year**, *Volume Number*, Page Range.

ISBN 978-3-7258-6212-2 (Hbk)
ISBN 978-3-7258-6213-9 (PDF)
https://doi.org/10.3390/books978-3-7258-6213-9

Contents

About the Editors

Kaiming Wang

Kaiming Wang, associate professor, received his PhD in Mechanical Engineering from the Department of Mechanical Engineering, Tsinghua University, Beijing, China, in 2021. He is currently working at the Changsha University of Science and Technology in China. He has authored more than 40 publications in the areas of his research interests, which include laser cladding, additive manufacturing and welding. He is a Guest Editor and Young Editorial Board Member for ten international journals such as *Advanced Powder Materials, Additive Manufacturing Frontiers, China Foundry* and *Coatings.* He has led over 10 projects, some of which are by the National Natural Science Foundation. Prof. Wang is the recipient of the 2025 Hunan Science Foundation for Excellent Young Scholars.

Dingding Xiang

Xiang Dingding obtained his PhD from Tsinghua University. He is an Associate Professor at Northeastern University. In 2019, he worked as a research assistant at the Singapore Centre for 3D Printing, School of Mechanical and Aerospace Engineering, Nanyang Technological University, Singapore. In 2023, he was a visiting scholar at the Research Institute for Advanced Manufacturing, The Hong Kong Polytechnic University. He serves as a council member and committee member of five national-level societies, including the Tribology Branch, Surface Engineering Branch, Additive Manufacturing Technology (3D Printing) Branch, and Extreme Manufacturing Branch of the Chinese Society of Mechanical Engineering, and the Additive Manufacturing Technology Specialized Committee of the Chinese Society for Nonferrous Metals. He is a Guest Editor for eight international SCI journals such as the *Journal of Functional Biomaterials* and *Lubricants*, and a Special Issue Editor, Editorial Board Member, and Young Editorial Board Member for over ten EI journals such as *China Surface Engineering, Surface Technology*, and *Materials Engineering*. He has published over 30 SCI papers and has led over 20 projects including several by the National Natural Science Foundation.

Jiang Ju

Jiang Ju received his Ph.D. in Materials Science and Engineering from Shanghai Jiao Tong University between 2017 and 2021. He is currently a postdoctoral fellow in the Faculty of Materials Science and Engineering at the City University of Hong Kong. He has authored numerous articles (more than 80 SCI papers) in prestigious peer-reviewed journals, including the *Chemical Engineering Journal, Acta Materialia, Nano Materials Science, Corrosion Science, Additive Manufacturing*, and *Composite Part B: Engineering*, among others. His research interests primarily focus on the design and microstructural regulation of advanced high-performance metal materials, the fine control of deformation mechanisms, high-temperature oxidation behavior, and wear mechanisms. He employs a range of cutting-edge techniques, including three-dimensional atomic probe tomography (3D-APT), high-resolution transmission electron microscopy (HR-TEM), on-axis transmission Kikuchi diffraction (on-axis TKD), and additive manufacturing (3D printing) technologies.

Zhenlin Zhang

Zhenlin Zhang, Ph.D., Associate Professor at Southwest Jiaotong University, and a Young Talent in National Defense of Sichuan Province; he mainly engages in research on laser additive manufacturing material development, innovative processes, and metallurgy. He has presided over 9 national and provincial projects and 4 enterprise projects, published 26 SCI papers as the first author and corresponding author, obtained 10 patents, and won 1 second prize of the Science and Technology Progress Award of the Ministry of Education in 2023.

coatings

MDPI

Editorial

Research and Progress of Laser Cladding: Process, Materials and Applications

Kaiming Wang [1,*], Zhenlin Zhang [2,*], Dingding Xiang [3] and Jiang Ju [4]

1 College of Automobile and Mechanical Engineering, Changsha University of Science and Technology, Changsha 410114, China
2 School of Materials Science and Engineering, Southwest Jiaotong University, Chengdu 610031, China
3 School of Mechanical Engineering and Automation, Northeastern University, Shenyang 110819, China
4 School of Materials Science and Engineering, Shanghai Jiao Tong University, Shanghai 200240, China
* Correspondence: kmwang@csust.edu.cn (K.W.); zzl21@swjtu.edu.cn (Z.Z.)

Laser cladding (LC) is a process in which materials with certain properties are added to the surface of a substrate [1–3]. Under a high-energy-density laser beam, the materials are melted and then solidified, forming a cladding layer which is metallurgically bonded with the substrate [4]. LC has the advantages of high energy density, small deformation, metallurgical bonding with the matrix and so on [5]. The addition of cladding materials mainly consists of the coaxial powder feeding method and the preset method; the coaxial powder feeding method can be divided into coaxial powder feeding and off-axis powder feeding [6]. The advantage of preset method is that the cladding material is not limited by powder size and fluidity; however, it is not suitable for large-scale industrial applications. The powder of the coaxial powder feeding method is sent out synchronously during the LC process, which is more conducive to industrial promotion [7]. In terms of the coaxial powder feeding method, the main process parameters of LC include laser power, laser scanning speed, powder feeding rate, spot diameter, defocus amount, overlap rate, shielding gas flow rate and so on. Process parameters have great effects on the macroscopic size, microstructures, defects and properties of the cladding layer. A variety of methods have been adopted to optimize the parameters of the LC process, such as the single variable method, the orthogonal test method and machine learning [8–12]. In order to avoid defects and reduce elemental segregation of the cladding layer, in recent years, some researchers have combined LC with other technologies and developed ultrasonic-assisted LC, electromagnetic-assisted LC and induction heating LC [13,14]. To further improve the production efficiency of LC, extreme high-speed LC technology has been developed, and its efficiency is three to five times higher than that of conventional LC. Extreme high-speed LC meets the requirements of green and pollution-free development and is expected to replace electroplating [15–17].

The LC process involves the interaction between the laser, powder and substrate. Analyzing the molten pool flow field, temperature field and stress field during the LC process is of great help in understanding the macroscopic morphology, microstructure evolution and properties control of LC. Many scholars simulated the molten pool flow behavior, temperature field, stress field and microstructure based on fluid dynamics and physical phase field process [18–22].

The cladding material exerts great influence on the performance of the cladding layers. Since the birth of LC technology, the research and development of cladding material has been of high concern for researchers. By selecting a specific cladding material, the wear resistance, corrosion resistance, high-temperature oxidation resistance and other properties of the substrate surface can be improved, and the consumption of precious metals can be reduced at the same time. To meet different components' working conditions, the flexible selection and design of cladding materials is an important issue. According to different material compositions, the cladding materials can be divided into self-fluxing alloy powders, ceramic powders, rare earth powders, amorphous alloys, high-entropy alloys and so on. The self-fluxing alloy powders mainly consist of iron-based, nickel-based and cobalt-based self-fluxing alloy powders, which are mixed with boron and silicon elements with deoxidation and self-fluxing properties [23–26]. Compared with metal alloy powders, ceramic powders exhibit the characteristics of high hardness, high melting point and low toughness and can be used as reinforcing phases in the LC layers. At present, the most commonly used ceramic powders are carbide powders, oxide powders and nitride powders, and the carbide powders include WC, TiC, VC, SiC, NbC, ZrC, and so on [27–32]. Rare earths and their oxides are mainly used as modified materials in LC, and the addition of less than 2% can significantly improve the microstructures and properties of the LC layers. At present, Ce, Y, La and their oxides are widely studied. The addition of rare earth to the cladding layers can increase the nucleation rate, refine the grains, and improve the high-temperature oxidation resistance and corrosion resistance of the cladding layers [33,34]. Due to the disordered structure of amorphous alloys, there are no grains or grain boundaries in amorphous alloys, and they generally possess high hardness and excellent wear resistance, as well as good corrosion resistance. The rapid heating and cooling of the LC process is very beneficial to the formation of amorphous alloys [35–37]. How to improve the amorphous ratio in LC layers is an important research topic in this field. Due to their multi-component and simple solid solution structures, such as FCC and BCC, high-entropy alloys display high hardness and excellent wear resistance, and have decent development prospects. At present, the most frequently studied LC high-entropy alloy coating systems include CoCrFeMnNi, CoCrCuFeNiTi, TiZrNbWMo, CoCrBFeNiSi and so on [38–42].

LC layers generally exhibit good wear resistance, corrosion resistance and high-temperature oxidation resistance. At present, the research on the properties of the LC layer is mainly focused on wear resistance, corrosion resistance, high-temperature oxidation resistance, self-lubrication and biocompatibility. One of the main methods to improve the wear resistance of the cladding layer is to directly add or in situ synthesize the hard phase, and the other is to improve the wear resistance of the coating by adding lubricants [43–49].

LC technology is mainly used in repairing and surface strengthening in the fields of the aerospace, automobile and petrochemical industries, as well as metallurgy and rail transit. It provides a new method for the repair of important damaged parts, thereby greatly reducing the cost and enhancing work efficiency. Using LC technology to strengthen the surface of important parts can improve the properties of the parts, thereby prolonging their service life and reducing costs [50–52].

Due to its unique processing characteristics, LC technology has great development potential in repairing and surface strengthening of engineering components. With the progress of computer hardware and software, the decline in laser price and the improvement in automation, LC technology has received extensive attention in recent years. However, the quality of the LC layer is affected by many factors, and there are still inevitable defects in the cladding layer that affect the performance of the components. It is necessary to conduct basic research to have a deeper understanding of the deposition process, reduce the formation of defects such as cracks and pores in the cladding layer, and improve the performance of the cladding layer, so that LC can achieve large-scale industrial production as soon as possible.

Author Contributions: Conceptualization, K.W., Z.Z., D.X. and J.J.; writing—original draft preparation, K.W. and Z.Z.; writing—review and editing, K.W., Z.Z., D.X. and J.J.; supervision, K.W. All authors have read and agreed to the published version of the manuscript.

Funding: The authors would like to acknowledge the financial support for this work from the Natural Science Foundation of Hunan Province, China (2022JJ40495).

Conflicts of Interest: The authors declare no conflict of interest.

References

1. Siddiqui, A.; Dubey, A. Recent trends in laser cladding and surface alloying. *Opt. Laser Technol.* **2021**, *134*, 106619. [CrossRef]
2. Liu, Y.; Ding, Y.; Yang, L.; Sun, R.; Zhang, T.; Yang, X. Research and progress of laser cladding on engineering alloys: A review. *J. Manuf. Process.* **2021**, *66*, 341–363. [CrossRef]
3. Cheng, J.; Xing, Y.; Dong, E.; Zhao, L.; Liu, H.; Chang, T.; Chen, M.; Wang, J.; Lu, J.; Wan, J. An Overview of Laser Metal Deposition for Cladding: Defect Formation Mechanisms, Defect Suppression Methods and Performance Improvements of Laser-Cladded Layers. *Materials* **2022**, *15*, 5522. [CrossRef]
4. Lu, P.; Lewis, S.; Fretwell-Smith, S.; Engelberg, D.; Fletcher, D.; Lewis, R. Laser cladding of rail; the effects of depositing material on lower rail grades. *Wear* **2019**, *438–439*, 203045. [CrossRef]
5. Yap, C.; Chua, C.; Dong, Z.; Liu, Z.; Zhang, D.; Loh, L.E.; Sing, S. Review of selective laser melting: Materials and applications. *Appl. Phys. Rev.* **2015**, *2*, 41101. [CrossRef]
6. Zhong, M.; Liu, W. Laser surface cladding: The state of the art and challenges. *Proc. Inst. Mech. Eng. Part C J. Mech. Eng. Sci.* **2009**, *224*, 1041–1060. [CrossRef]
7. Chen, W.; Yang, X.; Li, X.; Chai, C.; Liu, W. Study on Microstructure and Properties of Nickel-Based Self-Lubricating Coating by Laser Cladding. *Coatings* **2022**, *12*, 753. [CrossRef]
8. Huang, Y.; Hu, Y.; Zhang, M.; Mao, C.; Wang, K.; Tong, Y.; Zhang, J.; Li, K. Multi-Objective Optimization of Process Parameters in Laser Cladding CoCrCuFeNi High-Entropy Alloy Coating. *J. Therm. Spray Technol.* **2022**, *31*, 1985–2000. [CrossRef]
9. Li, Y.; Wang, K.; Fu, H.; Zhi, X.; Guo, X.; Lin, J. Prediction for Dilution Rate of AlCoCrFeNi Coatings by Laser Cladding Based on a BP Neural Network. *Coatings* **2021**, *11*, 1402. [CrossRef]
10. Saeedi, R.; Razavi, R.S.; Bakhshi, S.; Erfanmanesh, M.; Bani, A.A. Optimization and characterization of laser cladding of NiCr and NiCr-TiC composite coatings on AISI 420 stainless steel. *Ceram. Int.* **2021**, *47*, 4097–4110. [CrossRef]
11. Mohammed, S.; Zhang, Z.; Kovacevic, R. Optimization of processing parameters in fiber laser cladding. *Int. J. Adv. Manuf. Technol.* **2020**, *111*, 2553–2568. [CrossRef]
12. Shu, L.; Li, J.; Wu, H.; Heng, Z. Optimization of Multi-Track Laser-Cladding Process of Titanium Alloy Based on RSM and NSGA-II Algorithm. *Coatings* **2022**, *12*, 1301. [CrossRef]
13. Meng, L.; Sheng, P.; Zeng, X. Comparative studies on the Ni60 coatings deposited by conventional and induction heating assisted extreme-high-speed laser cladding technology: Formability, microstructure and hardness. *J. Mater. Res. Technol.* **2021**, *16*, 1732–1746. [CrossRef]
14. Farahmand, P.; Liu, S.; Zhang, Z.; Kovacevic, R. Laser cladding assisted by induction heating of Ni-WC composite enhanced by nano-WC and La_2O_3. *Ceram. Int.* **2014**, *40*, 15421–15438. [CrossRef]
15. Wang, K.; Du, D.; Liu, G.; Pu, Z.; Chang, B.; Ju, J. A study on the additive manufacturing of a high chromium Nickel-based superalloy by extreme high-speed laser metal deposition. *Opt. Laser Technol.* **2021**, *133*, 106504. [CrossRef]
16. Li, R.; Yuan, W.; Yue, H.; Zhu, Y. Study on microstructure and properties of Fe-based amorphous composite coating by high-speed laser cladding. *Opt. Laser Technol.* **2022**, *146*, 107574. [CrossRef]
17. Sommer, N.; Stredak, F.; Böhm, S. High-Speed Laser Cladding on Thin-Sheet-Substrates—Influence of Process Parameters on Clad Geometry and Dilution. *Coatings* **2021**, *11*, 952. [CrossRef]
18. Liu, G.; Du, D.; Wang, K.; Pu, Z.; Chang, B. Epitaxial growth behavior and stray grains formation mechanism during laser surface re-melting of directionally solidified nickel-based superalloys. *J. Alloy. Compd.* **2021**, *853*, 157325. [CrossRef]
19. Zhang, Z.; Kovacevic, R. A thermo-mechanical model for simulating the temperature and stress distribution during laser cladding process. *Int. J. Adv. Manuf. Technol.* **2019**, *102*, 457–472. [CrossRef]
20. Sun, S.; Fu, H.; Chen, S.; Ping, X.; Wang, K.; Guo, X.; Lin, J.; Lei, Y. A numerical-experimental investigation of heat distribution, stress field and crack susceptibility in Ni60A coatings. *Opt. Laser Technol.* **2019**, *117*, 175–185. [CrossRef]
21. Qiao, X.; Xia, T.; Chen, P. Numerical research on effect of overlap ratio on thermal-stress behaviors of the high-speed laser cladding coating. *Chin. Phys. B* **2021**, *30*, 18104. [CrossRef]
22. Chen, J.; Zhang, Z.; Zhang, Z.; Liu, Y.; Zhao, X.; Chen, J.; Chen, H. Effect of the Cooling Rate of Thermal Simulation on the Microstructure and Mechanical Properties of Low-Carbon Bainite Steel by Laser-Arc Hybrid Welding. *Coatings* **2022**, *12*, 1045. [CrossRef]
23. Yu, J.; Ho, H. Microstructure and Mechanical Properties of (Ti, Nb)C Ceramic-Reinforced 316L Stainless Steel Coating by Laser Cladding. *Appl. Sci.* **2022**, *12*, 6684. [CrossRef]
24. Han, B.; Li, G.; Chen, Z.; Zhang, G. Microstructure and Wear Property of Graphene Nanoplatelets Reinforced Nickel-Based Composite Coating by Laser Cladding. *Metals* **2022**, *12*, 1247. [CrossRef]
25. Qunshuang, M.; Haozhe, C.; Hui, Z.; Wei, M.; Lei, H.; Huadong, L.; Yitao, W.; Xiaohui, Y. The alloying effects of Cr on in-situ phase evolution and wear resistance of nickel composite coatings fabricated by wide-band laser deposition. *Surf. Coat. Technol.* **2020**, *397*, 126019. [CrossRef]
26. Ju, J.; Yang, C.; Ma, S.; Kang, M.; Wang, K.; Li, J.; Fu, H.; Wang, J. Effect of temperature on oxidation resistance and isothermal oxidation mechanism of novel wear-resistant Fe-Cr-B-Al-C-Mn-Si alloy. *Corros. Sci.* **2020**, *170*, 108620. [CrossRef]
27. Wei, A.; Tang, Y.; Tong, T.; Wan, F.; Yang, S.; Wang, K. Effect of WC on Microstructure and Wear Resistance of Fe-Based Coating Fabricated by Laser Cladding. *Coatings* **2022**, *12*, 1209. [CrossRef]

28. Qunshuang, M.; Yajiang, L.; Juan, W.; Kun, L. Microstructure evolution and growth control of ceramic particles in wide-band laser clad Ni60/WC composite coatings. *Mater. Des.* **2016**, *92*, 897–905. [CrossRef]
29. Chen, L.; Yu, T.; Xu, P.; Zhang, B. In-situ NbC reinforced Fe-based coating by laser cladding: Simulation and experiment. *Surf. Coat. Technol.* **2021**, *412*, 127027. [CrossRef]
30. Lu, S.; Zhou, J.; Wang, L.; Liang, J. Effect of V and Cr transition layers on microstructure and mechanical properties of Ni-based coating on titanium alloy fabricated by laser cladding. *Surf. Coat. Technol.* **2021**, *405*, 126734. [CrossRef]
31. Liu, Y.; Li, J.; Xu, Q.; Zhang, Y.; Yan, X.; Chen, Y.; He, H. Microstructure and Wear Behavior of TC_4 Laser Cladding Modified via SiC and MoS_2. *Coatings* **2022**, *12*, 792. [CrossRef]
32. Peng, Z.; Zhang, J.; Zhang, M.; Wang, K.; Peng, P. Laser in-situ preparation and mechanical properties of VC reinforced Fe-based wear-resistant composite cladding. *Ceram. Int.* **2022**, *48*, 28240–28249. [CrossRef]
33. Liu, L.; Qiao, Y.; Xu, P. Effect of La_2O_3 on Microstructure and Properties of Laser Cladding SMA Coating on AISI 304 Stainless Steel. *Coatings* **2022**, *12*, 1004. [CrossRef]
34. Sun, S.; Fu, H.; Ping, X.; Guo, X.; Lin, J.; Lei, Y.; Wu, W.; Zhou, J. Effect of CeO_2 addition on microstructure and mechanical properties of in-situ (Ti, Nb)C/Ni coating. *Surf. Coat. Technol.* **2019**, *359*, 300–313. [CrossRef]
35. Hou, X.; Du, D.; Chang, B.; Ma, N. Influence of Scanning Speed on Microstructure and Properties of Laser Cladded Fe-Based Amorphous Coatings. *Materials* **2019**, *12*, 1279. [CrossRef]
36. Hou, X.; Du, D.; Wang, K.; Hong, Y.; Chang, B. Microstructure and Wear Resistance of Fe-Cr-Mo-Co-C-B Amorphous Composite Coatings Synthesized by Laser Cladding. *Metals* **2018**, *8*, 622. [CrossRef]
37. Xiao, M.; Gao, H.; Sun, L.; Wang, Z.; Jiang, G.; Zhao, Q.; Guo, C.; Li, L.; Jiang, F. Microstructure and mechanical properties of Fe-based amorphous alloy coatings prepared by ultra-high speed laser cladding. *Mater. Lett.* **2021**, *297*, 130002. [CrossRef]
38. Shen, X.; Liu, C.; Wang, B.; Zhang, Y.; Su, G.; Li, A. Surface Properties of Medium-Entropy Alloy Coatings Prepared through a Combined Process of Laser Cladding and Ultrasonic Burnishing. *Materials* **2022**, *15*, 5576. [CrossRef]
39. Huang, Y.; Hu, Y.; Zhang, M.; Mao, C.; Tong, Y.; Zhang, J.; Li, K.; Wang, K. On the enhanced wear resistance of laser-clad CoCrCuFeNiTix high-entropy alloy coatings at elevated temperature. *Tribol. Int.* **2022**, *174*, 107767. [CrossRef]
40. Wei, X.; Zhang, P.; Yu, Z.; Yan, H.; Wu, D.; Shi, H.; Chen, J.; Lu, Q.; Tian, Y.; Ma, S.; et al. Effect of phase transformation on mechanical properties of $A_{l16.80}Co_{20.74}Cr_{20.49}Fe_{21.28}Ni_{20.70}$ high entropy alloy coatings processed by laser cladding. *J. Alloy. Compd.* **2021**, *862*, 158563. [CrossRef]
41. Zeng, X.; Liu, Z.; Wu, G.; Tong, X.; Xiong, Y.; Cheng, X.; Wang, X.; Yamaguchi, T. Microstructure and high-temperature properties of laser cladded $AlCoCrFeNiTi_{0.5}$ high-entropy coating on Ti 6Al-4V alloy. *Surf. Coat. Technol.* **2021**, *418*, 127243. [CrossRef]
42. Yutao, L.; Kaiming, W.; Hanguang, F.; Xingye, G.; Jian, L. Microstructure and wear resistance of in-situ TiC reinforced AlCoCrFeNi-based coatings by laser cladding. *Appl. Surf. Sci.* **2022**, *585*, 152703.
43. Sun, J.; Dai, S.; Zhang, D.; Si, W.; Jiang, B.; Shu, D.; Wu, L.; Zhang, C.; Zhang, M.; Xiong, X. Friction and Wear Properties of CoCrFeNiMnSnx High Entropy Alloy Coatings Prepared via Laser Cladding. *Metals* **2022**, *12*, 1230. [CrossRef]
44. Ben, Q.; Zhang, Y.; Sun, L.; Wang, L.; Wang, Y.; Zhan, X. Wear and Corrosion Resistance of FeCoCrxNiAl High-Entropy Alloy Coatings Fabricated by Laser Cladding on Q345 Welded Joint. *Metals* **2022**, *12*, 1428. [CrossRef]
45. Yao, Z.; Chen, J.; Qian, H.; Nie, Y.; Zhang, Q.; Yao, J. Microstructure and Tensile Property of Laser Cladding Assisted with Multidimensional High-Frequency Vibration. *Materials* **2022**, *15*, 4295. [CrossRef]
46. Gu, J.; Li, D.; Liu, S.; Si, J.; Cai, S.; Liu, X.; Liu, P.; Wang, K. Investigation on Microstructure and Dry Sliding Wear Behavior of a Novel Fe-Cr-B-C-Al-Si-Mn Composite Coatings on 2Cr13 Steel by Laser Cladding. *J. Mater. Eng. Perform.* **2021**, *31*, 2381–2390. [CrossRef]
47. Liu, S.; Zhang, M.; Zhao, G.; Wang, X.; Wang, J. Microstructure and properties of ceramic particle reinforced FeCoNiCrMnTi high entropy alloy laser cladding coating. *Intermetallics* **2022**, *140*, 107402. [CrossRef]
48. Wang, K.; Du, D.; Liu, G.; Pu, Z.; Chang, B.; Ju, J. High-temperature oxidation behaviour of high chromium superalloys additively manufactured by conventional or extreme high-speed laser metal deposition. *Corros. Sci.* **2020**, *176*, 108922. [CrossRef]
49. Ju, J.; Shen, Z.; Kang, M.; Zhang, J.; Wang, J. On the preferential grain boundary oxidation of a Ni-Co-based superalloy. *Corros. Sci.* **2022**, *199*, 110203. [CrossRef]
50. Chen, W.; Chen, Q.; Zhang, Z.; Tang, S.; Cai, Q. Microstructure and Fatigue of EA4T Steel in Laser Cladding Remanufacturing. *Coatings* **2022**, *12*, 1106. [CrossRef]
51. Kendall, O.; Fasihi, P.; Abrahams, R.; Paradowska, A.; Reid, M.; Lai, Q.; Qiu, C.; Mutton, P.; Soodi, M.; Yan, W. Application of a New Alloy and Post Processing Procedures for Laser Cladding Repairs on Hypereutectoid Rail Components. *Materials* **2022**, *15*, 5447. [CrossRef] [PubMed]
52. Wang, Y.; Yu, Z.; Zhang, C.; Kang, L.; Xu, C.; Liu, Y. Simultaneously Improving Microstructures and Wear Properties of Ni60 Coating by Heat Treatment. *Metals* **2022**, *12*, 1294. [CrossRef]

coatings

MDPI

Article

Prediction for Dilution Rate of AlCoCrFeNi Coatings by Laser Cladding Based on a BP Neural Network

Yutao Li [1], Kaiming Wang [2,*], Hanguang Fu [1,*], Xiaohui Zhi [3], Xingye Guo [1] and Jian Lin [1]

1 Key Laboratory of Advanced Functional Materials, Ministry of Education, Institute of Welding and Surface Technology, Beijing University of Technology, Beijing 100124, China; liyutao0123@outlook.com (Y.L.); xyguo@bjut.edu.cn (X.G.); linjian@bjut.edu.cn (J.L.)

2 College of Automobile and Mechanical Engineering, Changsha University of Science and Technology, Changsha 410114, China

3 School of Mechanical Engineering, Shijiazhuang Tiedao University, North 2nd Ring East Road, Changan District, Shijiazhuang 050043, China; mkmkzxh@hotmail.com

* Correspondence: kmwang@csust.edu.cn (K.W.); hgfu@bjut.edu.cn (H.F.); Tel./Fax: +86-010-67396244 (H.F.)

Abstract: The dilution rate has a significant impact on the composition and microstructure of the coatings, and the dilution rate and process parameters have a complex coupling relationship. In this study, three process parameters, namely laser power, powder feeding rate, and scanning speed, were selected as variables to design the orthogonal experiment. The dilution rate and hardness data were obtained from AlCoCrFeNi coatings based on orthogonal experiments. Then, a BP neural network was used to establish a prediction model of the process parameters on the dilution rate. The established BP neural network exhibited good prediction of the dilution rate of AlCoCrFeNi coatings, and the average relative error between the predicted value and the experimental value was only 5.89%. Subsequently, the AlCoCrFeNi coating was fabricated with the optimal process parameters. The results show that the coating was well-formed without defects, such as cracks and pores. The microhardness of the AlCoCrFeNi coating prepared with the optimal process parameters was 521.6 $HV_{0.3}$. The elements were uniformly distributed in the microstructure, and the grain size was about 20–60 μm. The microstructure of the AlCoCrFeNi coating was only composed of the BCC phase without the existence of the FCC phase and intermetallic compounds.

Keywords: laser cladding; AlCoCrFeNi coatings; BP neural network; dilution rate

1. Introduction

Laser cladding is an important surface modification technology that uses a high-energy laser beam to melt and deposit powder or wire on a substrate surface to form a single-layer or multi-layer coating. The coatings are firmly combined with the substrate by metallurgical bonding, which can significantly improve the wear resistance, corrosion resistance, and oxidation resistance of the substrate surface [1,2]. In addition, it also has the advantages of fast forming speed, low heat input, and low dilution rate [3–6].

High-entropy alloys (HEAs) are also called multi-principal element alloys. Their composition is composed of several main metallic elements. As a structural and functional material, it has unique properties and arouses widespread interest [7,8]. The AlCoCrFeNi alloy is a typical high-entropy alloy with an equal atomic ratio, and it possesses excellent mechanical properties, because its microstructure is mainly composed of the BCC phase [9]. The related work of preparing AlCoCrFeNi coatings by laser cladding technology has been extensively studied. The hardness of AlCoCrFeNi coatings prepared by laser cladding is significantly higher than that of as-cast AlCoCrFeNi, and the grain size is also significantly reduced [10–13]. The microstructure of the AlCoCrFeNi coatings mainly contains the BCC phase and a very small amount of the FCC phase. It is difficult for dislocations to slip in the BCC phase, which is also the main reason for the high strength of AlCoCrFeNi high-entropy alloys [9].

The dilution rate refers to the change of the cladding alloy composition caused by the mixing of molten matrix in the process of laser cladding. The composition of the coating changes under a high dilution rate, resulting in a decrease in the mechanical properties of the coating. However, under a too-small dilution rate, the bond between the substrate and the coating is poor. Therefore, the dilution rate is one of the most important process control parameters and the key to obtaining a high-quality coating.

The high entropy coating prepared by laser cladding usually has a high dilution rate of about 49.3% [14,15]. The hardness of the AlCoCrFeNi alloy can reach up to 500 HV. The substrate used in this study is soft; its hardness is only 240 HV. At a high dilution rate, a large amount of molten substrate enters into the AlCoCrFeNi coating, resulting in a sharp decrease of hardness. Moreover, the elements from the molten substrate have an important influence on the microstructure of the AlCoCrFeNi high-entropy alloy coatings. The excessive Fe and Cr elements from the substrate material promote the generation of the FCC phase with low strength and hardness [14], resulting in a decrease of strength and hardness. Therefore, it is of great significance to predict and control the dilution rate of the AlCoCrFeNi high-entropy alloy coating.

The dilution rate is mainly affected by process parameters, among which the most important process parameters are laser power, scanning speed, and powder feeding rate. In actual production, the relationship between dilution rate and process parameters is a complex non-linear relationship, and there are many influencing factors [13,16]. Traditional experimental methods are applied to explore the accurate relationship between process parameters and dilution rate, which consumes a lot of time and labor.

In solving complex nonlinear problems, neural networks have excellent nonlinear processing capabilities. Neural networks provide an effective method to construct the relationship between process parameters and the dilution rate of the cladding layer [17]. Some researchers employed the back propagation neural network (BPNN) to predict process parameters and cladding forming quality [18,19]. Deng et al. [20] adopted quantum-behaved particle swarm optimization and a BP neural network to build a model of the process parameters and microhardness of the Ti(C, N) cladding layer. The model has strong predictive ability, and the relative error can reach 9.12%. Guo et al. [21] used a BP neural network to optimize and predict the process parameters of laser cladding a Co-based alloy, and the average relative error between the predicted value and the experimental value was less than 10%.

In this paper, 18 groups of process parameters were designed by the orthogonal experiment method, and the corresponding AlCoCrFeNi coatings were prepared. Then, the effects of laser power, powder feeding rate, and scanning speed on the dilution rate of AlCoCrFeNi high-entropy alloy coatings were analyzed preliminarily. According to the obtained dilution rate data, the BP neural network was conducted to establish the prediction model of the process parameters and the dilution rate. Finally, the microstructure of the coating prepared with the optimal process parameters was analyzed.

2. Experimental Procedures and Methods

2.1. Experimental Materials

AlCoCrFeNi high entropy alloy powder is a spherical powder prepared by a gas atomization process, and the powder particle size is 45–105 μm. The specific morphology is shown in Figure 1, and the content of each element is shown in Table 1.

Table 1. Chemical composition of AlCoCrFeNi powder by XRF.

Element	Al	Co	Cr	Fe	Ni
Concentration (%)	9.8171	23.1098	20.5944	22.8544	23.5598
Atomic ratio (at.%)	18.5431	19.9760	20.1981	20.8136	20.4691

The material of substrate was 40CrNiMo alloy steel, with a size of 12 × 60 × 100 mm^3. Before laser cladding, the surface of substrate was ground and polished, then it was scrubbed with alcohol and acetone to remove the oil on the surface.

Figure 1. Morphology of AlCoCrFeNi powder.

2.2. Design of Orthogonal Experiment

The laser cladding was conducted using a continuous fiber laser machine (IPG, YLS-6000-S2, IPG Photonics, Oxford, MA, USA); the laser wavelength was 1070 nm, the maximum power was 6 kW, the laser focal length was 460 mm, and the square spot was 5 × 5 mm^2. The coaxial powder feeding method was adopted to transport the powder to the position near the focus of the beam under the argon carrier flow.

The orthogonal experiment selected the laser power (P), scanning speed (V), and powder feeding rate (R) as experimental factors, as shown in Table 2. Each factor of laser power (P) and scanning speed (V) contained 3 levels, and the powder feeding rate contained 2 levels. The parameters of the 18 times L_{18} ($3^3 \times 2^1$) orthogonal experiment are shown in Table 3. A cross-sectional view of the cladding layer is shown in Figure 2; h is the depth of the fusion zone, H is the height of the cladding layer, and W is the width of the cladding layer. The dilution rate (η) is expressed by the percentage of the substrate alloy molten into the cladding layer, and the calculation equation is $\eta = h/(H + h)$.

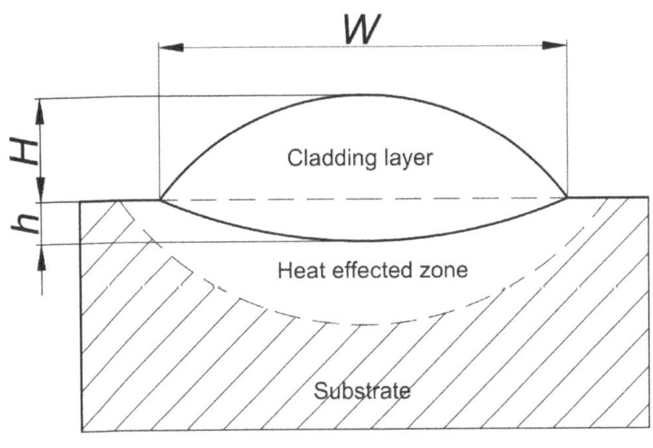

Figure 2. Schematic diagram of dilution rate (η).

Table 2. Experimental factors and levels.

Levels	Factors		
	P/(kW)	R/(g/min)	V/(mm/s)
Level 1	2.00	15.00	6.00
Level 2	2.50	10.00	4.00
Level 3	1.50	–	2.00

Table 3. Geometry sizes and laser cladding process parameters.

Sample	P (kW)	R (g/min)	V (mm/s)	H (mm)	W (mm)	h (mm)	η (%)	H_v ($HV_{0.3}$)
S-1	1.50	15.00	4.00	1.09	6.03	0.36	25	491.87
S-2	2.00	15.00	4.00	1.20	6.57	0.42	26	516.50
S-3	2.50	15.00	4.00	1.38	7.43	0.68	33	503.80
S-4	1.50	15.00	2.00	2.20	5.26	0.30	12	499.75
S-5	2.00	15.00	2.00	2.40	6.68	0.45	16	521.60
S-6	2.50	15.00	2.00	2.10	7.08	0.69	25	558.87
S-7	1.50	15.00	6.00	0.49	6.27	0.44	47	259.00
S-8	2.00	15.00	6.00	0.79	7.11	0.32	29	320.97
S-9	2.50	15.00	6.00	1.02	7.78	0.55	35	363.07
S-10	1.50	10.00	4.00	0.88	8.43	0.79	47	333.27
S-11	2.00	10.00	4.00	0.92	6.45	0.45	33	305.43
S-12	2.50	10.00	4.00	0.81	6.91	0.48	37	264.47
S-13	1.50	10.00	2.00	1.48	6.40	0.31	17	467.4
S-14	2.00	10.00	2.00	1.61	7.67	0.42	21	441.97
S-15	2.50	10.00	2.00	1.58	8.10	0.49	24	413.97
S-16	1.50	10.00	6.00	0.41	7.47	0.39	49	351.53
S-17	2.00	10.00	6.00	0.68	8.34	0.72	51	407.63
S-18	2.50	10.00	6.00	0.72	9.38	0.82	53	256.23

2.3. Microhardness Measurement

The microhardness can be used as an important reference index for the wear resistance of coatings, as well as a simple index for evaluating the quality of the coatings. Therefore, the microhardness of the coatings was also measured in this study. The microhardness was measured using a MICRO MET-5103 digital microhardness tester (Baihe, Shanghai, China); the test load was 300 g, and the indenter loading time was 15 s. Coatings were measured for five points at different positions in the upper middle area; the maximum and minimum values were removed, and the average of the remaining 3 values were taken as the microhardness value of the coatings.

The cross section of prepared coatings was ground and polished. In order to clearly distinguish the boundary between the coating and the substrate, aqua regia (HNO_3:HCl = 1:3) was used for etching, and the etching time was 30 s. The cross section corroded was observed under a stereo microscope; then, the geometric dimensions of each part of the coating were measured.

2.4. Microstructure Characterization

The phase qualitative analysis of the coatings was carried out by X-ray polycrystalline diffractometer (XRD, Bruker D8 Advance, Bruker, Madison, WI, USA), and the specific parameters were the following: Cu target X-ray, tube voltage 40 kV, tube current 40 mA, scanning speed 4°/min, scanning range 20–90°, and step size 0.02°.

The polished coating samples were etched for 6 s, and then, the morphology and microstructure were observed by a scanning electron microscope (ZEISS Gemini SEM 300, Carl Zeiss, Oberkohen, Germany). The composition of coating samples was examined by the attached energy dispersive spectroscopy (EDS, Carl Zeiss, Oberkohen, Germany). In order to prepare electron backscatter diffraction (EBSD, Carl Zeiss, Oberkohen, Germany) samples, the polished metallographic samples were placed in a perchlorate alcohol solution; then, the surface stress layer was removed by electrolytic polishing.

2.5. Design of BP Neural Network

A back-propagation neural network (BPNN version x) is an error back-propagation neural network model with a strong nonlinear approximation ability. In this study, a three-layer structure of BP neural network was established to build the relationship between the dilution rate and process parameters. The structure of BPNN is shown in Figure 3, including 1 input layer, 1 hidden layer, and 1 output layer. The input layer has 3 neurons (X_1, X_2, X_3), corresponding to the laser power (P), the powder feeding rate (R), and the scanning speed (V), respectively. The P, R, and V are the input parameters used to train the BP neural network. The output layer only has a neuron Y, which is the dilution rate (η). The η is the output parameter used to train the BP neural network. The hidden layer contains 6 neurons, the number of which can be determined according to empirical formulas. The activation function (transfer function) employs the Tan-Sigmoid function. The training process of the BP neural network is as follows:

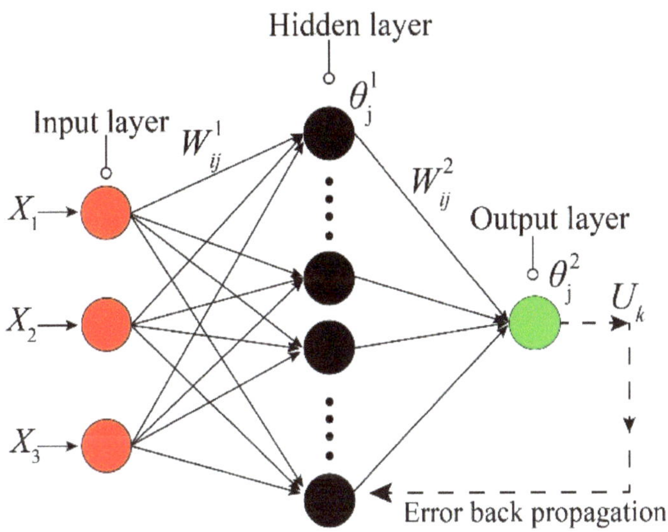

Figure 3. The structure of BP neural network.

Step 1: Input data and output data is an $m \times n$ two-dimensional array, defined as $(x_{ij})_{m\times n}$. The input data and output data are normalized to the interval [–1,1] according to Formula (1).

$$X_{ij} = 2*\frac{x_{ij} - x_{j\min}}{x_{j\max} - x_{j\min}} - 1 \quad (1)$$

$(X_{ij})_{m\times n}$ is the normalized $m \times n$ matrix, x_{jmax} and x_{jmin} are the maximum and minimum values of the jth column of the $(x_{ij})_{m\times n}$ matrix, respectively.

Step 2: The input layer neurons transmit the normalized data to the neurons of the hidden layer, or it transmits the value calculated by the neurons of the previous hidden layer to the next neurons of the hidden layer. The input of the lth layer neurons can be expressed by Formula (2):

$$U_j^l = \begin{cases} X_j = (X_{1j}, X_{2j}, \ldots X_{3j})^T, l = 0 \\ f\left(\sum_{i=1}^{h} w_{ij}^l U_i^{l-1} + \theta_j^l\right), l = 1,2,3\cdots P \end{cases} \quad (2)$$

where X_j is the jth column vector of the above two-dimensional matrix $(X_{ij})_{m\times n}$, and w^l_{ij} is the connection weight of the neuron (i) in the ($l-1$) layer and the neuron (j) in the l layer. U_i^{l-1} is the input of neuron (i) in the layer ($l-1$); θ_j^l is the threshold value of neuron (j) in the layer l, and the tansig function is adopted as the transfer function.

Step 3: The output neurons in the output layer (U_k) can be calculated by the following Formula (3):

$$U_k = \sum_{j=1}^{h} U_j^P w_{jk}^{P+1} + \theta_k^{P+1}, k = 1,2,3\ldots n_0 \tag{3}$$

where w_{jk}^{P+1} is the connection weight of neuron (j) in the p layer and neuron (k) in the ($p+1$) layer, and θ_k^{P+1} is the threshold value of the (p+1) layer neuron (k).

Step 4: The error of network training is evaluated by the Mean Square Error (MSE), which is calculated by Formula (4). The weight correction of each layer is calculated by the derivation rule of the differential chain in the following Formula (5):

$$\text{MSE} = \frac{1}{N_S}\sum_{k=1}^{N_s}\left[\frac{1}{2}\sum_{k\in\{C\}}(d_k - U_k)^2\right] \tag{4}$$

$$\Delta w_{ij}^l = -\lambda \frac{\partial \text{MSE}}{\partial y_k}\frac{\partial y_k}{\partial U_j^l}\frac{\partial U_j^l}{w_{ij}^l} \tag{5}$$

where d_k is the experimental response, N_s is the total number of samples, $\{C\}$ is the output data set, and λ is the learning rate usually in the interval [0,1].

3. Results and Discussion

The ultra-depth-of-field 3D microscope was used to observe the morphology of the single-track coatings, and the results are shown in Figure 4. The interface between coatings and the substrate presents a good metallurgical bond, and there was no crack in the coatings. The data, such as the geometry sizes, dilution rate, and hardness of the coatings, are shown in Table 3.

3.1. The Effect of Process Parameters on the Dilution Rate

The influence of laser cladding process parameters on the geometry and dilution rate of the coatings is shown in Figure 5. It could be found that when the powder feeding rate (R) and scanning speed (V) remained unchanged, as the laser power increased, the melting width, depth, and dilution rate of the coatings showed an increasing trend. However, the laser power (P) and scanning speed (V) remained unchanged; when the powder feeding rate increased from 10 g/min to 15 g/min, the melting width decreased, as shown in Figure 5a,b.

The depth rises with the increment of laser power, and the greater the powder feeding rate, the smaller the coating depth, as shown in Figure 5c,d. For coaxial powder feeding laser cladding, the substrate is heated by the laser energy that passes through the powder particle cloud [22]. The more powder particles in the laser beam, the stronger the shielding effect on the laser beam, resulting in forming of a smaller clad depth. When the powder feeding rate is 15 g/min, the scanning speed has the least influence on the clad depth, and the laser power has the greatest influence on the clad depth.

According to Figure 5e,f, the laser power has a minor effect on melting height, whereas scanning speed and powder feeding rate have a greater influence on coating height. At a powder feeding rate of 10 g/min, when the power increases from 2000 W to 2500 W, the height of coatings shows a slight downward trend, which indicates that the powder has been sufficiently melted at the power of 2000 W.

The variation in the dilution rate (η) is depicted in Figure 5g,h. As the laser power increases, the dilution rate rises correspondingly. Under the same laser power and scanning speed, the increase in powder feeding rate can significantly reduce the dilution rate of the coatings. It can be seen from Figure 5g that the scanning speed has the greatest influence on the dilution rate. A lower scanning speed can increase the amount of powder in the molten pool unit area, increasing the coatings height. In addition, the lower scanning speed has more heating time so that the powder can be melted more fully.

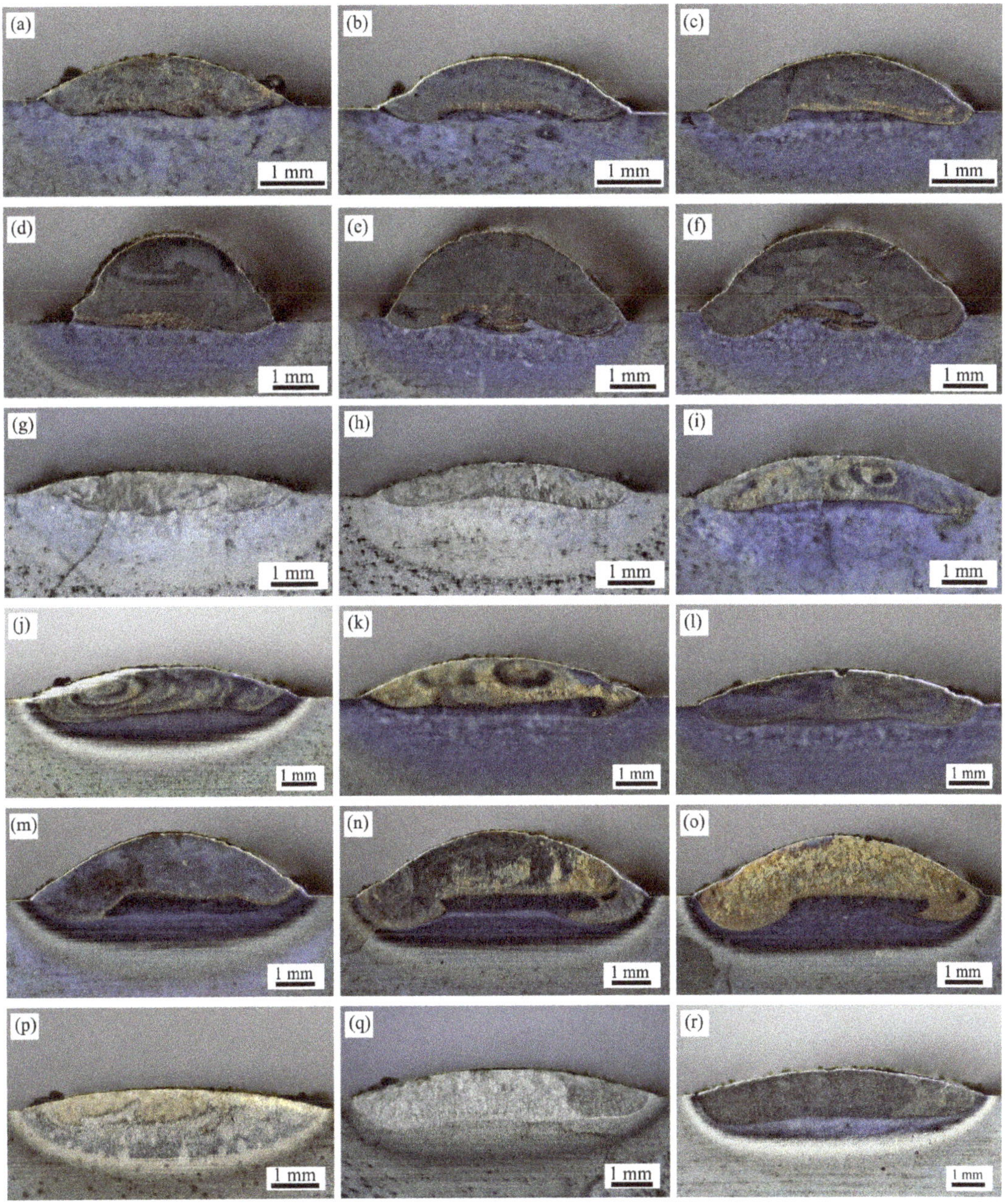

Figure 4. Cross-sections of coating with corrosion: (**a–c**) S-1, S-2, S-3; (**d–f**) S-4, S-5, S-6; (**g–i**) S-7, S-8, S-9; (**j–l**) S-10, S-11, S-12; (**m–o**) S-13, S-14, S-15; (**p–r**) S-16, S-17, S-18.

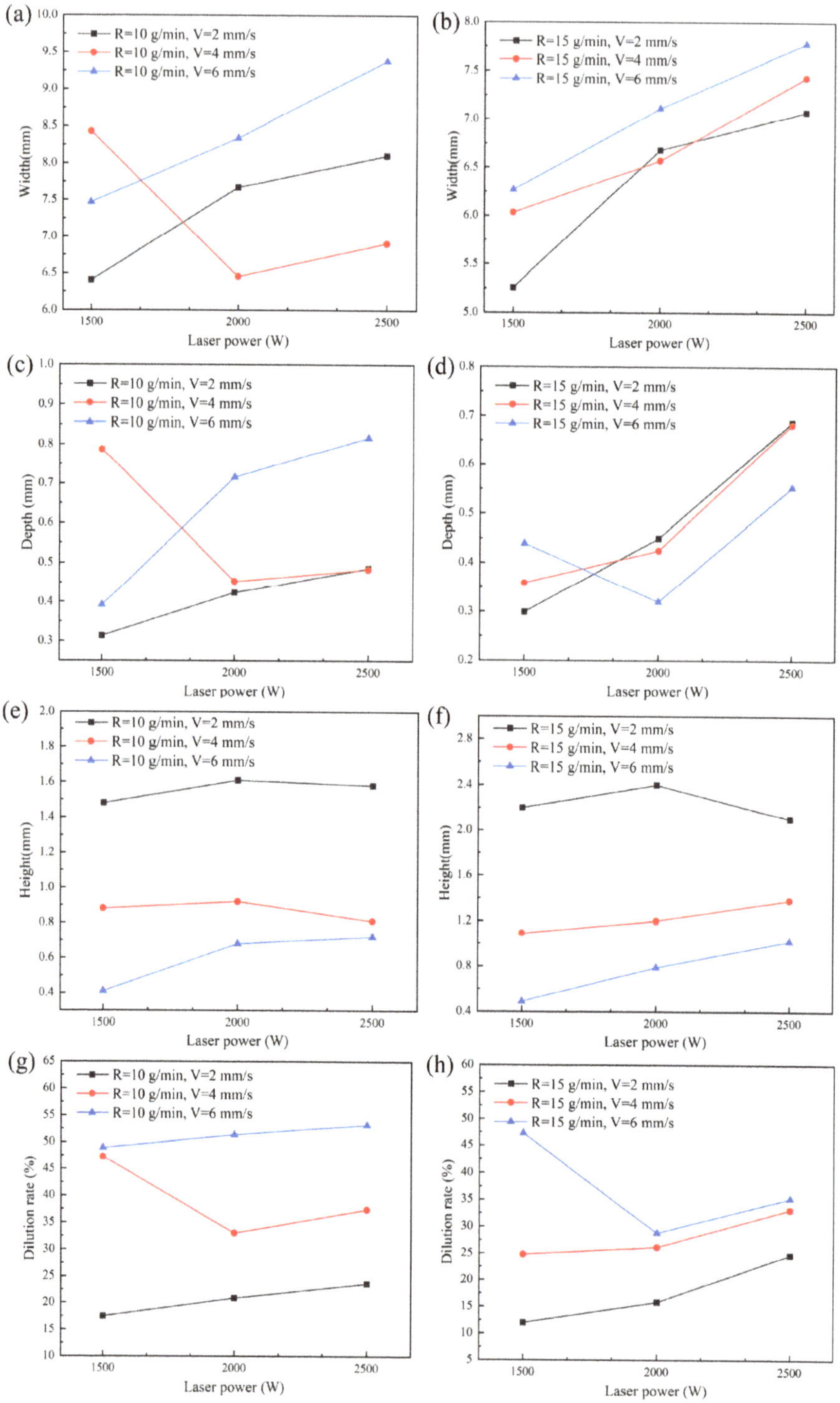

Figure 5. Effect of process parameters on the clad geometry: (**a**,**b**) width of coatings; (**c**,**d**) depth of coatings; (**e**,**f**) dilution rate of coatings; (**g**,**h**) height of coatings.

Moreover, there is an interesting phenomenon that the dilution rate can influence the microhardness of AlCoCrFeNi coatings. When the dilution rate is greater than the 40%, the microhardness is about 330 $HV_{0.3}$, such as in S-11 and S-16 samples. The microhardness is obviously lower than that of the sample with a dilution ratio of less than 25%.

The sample with the smallest dilution rate was S-4; the dilution rate was 12%; the microhardness value was 499.75 $HV_{0.3}$; and the process parameters were a laser power of 1500 W, scanning speed of 2 mm/s, and powder feeding rate of 15 g/min. For the S-5 sample, the dilution rate was 16%, but the microhardness value was 521.6 $HV_{0.3}$, and the coating height was 2.4 mm, which was higher than the 2.2 mm height of the S-4 sample. Obviously, the AlCoCrFeNi powder can be melted more abundantly under the power of 2000 W, so the overall mechanical properties were better than those of the S-4 sample.

The optimal process parameters were not directly given by the BP neural network. In this study, the dilution rate was the most important reference index. However, the laser cladding also needed to consider the hardness. If considering only the pursuit of low dilution rate, the coating may have defects, such as peeling and incomplete melting. Therefore, the optimal process parameter was the combination of hardness and dilution rate. The process parameters used for the S-5 sample, with a dilution rate of 16% and hardness of 521.6 $HV_{0.3}$, were optimal. The optimal process parameters were as follows: the laser power: 2 kW, the scanning speed: 2 mm/s, and the power feeding rate: 15 g/min.

3.2. *Analysis of Performance of BP Model*

The feedforwardnet function in the neural network training tool of MATLAB (MATLAB R2018b) was used to build and train a BP neural network. Due to the geometric size and the dilution rate data of samples, S-7 and S-10 were relatively abnormal; these two groups of data were eliminated, and a total of 16 groups of data were used for the establishment and training of the BP neural network. Twelve groups of data were randomly selected as the training data for the BP neural network. The sample data used for test and validation consisted of 4 groups, namely samples S-9, S-11, S-12, and S-15.

It can be seen from Figure 6 that the training of the BP neural network underwent 8 epochs, and it reached the best training performance at the 5th epoch, with a mean square error of 0.05%. Table 4 and Figure 7a show that the predicted value of the dilution rate in the training data was very close to the experimental value. As shown in Figure 7b, during the training process, the maximum value of the relative error between the predicted value and the training value was 1.70%, which proves that the BP neural network training effect was good. The predicted values of the test and validation samples were 32.99414, 30.54423, 38.58375, and 22.53957. Compared with the experimental values, the relative errors were 5.73%, 7.44%, 4.28%, and 6.13%, and the average relative error was 5.89%. It can be seen from the relative error of the test data that the BP neural network established in this paper had excellent predictive performance after training.

3.3. *Microstructure of the AlCoCrFeNi HEA Coating*

The AlCoCrFeNi HEA coating was fabricated by the optimal process parameters, and then, its microstructure was observed. The XRD diffraction pattern of the AlCoCrFeNi coating is shown in Figure 8. It could be found that the phase of the coating was mainly the BCC phase, and the BCC phase was divided into disordered phase A2 and ordered phase B2. The disordered A2 phase is generally an Fe–Cr-rich phase, and the ordered B2 phase is generally an Al–Ni-rich phase [23]. Due to the fact that three crystal structures of these two phases are a cubic orientation and the lattice parameters are similar, the formed X-ray diffraction was coherent. Therefore, most of the diffraction peaks on the XRD pattern overlapped, and it was difficult to distinguish [24–26]. However, there was a small diffraction peak near 31°, corresponding to the (100) plane of the B2 phase, which confirmed the existence of the ordered phase B2 [27,28].

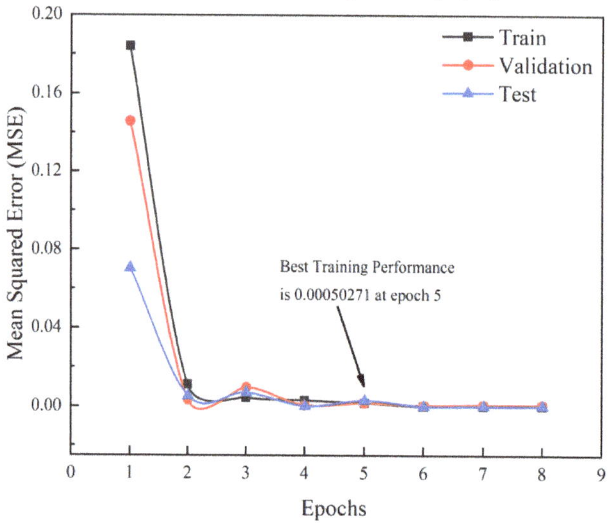

Figure 6. BPNN training process.

Table 4. Dilution rate of prediction.

No.	Sample	Experimental η (%)	Predicted η (%)
1	S-1	25	24.26575
2	S-2	26	26.06222
3	S-3	33	32.87267
4	S-4	12	11.67059
5	S-5	16	16.06459
6	S-6	25	25.05431
7	S-8	29	28.77535
8	S-9 (test)	35	32.99414
9	S-11 (test)	33	30.54423
10	S-12 (test)	37	38.58375
11	S-13	17	17.01862
12	S-14	21	21.08211
13	S-15 (test)	24	22.52957
14	S-16	49	48.90431
15	S-17	51	50.79725
16	S-18	53	52.09692

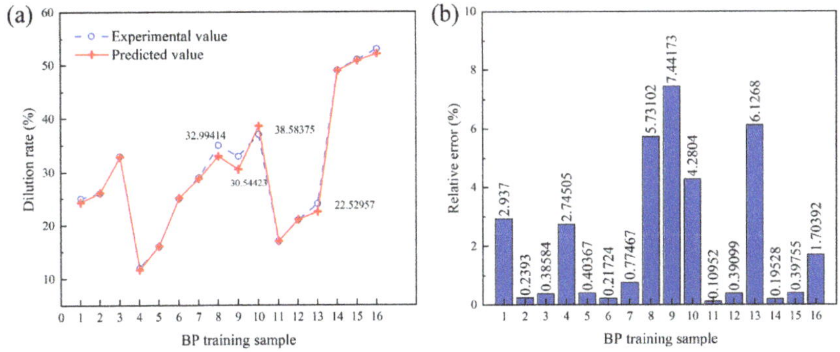

Figure 7. Comparison diagram of training values and experimental values: (**a**) dilution rate; (**b**) relative error.

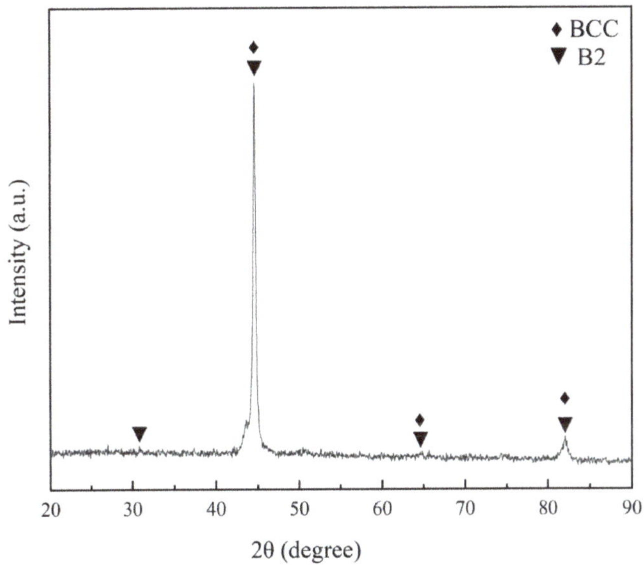

Figure 8. XRD pattern of AlCoCrFeNi coating (S-5 sample).

Clear grain boundaries could be observed at the cross-section of the AlCoCrFeNi coating corroded by aqua regia, as shown in Figure 9. The elements distributed homogeneously, and the microstructure uniformly consisted of a simple BCC solid solution phase, as shown in Figure 9a. The crystal grains were equiaxed with a size of about 20–60 μm. The high magnification of the microstructure is shown in Figure 9b; there was no precipitation of intermetallic compounds with an abnormal color contrast. There are two kinds of equiaxed grains with different contrasts in Figure 9a. However, according to EDS analysis, the two kinds of grains have very similar chemical compositions. This result is also consistent with the results obtained from other studies [11]. EDS point analysis was performed on the microstructure of the coatings in Figure 9b, and the results are shown in Table 5. At the region marked B in Figure 9b, the Fe element content was the highest, which was 30.40 at.%; and the Al element content was the lowest, which was 14.98 at.%. This is due to a small part of the molten substrate material entering into the molten pool, increasing the content of the Fe element in the region B.

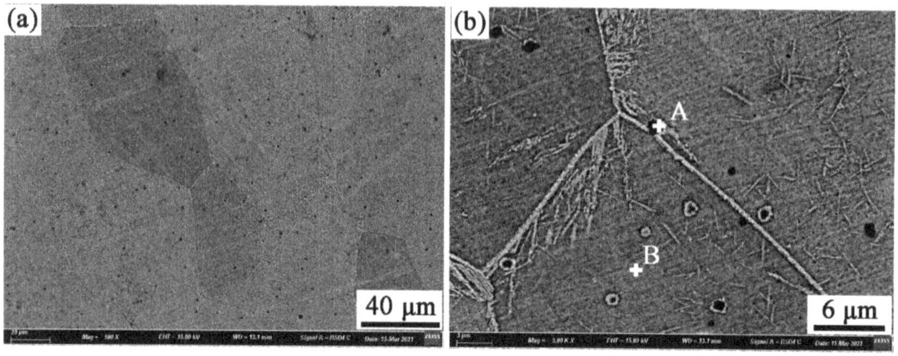

Figure 9. Back-scattered SEM images of the AlCoCrFeNi coating (S-5 sample) microstructure: (**a**) low magnification; (**b**) high magnification.

Table 5. EDS point of the AlCoCrFeNi coating (S-5 sample) (at.%).

Points	Al	Co	Cr	Fe	Ni	N
A	38.41	6.02	9.87	13.50	6.97	25.23
B	14.98	19.66	17.32	30.40	17.64	0

The black granular microstructure (marked A) had the highest content of Al and N elements, which were 38.41 at.% and 25.23 at.%, which indicates that the particles were the AlN phase. This kind of microstructure has also been found in the related research of AlCoCrFeNi high-entropy alloy, and it was confirmed that the structure was AlN particles, and the N element came from the impurities contained in the material [29]. Due to the small size and low content of the particles, they were not detected in XRD detection.

Figure 10 is the EDS element mapping of the AlCoCrFeNi coatings. The Al element aggregated at the black particles, resulting in the increasement of Al element concentration. This also explains why the Al element content in the other area of coatings was less. Except for the aggregation of partial Al elements, the other four elements were uniformly distributed, and no segregation phenomenon occurred, which also proves that the microstructure was composed of uniform solid solution.

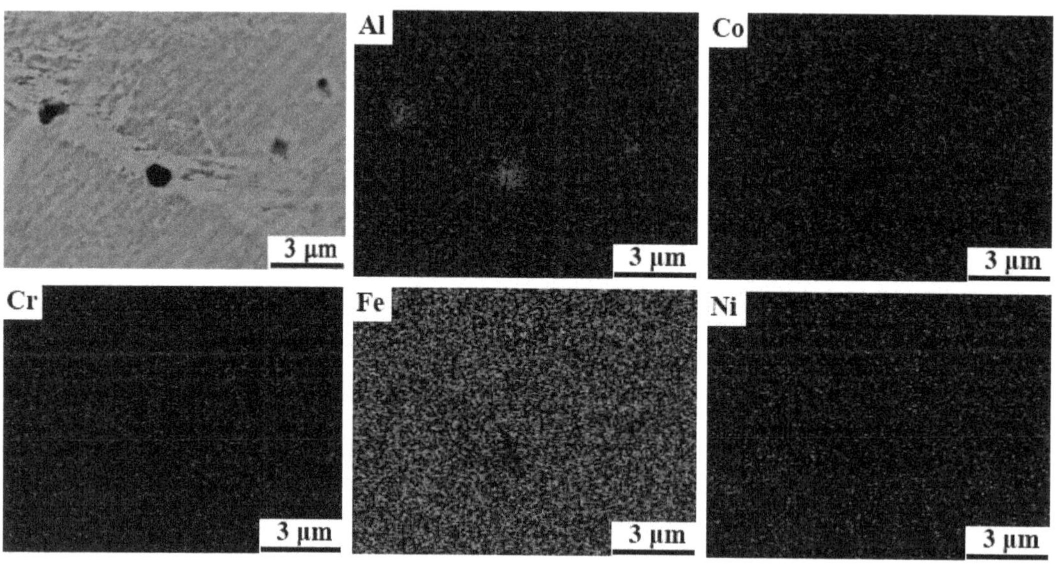

Figure 10. EDS maps of AlCoCrFeNi coating (S-5 sample).

Figure 11a is an IPF diagram of AlCoCrFeNi coatings. The high-angle boundaries are shown with the black line; the misorientation angle exceeds 10°. There were many different orientations of grains in the microstructure of coatings. It could be found that the AlCoCrFeNi coating possessed a texture index in the three pole figures (PFs), with an intensity range of 0–6.92, especially with a maximum texture index in the {100} PF up to 6.92, as depicted in Figure 11d.

The microstructure of the AlCoCrFeNi coating was composed of the BCC phase and almost contained no FCC phase, as shown in Figure 11b. This result is also consistent with the previous XRD and SEM analysis results. However, a small part of the substrate melted into the molten pool during the laser cladding, resulting in the Fe element content in the coatings exceeding 20 at.%. The microstructure of the coatings did not change and was still composed of the BCC solid solution phase. The dilution rate corresponding to the laser cladding process parameters did not cause the formation of the FCC phase.

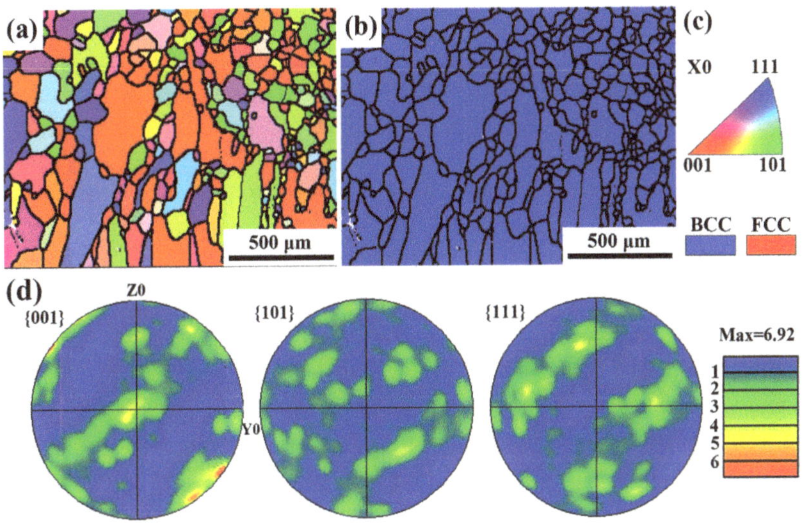

Figure 11. EBSD images of the AlCoCrFeNi coating (S-5 sample): (**a**) inverse pole figure (IPF); (**b**) phase map; (**c**) color coding for IPF and phase map; (**d**) pole figure.

4. Conclusions

In this study, a BP neural network was used to establish a prediction model of the laser cladding process parameters on the dilution rate, based on data obtained from an orthogonal experiment of AlCoCrFeNi high-entropy alloy coatings. The following conclusions could be drawn:

(1) The training performance of the established BP neural network was outstanding, and the maximum relative error between the training value and the predicted value was 1.70%. The prediction performance was excellent, with an average relative error of only 5.89% between the predicted value and the experimental value.

(2) The optimal process parameters were as follows: laser power of 2000 W, a scanning speed of 2 mm/s, and a powder feeding rate of 15 g/min. The dilution rate was 16%, and the microhardness value was 521.6 $HV_{0.3}$.

(3) The grains of the coatings under this process parameter were equiaxed, with uniform composition distribution, and no serious composition segregation occurred. The microstructure of coatings was still composed of the BCC phase solid solution. The FCC phase does not form in the microstructure due to the excessive Fe element content. These results also prove that the dilution rate under the process param-

eters does not have a major impact on the microstructure and properties of the AlCoCrFeNi coatings.

Author Contributions: Conceptualization, writing-original draft, Y.L.; date curation, K.W.; analysis, writing—review and editing, H.F.; funding acquisition, X.Z.; investigation, X.G.; methodology, J.L. All authors have read and agreed to the published version of the manuscript.

Funding: This research was funded by the National Natural Science Foundation of China (Grant No. 52075010) and the Natural Science Foundation of Hebei Province, China (Grant No. E2019210025).

Institutional Review Board Statement: Not applicable.

Informed Consent Statement: Not applicable.

Data Availability Statement: All data is contained within the article.

Conflicts of Interest: The authors declare no conflict of interest.

References

1. Zhou, Z.; Liu, X.; Zhuang, S.; Yang, X.; Wang, M.; Sun, C. Preparation and high temperature tribological properties of laser in-situ synthesized self-lubricating composite coatings containing metal sulfides on Ti6Al4V alloy. *Appl. Surf. Sci.* **2019**, *481*, 209–218. [CrossRef]
2. Zhu, Y.; Liu, X.; Liu, Y.; Wang, G.; Wang, Y.; Meng, Y.; Liang, J. Development and characterization of Co-Cu/Ti_3SiC_2 self-lubricating wear resistant composite coatings on Ti6Al4V alloy by laser cladding. *Surf. Coat. Technol.* **2021**, *424*, 127664. [CrossRef]
3. Oliveira, U.D.; Ocelík, V.; De Hosson, J.T.M. Analysis of coaxial laser cladding processing conditions. *Surf. Coat. Technol.* **2005**, *197*, 127–136. [CrossRef]
4. Weng, F.; Chen, C.; Yu, H. Research status of laser cladding on titanium and its alloys: A review. *Mater. Des.* **2014**, *58*, 412–425. [CrossRef]
5. Wang, K.; Du, D.; Liu, G.; Pu, Z.; Chang, B.; Ju, J. Microstructure and mechanical properties of high chromium nickel-based superalloy fabricated by laser metal deposition. *Mater. Sci. Eng. A* **2020**, *780*, 139185. [CrossRef]
6. Wang, K.; Du, D.; Liu, G.; Pu, Z.; Chang, B.; Ju, J. A study on the additive manufacturing of a high chromium Nickel-based superalloy by extreme high-speed laser metal deposition. *Opt. Laser Technol.* **2021**, *133*, 106504. [CrossRef]
7. Cantor, B.; Chang, I.T.H.; Knight, P.; Vincent, A.J.B. Microstructural development in equiatomic multicomponent alloys. *Mater. Sci. Eng. A* **2004**, *375*, 213–218. [CrossRef]
8. Yeh, J.; Chen, S.; Lin, S.; Gan, J.; Chin, T.; Shun, T.; Tsau, C.; Chang, S. Nanostructured high-entropy alloys with multiple principal elements: Novel alloy design concepts and outcomes. *Adv. Eng. Mater.* **2004**, *6*, 299–303. [CrossRef]
9. George, E.P.; Curtin, W.A.; Tasan, C.C. High entropy alloys: A focused review of mechanical properties and deformation mechanisms. *Acta Mater.* **2020**, *188*, 435–474. [CrossRef]
10. Qiu, X.; Zhang, Y.; He, L.; Liu, C. Microstructure and corrosion resistance of AlCrFeCuCo high entropy alloy. *J. Alloys Compd.* **2013**, *549*, 195–199. [CrossRef]
11. Zhang, S.; Wu, C.L.; Zhang, C.H.; Guan, M.; Tan, J.Z. Laser surface alloying of FeCoCrAlNi high-entropy alloy on 304 stainless steel to enhance corrosion and cavitation erosion resistance. *Opt. Laser Technol.* **2016**, *84*, 23–31. [CrossRef]
12. Kunce, I.; Polanski, M.; Karczewski, K.; Plocinski, T.; Kurzydlowski, K.J. Microstructural characterisation of high-entropy alloy AlCoCrFeNi fabricated by laser engineered net shaping. *J. Alloys Compd.* **2015**, *648*, 751–758. [CrossRef]
13. Juan, Y.; Li, J.; Jiang, Y.; Jia, W.; Lu, Z. Modified criterions for phase prediction in the multi-component laser-clad coatings and investigations into microstructural evolution/wear resistance of $FeCrCoNiAlMo_x$ laser-clad coatings. *Appl. Surf. Sci.* **2019**, *465*, 700–714. [CrossRef]
14. Guo, L.; Xiao, D.H.; Wu, W.Q.; Ni, S.; Song, M. Effect of Fe on microstructure, phase evolution and mechanical properties of $(AlCoCrFeNi)_{100-x}Fe_x$ high entropy alloys processed by spark plasma sintering. *Intermetallics* **2018**, *103*, 1–11. [CrossRef]
15. Cai, Y.; Chen, Y.; Manladan, S.M.; Luo, Z.; Gao, F.; Li, L. Influence of dilution rate on the microstructure and properties of FeCrCoNi high-entropy alloy coating. *Mater. Des.* **2018**, *142*, 124–137. [CrossRef]
16. Sun, Y.; Hao, M. Statistical analysis and optimization of process parameters in Ti6Al4V laser cladding using Nd:YAG laser. *Opt. Laser Eng.* **2012**, *50*, 985–995. [CrossRef]
17. Casalino, G. Computational intelligence for smart laser materials processing. *Opt. Laser Technol.* **2018**, *100*, 165–175. [CrossRef]
18. Rong, Y.; Zhang, Z.; Zhang, G.; Yue, C.; Gu, Y.; Huang, Y.; Wang, C.; Shao, X. Parameters optimization of laser brazing in crimping butt using taguchi and BPNN-GA. *Opt. Laser Eng.* **2015**, *67*, 94–104. [CrossRef]
19. Sathiya, P.; Panneerselvam, K.; Soundararajan, R. Optimal design for laser beam butt welding process parameter using artificial neural networks and genetic algorithm for super austenitic stainless steel. *Opt. Laser Technol.* **2012**, *44*, 1905–1914. [CrossRef]

20. Deng, Z.; Chen, T.; Wang, H.; Li, S.; Liu, D. Process parameter optimization when preparing Ti (C, N) ceramic coatings using laser cladding based on a neural network and quantum-behaved particle swarm optimization algorithm. *Appl. Sci.* **2020**, *10*, 6331. [CrossRef]
21. Guo, S.; Chen, Z.; Cai, D.; Zhang, Q.; Kovalenko, V.; Yao, J. Prediction of simulating and experiments for Co-based alloy laser cladding by HPDL. *Phys. Procedia* **2013**, *50*, 375–382. [CrossRef]
22. Xi, W.; Song, B.; Zhao, Y.; Yu, T.; Wang, J. Geometry and dilution rate analysis and prediction of laser cladding. *Int. J. Adv. Manuf. Technol.* **2019**, *103*, 4695–4702. [CrossRef]
23. Manzoni, A.; Daoud, H.; Völkl, R.; Glatzel, U.; Wanderka, N. Phase separation in equiatomic AlCoCrFeNi high-entropy alloy. *Ultramicroscopy* **2013**, *132*, 212–215. [CrossRef] [PubMed]
24. Butler, T.M.; Weaver, M.L. Oxidation behavior of arc melted AlCoCrFeNi multi-component high-entropy alloys. *J. Alloys Compd.* **2016**, *674*, 229–244. [CrossRef]
25. Xu, X.; Cheng, H.; Wu, W.; Liu, Z.; Li, X. Stress corrosion cracking behavior and mechanism of Fe-Mn-Al-C-Ni high specific strength steel in the marine atmospheric environment. *Corros. Sci.* **2021**, *191*, 109760. [CrossRef]
26. Li, Y.; Liu, Z.; Fan, E.; Huang, Y.; Fan, Y.; Zhao, B. Effect of cathodic potential on stress corrosion cracking behavior of different heat-affected zone microstructures of E690 steel in artificial seawater. *J. Mater. Sci. Technol.* **2021**, *64*, 141–152. [CrossRef]
27. Wang, R.; Zhang, K.; Davies, C.; Wu, X. Evolution of microstructure, mechanical and corrosion properties of AlCoCrFeNi high-entropy alloy prepared by direct laser fabrication. *J. Alloys Compd.* **2016**, *694*, 971–981. [CrossRef]
28. Niu, P.; Li, R.; Yuan, T.; Zhu, S.; Chen, C.; Wang, M.; Huang, L. Microstructures and properties of an equimolar AlCoCrFeNi high entropy alloy printed by selective laser melting. *Intermetallics* **2019**, *104*, 24–32. [CrossRef]
29. Liu, Z.; Zhang, X.; Xuan, F.; Wang, Z.; Tu, S. In situ synthesis of TiN/Ti_3Al intermetallic matrix composite coatings on Ti6Al4V alloy. *Mater. Des.* **2012**, *37*, 268–273. [CrossRef]

coatings

Article

Microstructure and Wear Behavior of TC4 Laser Cladding Modified via SiC and MoS_2

Yan Liu [1,*], Junjie Li [1], Qian Xu [2], Yunhua Zhang [1], Xiulin Yan [2], Yong Chen [2] and Huabing He [2]

1 School of Materials Science and Engineering, Southwest Jiaotong University, Chengdu 611756, China; 15084989475@163.com (J.L.); iszhangyunhua@163.com (Y.Z.)
2 AVIC Chengdu Aircraft Industrial (Group) Co., Ltd., Chengdu 610091, China; xuqian9654@163.com (Q.X.); xiulinjd@163.com (X.Y.); yognchen@my.swjtu.edu.cn (Y.C.); hhbzdx@163.com (H.H.)
* Correspondence: liuyanzt@163.com

Abstract: A TC4 composite coating reinforced by SiC ceramic phase and MoS_2 self-lubricating phase was prepared on a muzzle brake by laser cladding to improve its wear resistance properties. In this study, we investigated the microstructure and wear behavior of the composite coating. The results show that the composite coating consisted of equiaxed grains with grain sizes ranging from 102.39 to 255.31 μm on the surface and columnar grains on the bottom. The grains with mesh basket microstructure were mainly with α-Ti and β-Ti phases. When wearing against H70 brass, the main wear mechanism of the composite coating at room temperature was adhesive wear, while the wear mechanism was dominated by adhesive wear, oxidation wear, and slight abrasive wear at high temperature. Compared with TC4 coating without SiC and MoS_2, the wear rate of the composite coating was reduced by 15%–35% when the temperature was below 400 °C, and the wear rate was reduced by about 55% at 600 °C, resulting from the addition of SiC. Moreover, the friction coefficient of the composite coating was about 10%, which was 30% lower than that of the substrate and TC4 coating when the temperature was below 400 °C and at 600 °C due to the forming of the dense oxide film and the MoS_2 friction transfer film on the friction contact surface of the titanium-based composite coating.

Keywords: laser cladding; titanium alloy; composite coating; wear resistance

1. Introduction

Titanium alloys are widely used in military, aerospace, marine, chemical, and biomedical equipment industries [1] because of their low density, high specific strength, and corrosion resistance [2]. However, the limitations of titanium alloys, such as low wear resistance and oxidation at high temperature, significantly restrict their service environment and application fields [3–5]. In the military field, the high friction coefficient of titanium alloys usually causes damage to the equipment components, which seriously affects weapons accuracy. For example, titanium muzzle brakes and brass shells undergo adhesive wear under high-temperature friction. It damages the performance of the brake and affects the accuracy of the shells. To solve the above problems, surface treatment techniques such as nitriding, carburizing, oxidation, physical vapor deposition, chemical vapor deposition, and ion implantation [6–9] are often used to improve the surface properties of titanium alloys.

Laser cladding is also used to improve the surface properties of titanium alloys due to the customized powder composition [10]. Many scholars have been working on laser cladding of titanium alloys to obtain coatings with good compatibility, excellent wear resistance, and high-temperature oxidation resistance. N. Jeyaprakash et al. [11] prepared NiCrMoNb and NiCrBSiC coatings on titanium alloys to improve the wear resistance of the cladding layer as the γ-Ni in the coating was mixed with chromium (Cr), silicon (Si), molybdenum (Mo), and boron (B) to form a dendritic eutectic structure. Zhang et al. [12] found that when the Si content was 20%, the NiCrBSi coating on TC4 alloys had the best wear resistance. The increase in the Si content accelerated the transformation of the coating from violent oxidation to slow oxidation. At the same time, it delayed the increase rate of the content of the high hardness phase Ti_5Si_3 in the oxidation process. However, coatings with Si content higher than 20% had higher cracking susceptibility. Ceramic coating can significantly improve the hardness, wear resistance, and oxidation resistance of titanium alloys [13], but the ceramic coating is easy to crack due to its high brittleness. Therefore, research mainly focuses on composite ceramic coatings, and there has been much work using density functional theory (DFT) to understand wear and structural properties in composite ceramic materials and reveal mechanisms for the strength of structures [14,15]. Yoshitaka Umeno [15] et al. calculated the ideal strength of SiC crystals by first-principles DFT. The effect of the polytype SiC structures on the shear strength of some shear systems appears to be significant, while that on the tensile strength is marginal. Under multi-axial stress conditions, SiC can not only experience cleavage by tension, but also slide due to shear stress, although it is basically brittle. Anand M. Murmu [16] et al. prepared a ZrO_2-Al_2O_3-TiO_2 ceramic coating on the TC4 substrate, in which the nano-ZrO_2 of the tetragonal phase and monoclinic phase reduced the friction coefficient and improved the hardness and wear resistance. Zhao et al. [17] prepared a TiO_x coating, mainly composed of TiO_x and Ti, with a small amount of Al_2O_3, and showed that the cured TiO_x coating had extremely high fracture strength, hardness, and excellent wear resistance. Li et al. [18] studied the effect of Al_2O_3-13%TiO_2 on the structure and properties of TiAlSi composite coatings. They found that the stable α-Al_2O_3 improved the mechanical properties of the coating; a small amount of TiO_2 reduced thermal stress, curing shrinkage and cracking tendency; and the hardening phase Ti_3AlC_2 improved the strength and thermal shock resistance. Zhang et al. [19] used Ni60/20%WS_2 powder to prepare a self-lubricating coating on the surface of TC4. During the cladding process, WS_2 was decomposed to Cr_xS_y-type compounds and TiS_2, in which Cr_xS_y formed a transfer film on the friction surface to achieve the anti-friction effect and TiS_2 acted as a lubricant. Moreover, MoS_2 reduced the friction coefficient of the Ni-based self-lubricating composite coating and improved the wear resistance by forming a lubricating transfer film during the dry sliding wear process [20].

Great progress has been made in the research of wear-resistant coatings on the surface of laser cladding titanium alloys, but there are still many problems to overcome. As mentioned above, most researchers used laser cladding to improve the surface wear resistance of titanium alloys by adding or generating a ceramic hardening phase with high hardness and high wear resistance in the coating. However, under high temperatures or in an environment where the lubricating medium was not suitable, the coating surface still exhibited a high friction coefficient in wear, which affected the surface quality of the coating itself and the wear parts and reduced the service life of the parts in long-term wear. Therefore, the coating should improve the wear resistance of the surface of the part and have a certain anti-friction performance to satisfy the part's long-term working under special conditions such as high temperature and vacuum. Nowadays, as the application of titanium alloys becomes more and more widespread, it is necessary to prepare a self-lubricating and wear-resistant composite coating by laser cladding on titanium alloys to improve their wear resistance. However, the lubricating phase can reduce the friction coefficient of the coating, and the excess lubricating phase can reduce the strength and hardness of the coating. Moreover, increasing the hardness of the coating can improve the wear resistance of the coating, but if the content of the hard phase is too high, the brittleness of the coating increases, and fatigue cracking is likely to occur in the process of friction and wear. Many scholars have carried out research on self-lubricating wear-resistant coatings. Dai et al. [21] fabricated self-lubricating TiB_2-TiC_xN_y ceramic coatings on Ti-6Al-4V alloy by laser surface alloying with B4C and graphite powders in nitrogen atmosphere. Gao et al. [22] fabricated Ti-Ni/TiN/TiW + TiS/WS_2 self-lubricating wear resistant composite coating on Ti-6Al-4V alloy by laser cladding with NiCrBSi, TiN, and WS_2 powder mixtures. Zhu et al. [23] prepared Co-Cu/Ti_3SiC_2 self-lubricating wear resistant composite coatings on Ti6Al4V alloy by laser cladding. The average microhardness of these coatings has been significantly improved compared with substrate due to the reinforcing phases, and the wear resistance of the coating has also been effectively improved. However, these studies mainly focus on the wear performance and wear mechanism at room temperature, and there are few studies on the wear performance at high temperatures.

The commonly used self-fluxing alloy materials and metal-based ceramic materials have poor compatibility with titanium alloys, which cannot solve the problems of high friction coefficient and easy adhesive wear of titanium alloys. At the same time, the hardness and toughness of the coating cannot be balanced. Therefore, we designed a novel titanium-based composite coating on the muzzle brake by laser cladding to enhance the wear resistance of the coating by adding SiC ceramic phase and MoS_2 self-lubricating phase. We explored the high-temperature wear mechanism of the composite coating by analyzing the microstructure, phase structure, microhardness, friction reduction, and wear resistance.

2. Materials and Methods

The substrate was TC4 titanium alloy, and its chemical composition is shown in Table 1. The substrate is a duplex microstructure in the as-cast state, with a size of 170 × 130 × 13 mm. A mixed powder consisting of TC4 powder for suitable compatibility with the substrate, SiC powder for hardening, and Ni-coated MoS_2 powder for lubricating was prepared, and its morphologies are shown in Figure 1. The TC4 powder particles have a uniform size of 100–150 μm high sphericity and smooth surface, which provides good fluidity during the cladding process. SiC powder particles with the purity of 99.9% have excellent high-temperature strength, good wear, corrosion resistance, and low cost, and are used to improve the wear resistance of the coating. Because MoS_2 was easy to burn during the cladding process, Ni-coated MoS_2 was selected as a lubricant phase. The powder contains 25.41% MoS_2, and the rest is Ni. The microscopic morphologies of powder particles are shown in Figure 1.

The effect of powder composition on the coating performance was explored through preliminary tests. We found that too high SiC content easily led to the cladding cracking or even peeling off. Excessive MoS_2 content caused the formation of large pores, and the presence of Mo reduced the elastic modulus of the Ti-Mo system. Therefore, the powder ratio was determined to be TC_4:SiC:Ni-coated MoS_2 = 98.5:1:0.5.

The laser cladding system consisted of a YLS-4000 fiber laser, powder feeding system, walking mechanism, and integrated control. The powder feeding system consisted of a TD03 laser cladding synchronous turntable double-cylinder powder feeder, an anti-static powder feeding tube, and a coaxial four-point nozzle. The cladding shielding gas was 99.999% high-purity argon. In the early stage of the experiment, we conducted single-factor experiments and multi-factor orthogonal experiments, through which the optimized process parameters were determined, as shown in Table 2. During the laser cladding process, the substrate was kept at 250 °C. TC4 coatings were prepared using the same process parameters for comparison. The schematic diagram of laser cladding is shown in Figure 2.

Table 1. Chemical composition of TC4 substrate and TC4 powder (wt %).

Materials	Al	V	Fe	C	N	O	H	Ti
Substrate	6.30	4.05	0.16	0.005	0.015	0.17	0.00388	Bal.
Powder	5.95	4.12	0.10	0.012	0.012	0.084	0.0032	Bal.

Figure 1. The morphology of powder: (**a**,**b**) TC4; (**c**,**d**) SiC; (**e**,**f**) nickel-coated MoS_2.

Table 2. Optimized process parameter table.

Laser Power (W)	Scanning Speed (mm/min)	Powder Feeding rate (g/min)	Shield Airflow (L/min)	Carrier Powder Gas Flow (L/min)	Preheating Temperature (°C)	Focused Beam Diameter (mm)
1500	300	9	35	8	250	0.2

Figure 2. Schematic diagram of laser cladding.

The as-deposited composite coating was evaluated for its microstructural characteristics, hardness, tensile properties, and high-temperature wear performance. The metallographic specimen was obtained with 10 mm × 10 mm × 10 mm via a wire-cutting machine, then polished by a series of abrasive paper, and etched by a mixed reagent with HNO_3:HF:H_2O = 2:3:5 for 10 s. The AIM Zeiss microscope was used to observe the metallographic structure of the cladding layer, and the FEI Nova Nano SEM 400 field emission scanning electron microscope (SEM) was used to observe the microscopic morphology. An energy dispersive spectrometer (EDS) in SEM was used to analyze the types and contents of constituent elements in the selected micro-zones on the surface of the coating and friction and wear samples. The X Pert Pro MPD X-ray diffractometer (XRD) with a Cu target, the voltage being 40 kV, the current being 40 mA, and the scanning step of 0.02°/min was used for the phase analysis of the laser cladding layer. The DHV-1000ZTEST micro-Vickers hardness tester was used to test the microhardness of the coating. The test area was a straight line from the near-surface of the cladding layer to the substrate area. The test load was 1.96 N, lasting for 15 s, and the test point interval was 0.1 mm. In this paper, the thicknesses of the coating, bonding interface, and heat-affected zone (HAZ) were relatively narrow, being only 1–2 mm. The micro-shear method was used to measure the shear strength of each area. The sample size was 1.5 mm × 1.5 mm × 20 mm, and the shearing speed was 1 mm/min. The micro-shear tester was developed by The Welding Laboratory of Southwest Jiaotong University. The high-temperature wear performance was performed by the HT-1000 high-temperature friction and wear tester, which is composed of four parts: a computer control system, a high-temperature furnace, a mechanical friction system, and a loading mechanism. The required load was applied after heating the furnace to the target temperature during the test. Meanwhile, the friction sample fixed on the plate was driven to rotate against the counter surface (ball). In this paper, the diameter of the friction sample was 30 mm and the diameter of the friction pair was 6 mm. The chemical composition of the friction pair was 69.7%–71% Cu, 0.085% Fe, 0.0029% Pb, and the rest was Zn. The test load was 2.94 N, the sliding speed was 6 m/s, and the friction radius was 6 mm. The experimental temperatures were room temperature, 200 °C, 400 °C, and 600 °C, and the experimental time was 10–30 min. The wear volume was obtained by the Bruker Contour GT white light interference three-dimensional profiler. The apparatus information used in this work is listed in Table 3.

Table 3. The information of the apparatus used.

Device Designation	Company	City	Country
AIM Zeiss microscope	Carl Zeiss AG	Oberkochen	Germany
FEI Nova Nano SEM 400 field emission scanning electron microscope	Thermo Fisher Scientific	Hillsboro, OR	USA
X Pert Pro MPD X-ray diffractometer	PANalytical B.V.	Almelo	The Netherlands
DHV-1000ZTEST micro-Vickers hardness tester	Chengdu Coret Intelligent Technology Co., Ltd.	Chengdu	China
Micro-shear tester	The Welding Laboratory of Southwest Jiaotong University	Chengdu	China
HT-1000 high-temperature friction and wear tester	Lanzhou Zhongke Kaihua Technology Development Co., Ltd.	Lanzhou	China
Bruker Contour GT white light interference three-dimensional profiler	Bruker Corporation	Billerica, MA	USA

3. Results and Discussion

3.1. Microstructural Characteristics of Composite Coating

The XRD phase analysis results of the composite coating prepared with optimized process parameters are shown in Figure 3. The XRD spectra was coincident with phase α-hexagonal-crystal-system Ti (a = b = 2.944 nm, c = 4.678) and Ti_3Al (a = b = 5.793 nm. c = 4.649 nm); it exhibited a prominent (101) and (201) peak. The diffraction peaks occurred at 2θ = 35°, 38°, 40°, 53°, 71°, 77°, and 78°, indicating the diffraction of (100), (002), (101), (102), (110), (103), (112), and (201) planes of the α matrix and the diffraction of (200), (002), (201), (202), (212), (311), (222), and (401) planes of the Ti_3Al matrix. The major peaks at around 38° indicated a strong presence of β-Ti in the sample, as shown in Figure 3, which indicated the diffraction of (110) planes. Therefore, the main phases in the coating were α-Ti, β-Ti, and Ti_3Al. SiC and MoS_2 were not detected due to the low content.

The cross-sectional morphology of samples is shown in Figure 4a. The cladding layer was roughly divided into four areas from top to bottom: the coating, the bonding interface, the heat-affected zone (HAZ), and the substrate. There were apparent differences in the microstructures of each area. The reason was that the cladding layer had different temperature gradients and solidification rates at different depths. Because the surface of the cladding layer was in contact with the air, there was convection heat dissipation between them, which led to a high solidification rate, resulting in equiaxed grains. On the contrary, the cladding layer's heat conduction at the bottom was relatively slow, so it was easy to form columnar grains, and these columnar grains grow directionally [24]. Consequently, from the surface layer to HAZ, the microstructure of each area was different. The microstructure of the surface of the coating exhibited equiaxed grains with the grain sizes ranging from 102.39 to 255.31 μm. The columnar grains were observed at the bottom with the grain length up to 910 μm, and the width ranged from 208.78 to 321.81 μm. The bonding interface (the junction of the cladding material and the substrate) was an excellent metallurgical bond. Figure 4b,c shows the microstructure of the cladding layer under optimized process parameters. It can be seen from Figure 4b that the β phase displayed a fine mesh basket microstructure with clear grain boundary. In Figure 4c, a large number of α-laths and fine white fragments and spheres are present. The EDS detection result in Table 4 showed that the spheres were SiC particles, and the white fragments were Ti_3Al. Due to the high cooling rate of laser cladding, the β phase was not totally transformed into the α phase. Some untransformed β phase transformed into a metastable phase α′, and others remained in the β phase structure. The basket-weave and sheet-like α phase with small ratio and staggered arrangement were distributed on the transformed β substrate. Acicular hexagonal martensite was not found in the metallographic structure, which might be attributed to the fact that it was a metastable phase and decomposed into α + β phase at a high temperature as the sample had been kept at 250 °C during the laser cladding process [25]. Ti_3Al intermetallic compounds were formed during the phase transformation when the Al content was 5%–25% [26]. In this work, the Al content in the cladding powder system was about 6%, and as a result, Ti_3Al precipitated from the molten pool during the solidification of the laser cladding. Zhang et al. [27] generated ultra-fine nanocrystalline Ti_3Al in the TC4 coating with the same feature as the fragmented white structure in Figure 4c. The size of Ti_3Al in Figure 4c was slightly larger than that due to the larger laser power used in this work. The precipitated Ti_3Al was dispersed and distributed in the coating to cause the precipitation strengthening and dispersion strengthening.

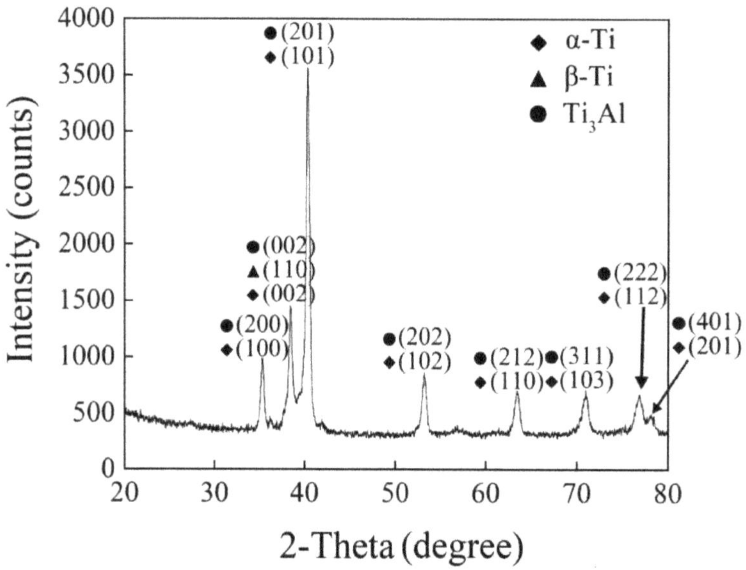

Figure 3. X-ray diffraction pattern of the composite coating.

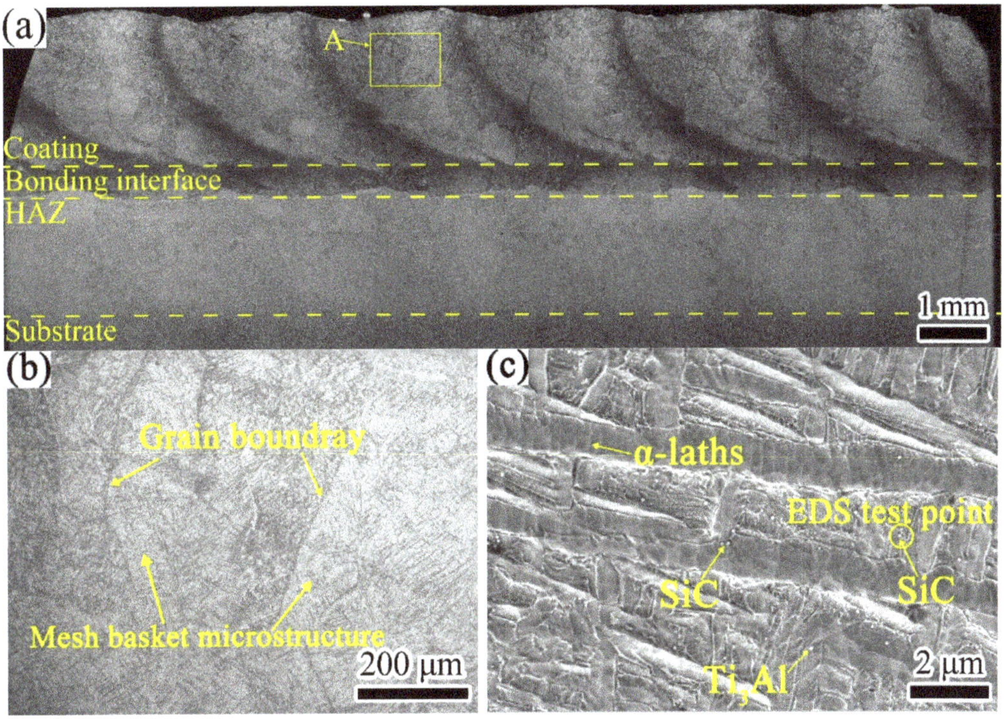

Figure 4. Laser cladding sample: (**a**) cross-sectional morphology; (**b**) enlarged area A in (**a**); (**c**) microstructure in (**b**) obtained with SEM.

Table 4. EDS detection result of the test point in Figure 4c (at %).

Element	Al	Ti	V	C	Si
Content	7.94	83.79	3.03	3.53	1.71

3.2. Microhardness and Shear Properties

Figure 5 shows the microhardness of the samples. The dispersion of SiC increased the microhardness to 466.75 HV, which is 1.5 times as high as the substrate. The uneven distribution of SiC and a small number of pores in the coating caused slight fluctuations in the microhardness. The average microhardness of the bonding interface was not much different from the microhardness of the substrate, which was only increased by 10%.

Figure 6 shows the shear curve (load–displacement curve) of each area. The maximum shear stress at fracture was arranged in order of coating, bonding interface, HAZ, and substrate. The fracture of the coating occurred immediately after the maximum load was applied without plastic deformation, indicating that the coating was brittle. On the contrary, both the bonding interface and the HAZ experienced a plastic deformation, crack formation, and propagation.

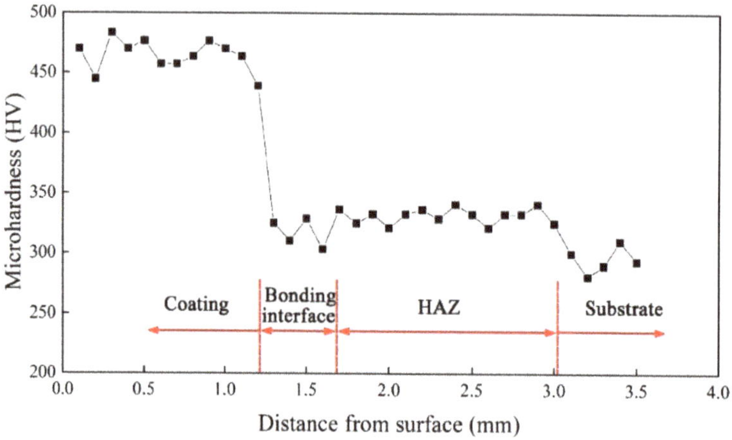

Figure 5. Microhardness of the titanium-based composite coating.

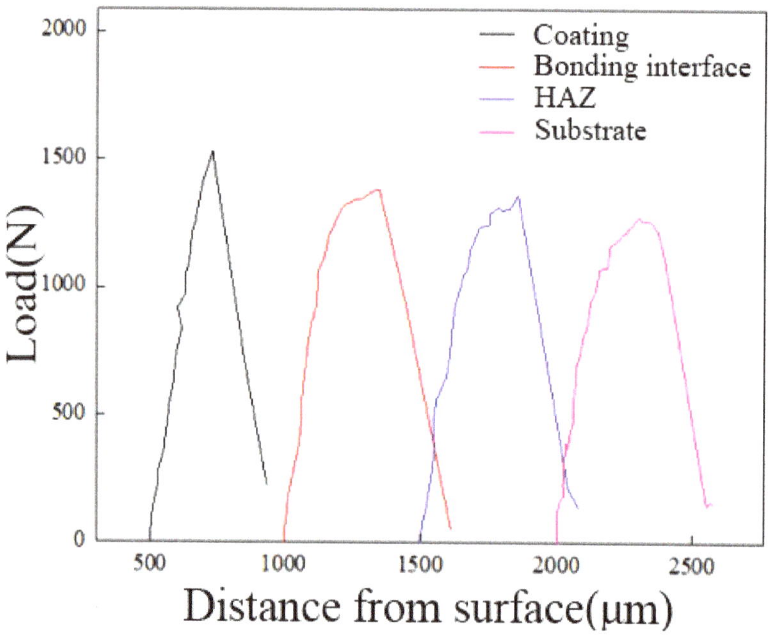

Figure 6. Load–displacement curve of each area.

In the shear test, the deformation of the shear surface varied with the cutting depth into the sample. The brittle and hard sample led to shallow cutting depth and small shear surface deformation. Therefore, the cutting depth and the deformation of the shear surface can reflect the plasticity of the material. In this paper, the indentation rate α was introduced as an index of plasticity. The calculation method was given by the following equation:

$$\alpha = \frac{(A_0 - A_k)}{A_0} \times 100\%, \tag{1}$$

where A_0 is the cross-sectional area of the sample and A_k is the minimum cross-sectional area after the sample was broken. The smaller the A_k is, the higher the indentation rate α is, which indicates that the material has strong deformability and good plasticity.

Figure 7 shows the shear strength and indentation rate of the prepared cladding samples. The shear strength gradually decreases from coating, bonding interface, HAZ, to substrate. The indentation rate arranged in order from large to small was bonding interface, HAZ, coating, and substrate. Ti_3Al was dispersedly distributed in the coating. Ti_3Al has a hexagonal close-packed crystal structure, and only one slip system at room temperature, which results in the plasticity of the coating being lower than that of the bonding interface. The strength and plasticity of the coating were better than those of the substrate, which indicated that the coating and the substrate were well bonded. Moreover, the bonding interface had a higher plastic deformation capacity than that of the substrate.

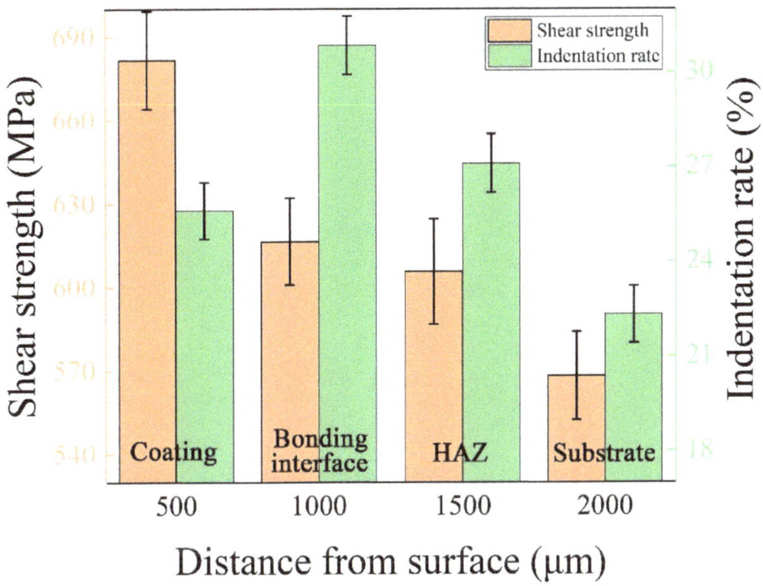

Figure 7. Shear strength and indentation rate of each area.

The shear fracture morphology of the coating is shown in Figure 8. The macroscopic morphology of the fracture was relatively smooth with evident shear bands and parabolic ductile dimples. The macroscopic morphology exhibited a rock-like pattern with a strong three-dimensional effect in area A, where the grain boundary and lots of pores in the fracture could be observed. According to area B in Figure 8a, a large number of small planes on the grain boundary of the fracture surface were found, suggesting that the fracture of the coating was a brittle intergranular fracture.

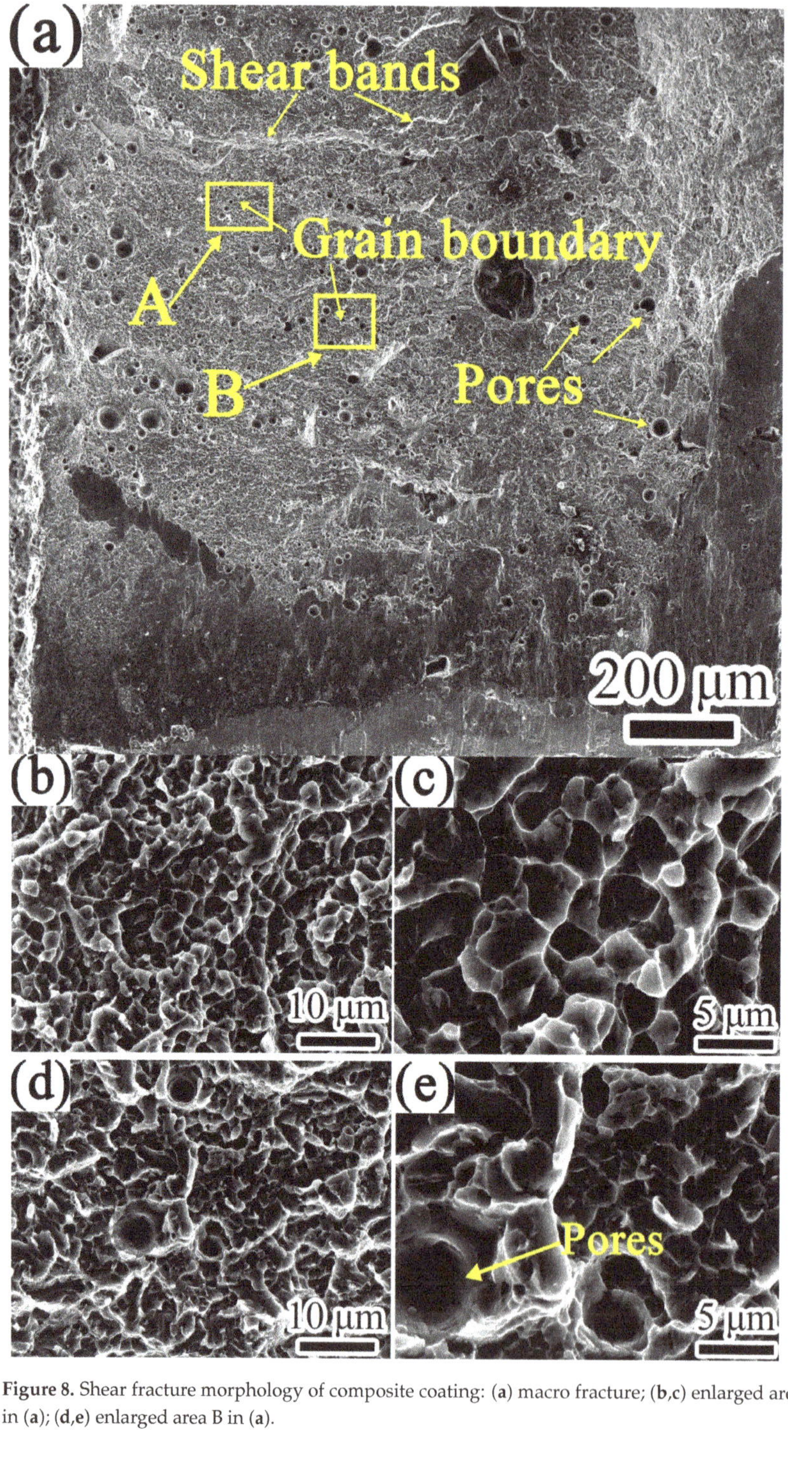

Figure 8. Shear fracture morphology of composite coating: (**a**) macro fracture; (**b**,**c**) enlarged area A in (**a**); (**d**,**e**) enlarged area B in (**a**).

3.3. Formatting of Mathematical Components

The friction coefficient, wear rate, and wear scar morphology of composite coatings, TC4 coating, and substrate were investigated by wear against H70 brass, which is the main component of the outer wall of the shell. The wear mechanism and the high temperature wear resistance of the composite coating were explored.

3.3.1. The Influence of Temperature on Friction Coefficient

The friction coefficients of three samples' wear against H70 brass at different temperatures are shown in Figure 9. At the beginning of wear, the contact surface between the worn ball and the sample was a tiny convex peak composed of the end surface and the cylindrical surface [28]. At this time, the contact area was small and constantly changing, resulting in unstable load and large fluctuation in the wear coefficient. Therefore, the unstable data of the wear process at the initial stage (about 3 min) of the wear test are ignored in this section. As the wear proceeded, the contact area increased so that the device rotated stably, and the friction coefficient gradually stabilized. At all temperatures, the friction coefficient of the composite coating were more stable than that of the substrate and the TC4 coating.

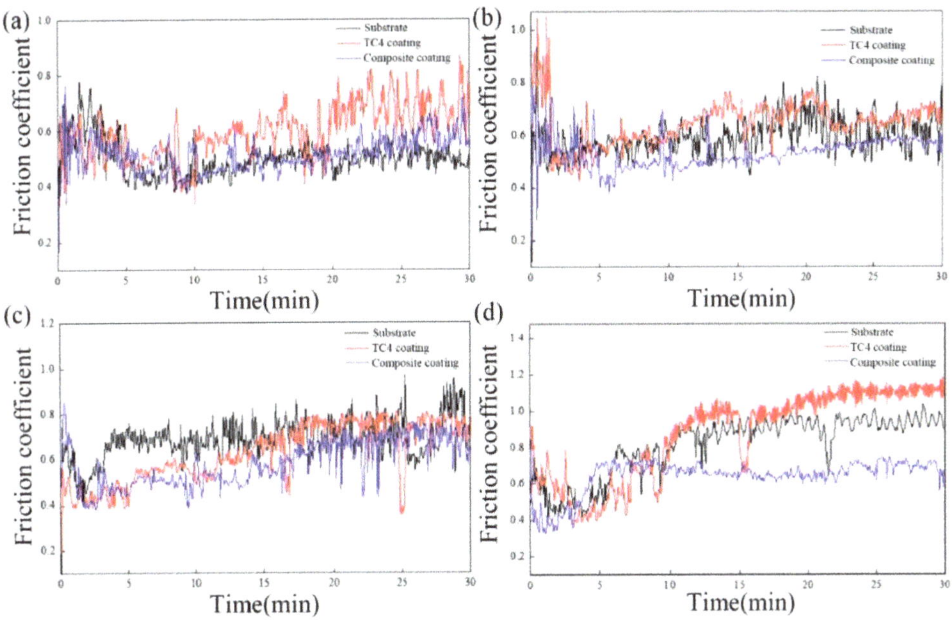

Figure 9. Friction coefficients of three samples' wear against H70 brass at different temperatures: (**a**) room temperature; (**b**) 200 °C; (**c**) 400 °C; (**d**) 600 °C.

At room temperature, the friction coefficient of the composite coating and the substrate gradually stabilized at a similar value over time. The friction coefficient of TC4 coating fluctuated greatly, and the value was high. After several minutes of wear at 200 °C and 400 °C, the friction coefficient of the composite coating was stabilized at about 0.6, and the friction coefficient of the substrate and the TC4 coating was higher, but neither exceeded 1.0. At 600 °C, the friction coefficients of the three samples for the first 5 min were stable at about 0.4–0.5. As the wear time increased, the friction coefficient of the composite coating increased to about 0.6, and it was relatively stable. However, due to the softening of H70 brass at high temperatures, adhesive wear occurred in the substrate and the TC4 coating, which rapidly increased the friction coefficient of them, and the friction coefficient of the TC4 coating even exceeded 1.0.

Figure 10 shows the average friction coefficient of the three samples' wear against H70 brass at different temperatures. The average friction coefficients of the three samples all increased with the temperature increasing. Among them, the average friction coefficient of the composite coating increased the slowest with temperature, while that of the TC4 coating was first gentle and then sharply increased. At all temperatures, the coefficient of friction of the TC4 coating was the largest, followed by the substrate and the composite coating. From room temperature to 400 °C, the friction coefficient of the composite coating was 2%–10% lower than that of the substrate and 10%–15% lower than that of the TC4 coating, while at 600 °C, it was 23% lower than that of the substrate and 31% lower than that of the TC4 coating. It could be concluded that the composite coating exhibited better anti-friction properties at high temperatures. The MoS_2 in the composite coating was a layered structure and prone to relative sliding between the atomic layers. There was only a small van der Waals force between the layers of MoS_2, which was easy to form a lubricating transfer film on the contact surface during the friction process to reduce the friction coefficient. MoS_2 oxidized above 400 °C, part of which generated MoO_3. MoO_3 was a layered structure with multiple sliding surfaces, which also played a role in reducing the friction coefficient [29].

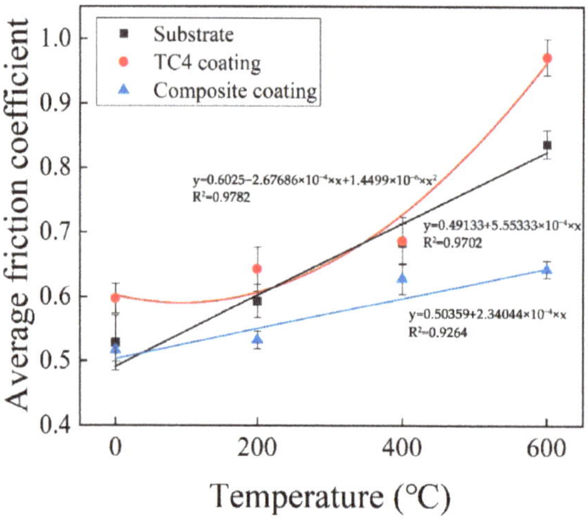

Figure 10. The average friction coefficient of three samples' wear against H70 brass at different temperatures.

3.3.2. The Influence of Temperature on Wear Volume and Wear Rate

The wear volume was selected as an index to measure the wear resistance of the coating. Figure 11 shows the wear volume of the three samples' wear against H70 brass at different temperatures and times. As the temperature increased, the wear volume of the substrate gradually increased. After 30 min, at room temperature, 200 °C, and 400 °C, the wear volume of the TC4 coating was not much different, but it became larger at 600 °C. When the temperature was below 400 °C, the wear volume of the composite coating gradually decreased as the temperature increased, but at 600 °C, it increased significantly, as seen in Figure 11c.

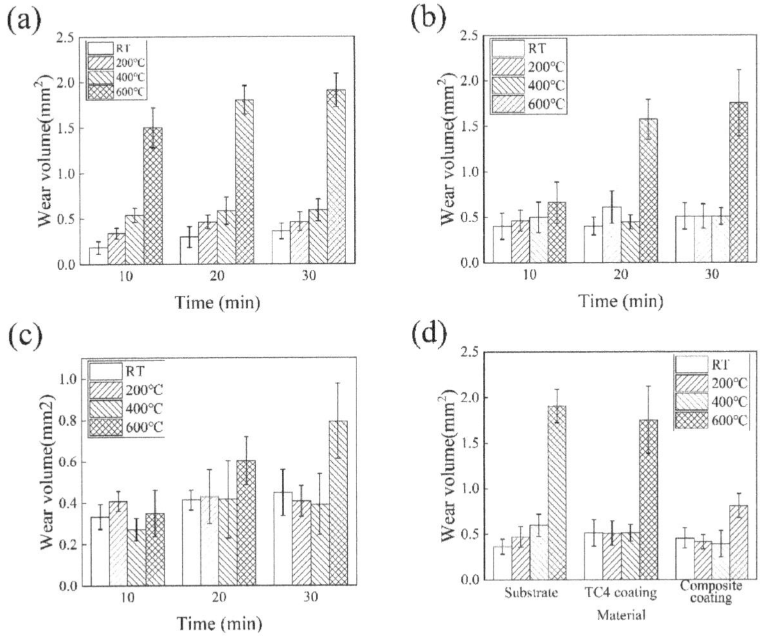

Figure 11. Wear volume of substrate, TC4 coating, and composite coating wear against H70 brass: (**a**) substrate; (**b**) TC4 coating; (**c**) composite coating; (**d**) comparison of wear for 30 min.

To reduce the test error, the wear rate K (mm^3/N·mm) was introduced to more accurately measure the wear resistance of the coating. Figure 12 shows the wear rate of the three samples' wear against H70 brass at different temperatures and different times. The wear rate of the three samples decreased with the increase in time. The wear rate reached the peak value in 10 min, and it almost remained the same in 20 min and 30 min. The friction and wear process could be divided into the "run-in" stage and the stable wear stage [30]. In the "run-in" stage, there were some tiny peaks on the wear contact surface, with large undulations and unevenness. The sample's surface was likely to flake off due to fatigue or plowing, which caused the wear to become more serious. In the stable wear stage, the wear rate decreased because the convex peaks of the contact surface were gradually smoothed, the large topographical fluctuations disappeared, the wear contact area increased, and the friction coefficient tended to be stable. At room temperature, the wear rate of the composite coating was 20% higher than that of the substrate and 10% lower than that of TC4 coating and did not exhibit wear resistance, which was consistent with the analysis result of the friction coefficient in 3.3.1. At 200 °C and 400 °C, the wear rate of the composite coating was 15%–35% lower than that of the substrate and about 20% lower than that of the TC4 coating. At 600 °C, the wear rate of the composite coating was 58% lower than that of the substrate and 54% lower than that of the TC4 coating. It can be concluded that the composite coating exhibited better wear resistance at high temperatures.

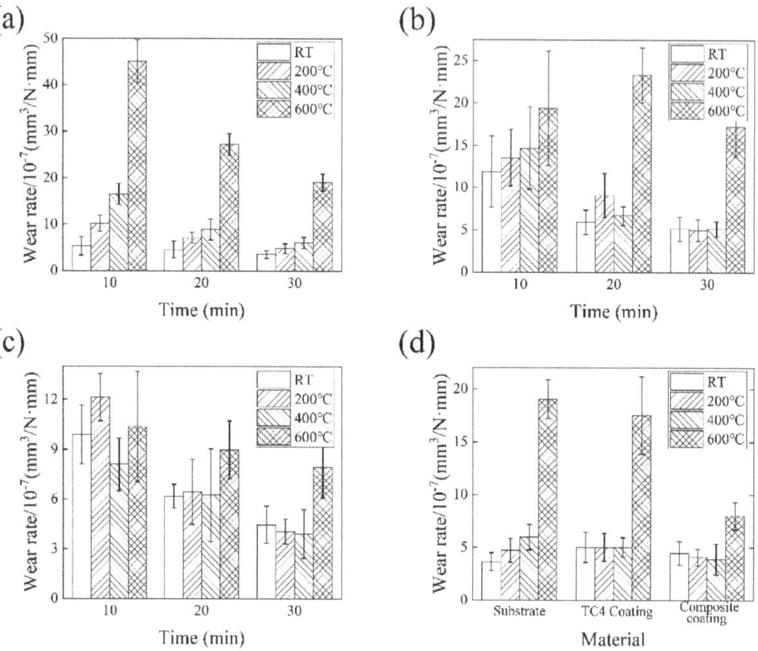

Figure 12. Wear rate of substrate, TC4 coating, and composite coating wear against H70 brass: (**a**) substrate; (**b**) TC4 coating; (**c**) composite coating; (**d**) comparison of wear for 30 min.

3.3.3. Wear Mechanism of Composite Coating

The wear morphology of the three samples' wear against H70 brass at different temperatures for 30 min is shown in Figure 13.

At room temperature, the surface of the substrate and the TC4 coating underwent severe plastic deformation and fracture, and the coating peeled off, which caused many wear particles to appear between the surface and the worn ball and caused abrasive wear to occur, resulting in their wear coefficient fluctuating greatly. At the same time, some brass adhered to the wear scar surface under the load. The composite coating only had slight scaly peeling and wear debris.

At 200 °C, the wear debris plowed grooves on the surfaces of the three samples under the action of normal stress. At high temperatures, the material softened, and its shear resistance was low. Meanwhile, because the TC4 titanium alloy material was relatively viscous [31], the wear surface was prone to adhesive force, which made it accessible to adhesive tearing under the action of external force [32]. Adhesive tearing of the composite coating was relatively low.

At 400 °C, the wear debris that accumulated on the surface of the substrate was filled into the furrow, and the furrow became shallow. Large pieces of brass adhered to the surface of the wear scar. The peeling area of the TC4 coating was enlarged, and more alloys were oxidized. Due to the lower strength of TiO_2, the oxide film was broken during continuous wear and mixed with the adhered brass to form a large area of adhesive wear. Coupled with the adhesion effect, the two caused the lump debris to become larger at the wear scar. The surface of the composite coating was relatively flat, and the furrow was not obvious. The primary wear mechanism was adhesive wear. Compared with the substrate and the TC4 coating, the adhesive wear of the composite coating was lower. This was because MoS_2 had good ductility and was prone to slippage. The formed MoS_2 friction film increased the friction contact area [33].

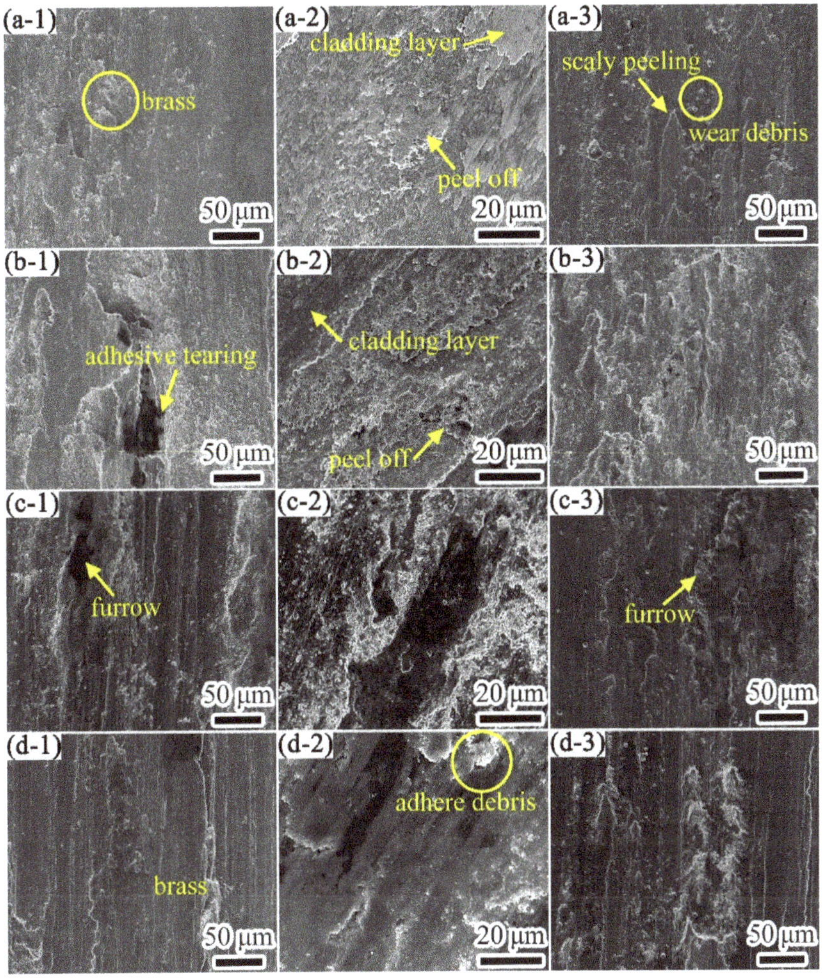

Figure 13. The wear morphology of the three types of samples' wear against H70 brass for 30 min at different temperatures: (**a-1**) substrate with RT, (**a-2**) TC4 coating with RT, (**a-3**) composite coating with RT, (**b-1**) substrate with 200 °C, (**b-2**) TC4 coating with 200 °C, (**b-3**) composite coating with 200 °C, (**c-1**) substrate with 400 °C, (**c-2**) TC4 coating with 400 °C, (**c-3**) composite coating with 400 °C, (**d-1**) substrate with 600 °C, (**d-2**) TC4 coating with 600 °C, (**d-3**) composite coating with 600 °C.

At 600 °C, the furrows on the surface of the wear scar on the substrate became wider, and a lot of brass attached to the surface was broken. It was difficult to observe furrows on the wear scar surface of the TC4 coating. A large amount of brass adhered to the wear scar surface. The main wear mechanisms were adhesive wear and oxidative wear for the substrate and the TC4 coating. The wear morphology of the composite coating was similar to that at 400 °C. For the composite coating, there was less adhesive wear than the substrate and the TC4 coating at 600 °C.

Titanium alloys have high lattice energy at high temperatures and are prone to dislocations and slippage, resulting in plastic deformation. Moreover, alloying elements in TC4 are easily oxidized at high temperatures and produce many oxidation products, such as TiO, TiO_2, Al_2O_3, and V_2O_3 [34]. Therefore, the wear mechanism at high temperatures is different from that at room temperature. We analyzed the wear mechanism of the composite coating at different temperatures.

The wear morphology of the composite coating wear against H70 brass for 10–30 min at different temperatures is shown in Figure 14, and the EDS detection results of each point in Figure 14 are shown in Table 5. As the wear time increased at room temperature, few scaly and round peak-like peeling layers and oxide abrasive grains gradually appeared on the friction surface [35]. The furrow was shallow, and the peeling was not obvious, but the wear scar surface was attached with large flakes of brass. Therefore, the wear mechanism at room temperature was mainly adhesive wear, accompanied by slight abrasive wear and oxidative wear.

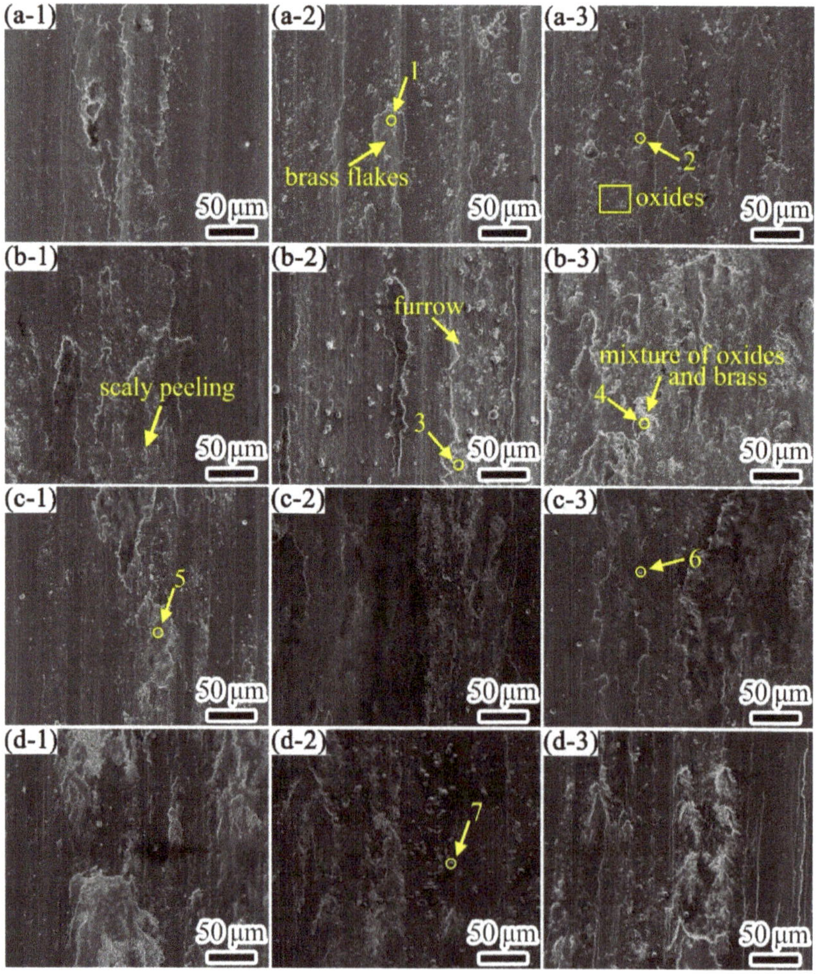

Figure 14. The morphology of the wear scar of composite coating wear against H70 brass at different temperatures and times: (**a-1**) RT with 10min, (**a-2**) RT with 20min, (**a-3**) RT with 30min, (**b-1**) 200 °C with 10min, (**b-2**) 200 °C with 20min, (**b-3**)200 °C with 30min, (**c-1**) 400 °C with 10min, (**c-2**) 400 °C with 20min, (**c-3**) 400 °C with 30min, (**d-1**) 600 °C with 10min, (**d-2**) 600 °C with 20min, (**d-3**) 600 °C with 30min.

Table 5. EDS detection results of each point in Figure 14 (at %).

Point	Ti	Al	V	Mo	Si	O	Cu	Zn
1	3.15	—	0.20	—	—	16.36	49.28	31.01
2	49.91	2.66	2.42	0.35	0.64	42.40	0.86	0.76
3	32.02	3.76	1.58	—	0.51	34.44	18.08	9.61
4	15.39	2.03	0.88	—	—	55.74	16.35	9.62
5	25.34	1.97	1.37	—	0.41	48.48	16.04	6.39
6	18.96	1.90	1.08	20.62	0.60	56.84	—	—
7	6.73	1.03	0.48	18.67	—	73.09	—	—

At 200 °C, some shallow furrows and scaly peeling appeared on the surface to which the oxide film and little brass adhered after wearing 10 min. As the wear time increased, the furrow was deepened and a large amount of abrasive debris was produced, and the brass that adhered to the surface was rolled and smoothed. After 30 min of wear, the abrasive grains on the surface disappeared and the mixture of oxides and brass increased. The mixed oxide film had a strong bearing capacity and compactness, which protected the coating and prevented abrasive wear [36]. Therefore, the wear mechanisms at 200 °C were adhesive wear, oxidative wear, and slight abrasive wear.

In the early stage of wear at 400 °C, as the brass was very soft, the furrow on the worn surface was very shallow, and the surface layer did not peel off. The main wear mechanisms were adhesive wear and oxidative wear. As the wear time increased, the adhesive wear increased. However, due to the combined action of the MoS_2 transfer film, and MoO_3 in the Magnéli phase, the aggravation of the adhesive wear on the coating was delayed. When the wear time increased to 30 min, the coating produced wear debris, which caused secondary wear on the surface, resulting in delamination and furrows. The main wear mechanisms were still adhesive wear and oxidative wear.

At 600 °C, the adhesive wear on the surface of composite coating was severe after 10 min of wear, showing shallow furrows and fine wear debris distributed on it. As the wear time increased, the furrow deepened, and at the same time, the number of abrasive particles increased and an oxide film formed on the wear scar's surface. The MoO_3 with a spherical shape, formed by the oxidation of MoS_2, was distributed on the surface. When the wear time increased to 30 min, the abrasive particles decreased, the wear system tended to be stable, and the wear volume increased slightly. The main wear mechanisms were oxidative wear, adhesive wear, and abrasive wear. The high-temperature environment and wear heat caused different degrees of oxidation on the wear contact surface. When the temperature was low, the MoS_2 transfer film on the wear contact surface and a small amount of dense Al_2O_3 oxide film with high load-bearing capacity played a role in reducing adhesive wear. As the temperature increased, the oxidation intensified, which produced a large amount of TiO and TiO_2. The oxidation wear occurred, and the loose oxide film reduced the wear resistance of the coating [37]. This resulted in a significant increase in the wear amount of the composite coating at 600 °C.

4. Conclusions

In this study, we prepared a TC4 composite coating reinforced by SiC and MoS_2 on a muzzle brake by laser cladding to improve its wear resistance properties and reduce the wear rate. The following results were obtained.

(1) The microstructure of the composite coating was examined. The composite coating consisted of equiaxed grains with grain sizes ranging from 102.39 to 255.31 μm on the surface, and columnar grains with a maximum length of 910 μm and width ranging from 208.78 to 321.81 μm on the bottom. The grains with mesh basket microstructure were mainly α-Ti and β-Ti phases. The strengthening phase SiC with a fine spherical shape was diffusely distributed in the coating.

(2) The microhardness and the shear properties of the composite coating were tested. The microhardness of the composite coating was about 466.75HV, which was almost

1.5 times that of the substrate due to the dispersion strengthening of SiC in the composite coating. Compared with the substrate, the plasticity of the coating, bonding interface, and HAZ increased by 14%, 25%, and 6%, respectively. The fracture mechanism was a brittle intergranular fracture. The high plasticity of the bonding interface indicated that the bonding of the coating is quite strong.

(3) The influence of temperature on wear properties was investigated, and the mechanisms were studied. When wearing against H70 brass, the friction coefficient of the composite coating gradually increased with the increase in temperature. The friction coefficient of the composite coating was about 10% lower, and the wear rate was 15%–35% lower than that of the substrate and TC4 coating when the temperature is below 400 °C. At 600 °C, they were reduced to about 30% and 55%, respectively. Compared with the substrate and TC4 coating, the wear rate of the composite coating was reduced by 15%–35% below 400 °C, and by 55% at 600 °C. The primary wear mechanism of composite coatings at room temperature was adhesive wear. At high temperatures, the main wear mechanisms were adhesive wear, oxidative wear, and slight abrasive wear. Compared with the substrate and the TC4 coating, the dense oxide film and the MoS_2 friction transfer film formed on the friction contact surface of the composite coating significantly reduced the degree of adhesive wear. We thus concluded that the composite coating exhibited enhanced anti-friction and wear resistance at high temperatures.

Author Contributions: Methodology, Q.X. and Y.Z.; Project administration, Y.L.; Supervision, H.H.; Validation, Q.X. and X.Y.; Writing—original draft, J.L.; Writing—review & editing, Y.L. and Y.C. All authors have read and agreed to the published version of the manuscript.

Funding: This research received no external funding.

Institutional Review Board Statement: Not applicable.

Informed Consent Statement: Not applicable.

Data Availability Statement: Data is contained within the article.

Conflicts of Interest: The authors declare no conflict of interest.

References

1. Tchein, G.J.; Jacquin, D.; Aldanondo, E.; Coupard, D.; Gutierrez-Orrantia, E.; Mata, F.G.; Lacoste, E. Analytical modeling of hot behavior of Ti-6Al-4V alloy at large strain. *Mater. Des.* **2019**, *161*, 114–123. [CrossRef]
2. Zhang, Z.X.; Qu, S.J.; Feng, A.H.; Shen, J. Achieving grain refinement and enhanced mechanical properties in Ti-6Al-4V alloy produced by multidirectional isothermal forging. *Mater. Sci. Eng. A* **2017**, *692*, 127–138. [CrossRef]
3. Kanyane, L.R.; Adesina, O.S.; Popoola, A.P.I.; Farotade, G.A.; Malatji, N. Microstructural evolution and corrosion properties of laser clad Ti-Ni on titanium alloy (Ti6Al4V). *Procedia Manuf.* **2019**, *35*, 1267–1272. [CrossRef]
4. Fei, W.; Chuanzhong, C.; Huijun, Y. Research status of laser cladding on titanium and its alloys: A review. *Mater. Des.* **2014**, *58*, 412–425.
5. Adebiyi, D.I.; Popoola, A.P.I.; Pityana, S.L. Phase constituents and microhardness of laser alloyed Ti-6Al-4 V alloy. *J. Laser Appl.* **2015**, *27*, S29104. [CrossRef]
6. Kim, T.S.; Park, Y.G.; Wey, M.Y. Characterization of Ti-6Al-4V alloy modified by plasma carburizing process. *Mater. Sci. Eng. A* **2003**, *361*, 275–280. [CrossRef]
7. El-Hossary, F.M.; Negm, N.Z.; Khalil, S.M.; Raaif, M. Surface modification of titanium by radio frequency plasma nitriding. *Thin Solid Films* **2006**, *497*, 196–202. [CrossRef]
8. Costa, M.Y.P.; Venditti, M.L.R.; Cioffi, M.O.H. Fatigue behavior of PVD coated Ti-6Al-4V alloy. *Int. J. Fatigue* **2011**, *33*, 759–765. [CrossRef]
9. Veronica, M.C.; Aguiara, C.; Vazquez, A.M.; Robin, A.L.M.; Barboza, M.J.R. Corrosion Behavior Analysis of Plasma-assited PVD Coated Ti-6Al-4V alloy in 2 M NaOH Solution. *Mater. Res.* **2017**, *20*, 436–444.
10. Watanabe, I.; Mcbride, M.; Newton, P.; Kurtz, K.S. Laser surface treatment to improve mechanical properties of cast titanium. *Dent. Mater.* **2009**, *25*, 629–633. [CrossRef]
11. Jeyaprakash, N.; Chehua, Y.; Tseng, S.P. Characterization and tribological evaluation of NiCrMoNb and NiCrBSiC laser cladding on near-α titanium alloy. *Int. J. Adv. Manuf. Technol.* **2020**, *106*, 2347–2361. [CrossRef]
12. Zhang, Y.L.; Li, J.; Zhang, Y.Y.; Kang, D.N. Evolution in microstructure and high-temperature oxidation behaviors of the laser-cladding coatings with the Si addition contents. *J. Alloys Compd.* **2020**, *827*, 154131. [CrossRef]

13. Maswuma, Z.; Popoola, A.P. The effect of process parameters on the hardness and wear resistance performance of laser cladded Ti-Si coatings on Ti-6Al-4V alloy. *Int. J. Microstr. Mater. Prop.* **2019**, *14*, 1–13. [CrossRef]
14. Ilawe Niranjan, V.; Zimmerman Jonathan, A.; Wong Bryan, M. Breaking Badly: DFT-D2 Gives Sizeable Errors for Tensile Strengths in Palladium-Hydride Solids. *J. Chem. Theory Comput.* **2015**, *11*, 5426–5435. [CrossRef] [PubMed]
15. Umeno, Y.; Kubo, A.; Nagao, S. Density functional theory calculation of ideal strength of SiC and GaN: Effect of multi-axial stress. *Comput. Mater. Sci.* **2015**, *109*, 105–110. [CrossRef]
16. Anand, M.M.; Shambhu, K.; Parida, S.K.; Das, A.K. Evaluation of hardness and wear behavior of laser cladding zirconia-alumina-titania ceramic top coating on pure Ti6Al4V. *Mater. Today Proc.* **2020**, *26*, 1103–1107.
17. Yitian, Z.; Mingyuan, L.; Zhiqi, F.; McCormick, P.; Tan, Q.; Mo, N.; Huang, H. Microstructures and mechanical properties of wear-resistant titanium oxide coatings deposited on Ti-6Al-4V alloy using laser cladding. *J. Eur. Ceram. Soc.* **2020**, *40*, 798–810.
18. Li, C.; Zeng, M.; Liu, C.; Wang, F.; Guo, F.; Wang, J.; Yang, Y.; Li, W.; Wang, Y. Microstructure and tribological behavior of laser cladding TiAlSi composite coatings reinforced by alumina-titania ceramics on Ti-6Al-4V alloys. *Mater. Chem. Phys.* **2020**, *240*, 122271. [CrossRef]
19. Liu, X.-B.; Zheng, C.; Liu, Y.-F.; Fan, J.-W.; Yang, M.-S.; He, X.-M.; Wang, M.-D.; Yang, H.-B.; Qi, L.-H. A comparative study of laser cladding high temperature wear-resistant composite coating with the addition of self-lubricating WS2 and WS2/(Ni-P) encapsulation. *J. Mater. Process. Technol.* **2013**, *213*, 51–58. [CrossRef]
20. Gao, Q.; Yan, H.; Qin, Y.; Zhang, P.; Chen, Z.; Guo, J.; Yu, Z. Self-lubricating Wear Resistant Composite Coating Ti-Ni+TiN+MoS_2/TiS Prepared on Ti-6Al-4V Alloy by Laser Cladding. *Chin. J. Mater. Res.* **2018**, *32*, 921–928.
21. Dai, J.; Li, S.; Zhang, H. Microstructure and wear properties of self-lubricating TiB_2-TiC_xN_y ceramic coatings on Ti-6Al-4V alloy fabricated by laser surface alloying. *Surf. Coat. Technol.* **2019**, *369*, 269–279. [CrossRef]
22. Gao, Q.; Yan, H.; Qin, Y.; Zhang, P.; Guo, J.; Chen, Z.; Yu, Z. Laser cladding Ti-Ni/TiN/TiW+TiS/WS_2 self-lubricating wear resistant composite coating on Ti-6Al-4V alloy. *Opt. Laser Technol.* **2019**, *113*, 182–191. [CrossRef]
23. Zhu, Y.; Liu, X.-B.; Liu, Y.-F.; Wang, G.; Wang, Y.; Meng, Y.; Liang, J. Development and characterization of Co-Cu/Ti_3SiC_2 self-lubricating wear resistant composite coatings on Ti6Al4V alloy by laser cladding. *Surf. Coat. Technol.* **2021**, *424*, 127664. [CrossRef]
24. Wang, K.; Du, D.; Liu, G.; Pu, Z.; Chang, B.; Ju, J. A study on the additive manufacturing of a high chromium Nickel-based superalloy by extreme high-speed laser metal deposition. *Opt. Laser Technol.* **2021**, *133*, 182–191. [CrossRef]
25. Zhu, Z. *Research and Development of New-Brand Titanium Alloys of High Performance for Aeronautical Applicaions*; Aviation Industry Press: Beijing, China, 2013.
26. Xiyan, Z.; Yongqing, Z.; Chenguang, B. *Titanium Alloys and Applications*; Chemical Industry Press: Beijing, China, 2005.
27. Zhang, T.G.; Sun, R.L. Microstructure and Properties of TC4 Layers by Laser Cladding on Ti811 Alloy Surface. *Trans. Chin. Weld. Inst.* **2019**, *40*, 41–45, 162.
28. Shizhu, W.; Ping, H. *Principles of Tribology*; Tsinghua University Publishing House Co., Ltd.: Beijing, China, 2008.
29. Ju, H.; Wang, R.; Yu, L.; Xu, J. Influence of Environmental Temperature on the Friction and Wear Properties of Ti-Mo-N Films. *J. Jiangsu Univ. Sci. Technol.* **2019**, *33*, 12–17.
30. Ding, Z.X.; Chen, W.; Wang, Q.; Cao, Z.H. Friction and Wear Characteristics of Multimodal WC-12Co Coatings Deposited by HVOF. *Tribology* **2011**, *31*, 425–430.
31. Shi, C.X.; Zhong, Q.P.; Li, C.G. *China Materials Engineering Canon*; Chemical Industry Press: Beijing, China, 2006; Volume 1.
32. Xiaofei, Y.; Faqin, X.; Yong, H.; Guoxian, Z.; Xiangqing, W. Effects of Temperature on Wear Properties and Friction Coefficient of TC4 Alloy. *Rare Metal Mater. Eng.* **2012**, *41*, 1463–1466.
33. Zehui, Y.; Nan, W.; Yongnan, C.; Long, Z.; Hong, C.; Jianmin, H.; Wenlong, S. Study on Self-lubricating MoS2/TiO2 Coating Synthesized on TC4 Surface by micro-arc oxidation. *Rare Metal Mater. Eng.* **2020**, *49*, 3195–3202.
34. Wang, K.; Du, D.; Liu, G.; Pu, Z.; Chang, B.; Ju, J. High-temperature Oxidation Behaviour of High Chromium Superalloys Additively Manufactured by Conventional or Extreme High-speed Laser Metal Deposition. *Corros. Sci.* **2020**, *176*, 108922. [CrossRef]
35. Wanzhen, G.; Zuomin, L.; Xinlei, G. *Surface Wear Resistance and Tribological Material Design*; Chemical Industry Press: Beijing, China, 2014.
36. Zhang, H.X.; Yu, H.J.; Chen, C.Z. Microstructure and Wear Resistance of Composite Coating by Laser Cladding Al/TiN on Ti-6Al-4V Substrate. *Surf. Rev. Lett.* **2015**, *22*, 1550044. [CrossRef]
37. Alam, M.O.; Haseeb, A.S.M.A. Response of Ti-6Al-4V and Ti-24Al-11Nb alloys to dry sliding wear against hardened steel. *Tribol. Int.* **2002**, *35*, 357–362. [CrossRef]

coatings

MDPI

Article

The Effect of Process-Induced Porosity on Fatigue Properties of Ti6Al4V Alloy via High-Power Direct Energy Deposition

Hang Lv [1], Zhenlin Zhang [1], Junjie Li [1], Yan Liu [1,*], Hui Chen [1,*], Huabing He [2], Jing Cheng [2] and Yong Chen [2]

1 Key Laboratory of Advanced Technologies of Materials, Ministry of Education, School of Materials Science and Engineering, Southwest Jiaotong University, Chengdu 610031, China; 13688061311@163.com (H.L.); zzl21@swjtu.edu.cn (Z.Z.); 15084989475@163.com (J.L.)

2 AVIC Chengdu Aircraft Industrial (Group) Co., Ltd., Chengdu 610073, China; hhbzdx@163.com (H.H.); chengjing_ustb@163.com (J.C.); yognchen@my.swjtu.edu.cn (Y.C.)

* Correspondence: liuyanzt@163.com (Y.L.); xnrpt@swjtu.edu.cn (H.C.)

Abstract: Titanium alloy is widely used in the aviation sector and has become the most important structural material in aircraft manufacturing. However, manufacturing a large-scale titanium component owns a high buy-to-fly ratio due to its poor machinability and expensive price. Over the last decade, the additive manufacturing (AM) technology has developed rapidly and has become a promising processing method for titanium alloys. In the future, in order to enhance processing efficiency and material utilization, a higher laser energy source is supposed to be applied in AM processes. Nevertheless, porosity within the AM fabricated part is the most important issue that restricts the application of AM technology. In the present work, two bulks with different porosities were fabricated using high-power direct energy deposition (HP-DED), and the high cycle fatigue (HCF) performance of the as-build part was tested and compared. The result shows that a lack of fusion (LOF), spherical pores and un-melted particles are the main porosity defects in the as-build part. The shape, size and location of the defect will have a synthetic effect on HCF performance. In addition, the unstable key-hole during the process will facilitate the formation of a pore, which consequently increases the porosity. Online monitoring and closed-loop feedback systems should be established for enhancing the process stability.

Keywords: high-power laser; Ti6Al4V; additive manufacturing; porosity; high cycle fatigue

1. Introduction:

Titanium alloys are widely used in the aerospace industry due to their high specific strength, superior corrosion resistance and excellent mechanical properties at elevated temperatures. Ti6Al4V alloy is extensively used to manufacture critical structural parts in the aviation sector, such as aircraft frame beam structures, compressor blades, turbine blades, undercarriages and so forth [1]. In recent years, with the rapid development of high-speed transport equipment, the demand for structural components with a higher strength, a higher reliability and a longer service life has increasingly risen.

During the last decade, additive manufacturing (AM) technologies have received extensive interest among researchers since they have various advantages against traditional processing methods in consideration of energy consumption, material utilization and design flexibility [2]. Hence, an increasing number of institutes and enterprises have established their own AM processing systems. However, there are still several issues that restrict the application of AM technologies into actual mass production, such as limited productivity, anisotropic performance, high porosity and poor plasticity.

For metal materials, there are four typical AM technologies: wire arc additive manufactured (WAAM), selective laser melting (SLM), electron beam melting (EBM) and direct energy deposition (DED). Each of them has several nomenclatures by different scholars but share an identical processing principle. It is well known that there are many large-scale titanium components in an aircraft structure. However, manufacturing large-scale titanium parts using traditional methods involves a high buy-to-fly ratio due to the costly price and poor machinability of titanium alloys. In the future, manufacturing large-scale titanium components using AM technologies is of great practical significance in terms of both ecological and economic purpose. Among the common AM technologies, direct energy deposition (DED) is the best candidate, which owns a moderate deposition efficiency and controllable processing accuracy. In the laser cladding process, it is well known that the width and height of a single-track increase with the increase of the applied laser power. Therefore, a higher laser power in the DED process will consequently lead to an enhanced processing efficiency, thus shortening the processing time.

Hitherto, there are limited open literatures in relation to fabricating titanium components using AM technologies with a laser power greater than 4 kw. Ren et al. successfully fabricated a Ti6Al4V part using direct energy deposition with a laser power of 7.6 kw. Afterwards, he conducted a series of investigations regarding fatigue crack growth (FCG) behavior [3] and low cycle fatigue (LCF) performance [4,5]. Wang et al. [6] fabricated a Ti-6.5Al-3.5Mo-1.5Zr part using laser additive manufacturing (LAM) with a laser power of 4–6 kw, and clarified the relationship between microstructural anisotropy and FCG behavior. Brandl fabricated a Ti6Al4V part using wire-feed direct energy deposition with a laser power of 3.5 kw, and systematically studied the microstructure feature [7] and tensile behavior [8] of as-fabricated alloy.

In summary, limited investigations have been carried out regarding the microstructure characterization and fatigue performance of Ti6Al4V as it is part fabricated by HP-DED. Therefore, a deep understanding for the HP-DED process is still needed in terms of microstructure evolution, defect characterization and service performance. In the process of HP-DED, the formation of a keyhole is inevitable due to a higher heat input and subsequent recoil pressure caused by metallic vapor, which is the same as laser welding. The unstable molten pool and keyhole will consequently lead to greater entrapments of gas bubbles and a higher porosity, and could thus deteriorate the fatigue performance.

Processing-induced porosity is the main issue that restricts the application of AM fabricated components [9]. There are three main porosity defects within AM fabricated parts.

A lack of fusion (LOF) occurs due to the insufficient laser energy input between layers, un-melted particles and pores with different shapes due to the entrapped gas bubbles during solidification [10]. The effect of porosity on the fatigue life has been well established through statistical analysis methods in recent years [11–17], and a corresponding modified Kitagawa-Takahashi diagram has been proposed to predict the fatigue life. In addition, Romali Biswal studied the effect of the pore location on the fatigue performance of AM fabricated Ti6Al4V through finite element analysis [18], and established the relationship between the stress concentration factor and pore shape. The result shows that the oblate spherical pore has the highest stress concentration factor. Moreover, machine learning is also a feasible approach for predicting the fatigue life of AM fabricated parts [19,20].

To the best knowledge of the authors, no investigation has been published to date in relation to the effect of processing-induced porosity on fatigue performance of Ti6Al4V parts fabricated by HP-DED. In the present work, the high cycle fatigue (HCF) performance of HP-DED fabricated Ti6Al4V parts was investigated. The influence of porosity on HCF performance is thoroughly discussed.

2. Experimental Material and Methods

2.1. Manufacturing Procedures

In an enclosed environmental chamber, an additive manufacturing (AM) system was established based on coaxially powder-blown direct energy deposition. Specifically, the AM system consisted of an argon atmosphere protecting chamber with a volume of approximately 20 cubic meters, a laser generator with a maximum output power of 6 kw, a three-axis CNC controlling mechanism, a powder feeder, an argon circulatory system, and an integrated controlling computer. The layout of the whole AM system is schematically shown in Figure 1.

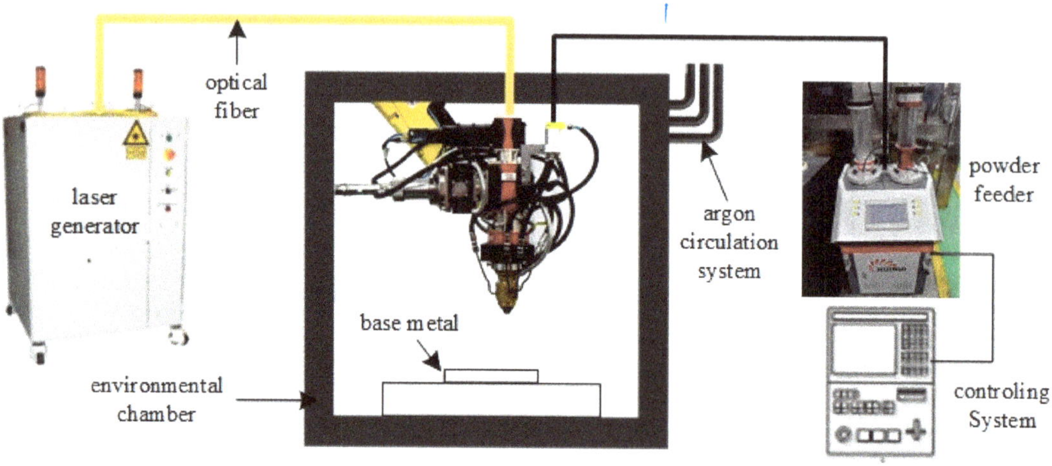

Figure 1. Diagrammatic sketch of the HP-DED system.

Ti6Al4V wrought plate and Ti6Al4V commercial powder (provided by Xi'an Bright Laser Technologies Co., Ltd., Xi'an, China) were used as a base metal and a feedstock material, respectively. Chemical composition of the powder is characterized by a high-frequency combustion infrared absorption method (C element), inert gas melting thermal conductivity method (O, N, H elements) and an inductively coupled plasma optical emission spectrometer method (other elements), which is listed in Table 1. Prior to the high-power direct energy deposition (HP-DED) process, the circulatory system was activated to pump argon gas into the chamber until the content of oxygen is below 30 ppm. The inner pressure of the chamber was kept slightly higher than the ambient pressure during the manufacturing process. The optimum processing parameters have been determined through previous single-track experiments, which are listed in Table 2. In addition, the scanning direction was kept at 90 degrees between the two successive layers for diminishing the residual stress and deformation, which is shown in Figure 2a.

Table 1. Chemical composition of Ti6Al4V feedstock powder.

Element	C	O	N	H	Al	V	Fe	Ti
Content	0.005	0.062	0.006	0.0028	6.11	3.97	0.194	Bal.

Table 2. Optimum processing parameters of the HP-DED process.

Laser Power (W)	Scanning Speed (mm/min)	Powder Feeding Rate (g/min)	Layer Thickness (mm)	Overlapping Ratio
4000	1200 (inner hatching) 800 (Contour scanning)	31	0.9	50%

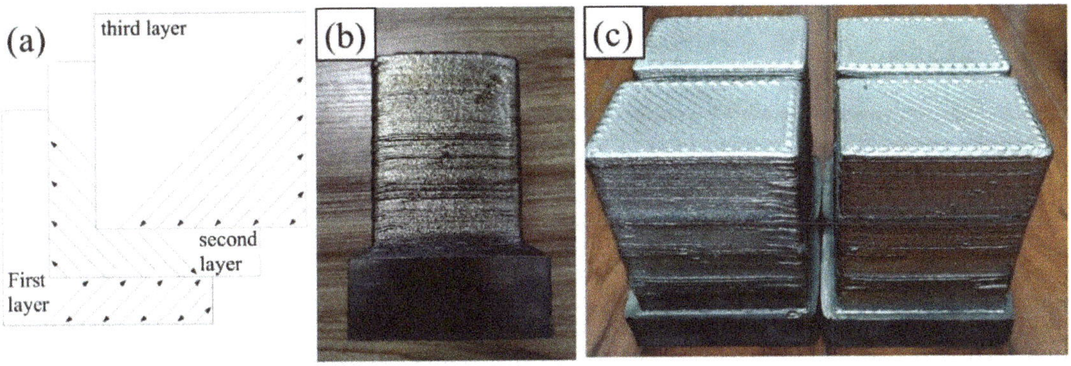

Figure 2. (**a**) Scanning strategy between layers; macroscopic morphology of (**b**) bulk1; (**c**) bulk2.

In order to study the effect of processing-induced porosity on fatigue performance, two batches of powder materials were prepared and one of them was deliberately contaminated. The powder was contaminated by being placed in the ambient environment for several hours without protection. Afterwards, two Ti6Al4V bulks were deposited using original powder and contaminated powder, respectively. For simplicity, the fabricated parts with the contaminated powder and the original powder were denoted as bulk1 and bulk2 in the remaining parts, respectively. The macroscopic morphology of the two bulks in their as-deposited state are shown in Figure 2b,c.

2.2. Microstructure Characterization, Tensile Testing and Fatigue Testing

The fabricated bulks were taken from the chamber after the HP-DED process was finished. Subsequently, the as-deposited bulks were separated from the base metal, and the additional material on the rough surface was removed by milling. The middle section in the height of the fabricated parts was chosen as the representative zone for subsequent characterization.

Specimens used for metallography characterization, tensile testing and high cycle fatigue (HCF) testing were cut by electric discharging machining. Then, the specimen for microstructure observation was prepared according to standard metallographic procedures, including mounting with resin and polishing with silica suspension. Subsequently, the scratches in the specimen were removed with a MIR-EPLAB-01 electrolytic polishing instrument, and then a solution of 5 mL HF, 12 mL HN_3O, and 83 mL H_2O was used to etch the specimen. The macroscopic morphology and microstructure were observed using a stereo microscope and an optical microscope (ZEISS Observer.A1m, Gottingen, Germany), respectively.

Axial tensile testing was conducted using an electronic universal tester (MTS-CMT5105, MTS, Eden Prairie, MN, USA) at room temperature. During the testing, an extensometer was fixed on the specimen to obtain a more accurate value of deformation. The ultimate tensile strength (UTS), yield strength (YS) and elongation (EI) were calculated by the average value of three specimens. The specimens for HCF testing were cut out along the deposition direction (i.e., height direction), with the sampling configuration and the dimension of the specimen as shown in Figure 3.

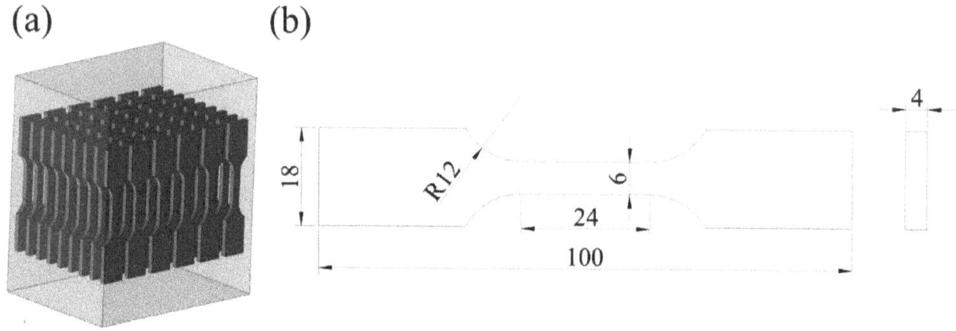

Figure 3. (**a**) Sampling configuration for HCF specimens; (**b**) dimension of the HCF specimen.

Load-controlled axial fatigue tests were conducted using a QBG-100 fatigue tester in air and ambient temperature, with a stress ratio of 0.1 and a frequency range from 80 to 90 Hz. The testing frequency was controlled by the fatigue tester automatically, which is related to the intrinsic property of the material. Referring to ISO 12107-2003, a staircase method and a group method were employed to obtain the fatigue limit and the S-N curve of the HP-DED fabricated Ti6Al4V part, respectively. The mean value and the standard deviation of the fatigue strength were calculated using the following equations.

$$\hat{\mu}^y = S_0 + d\left(\frac{A}{C} \pm \frac{1}{2}\right) \quad (1)$$

$$\hat{\sigma}_y = 1.62d(D + 0.029) \quad (2)$$

where S_0 is the minimum stress level in the fracture case, d is the stress spacing10 MPa for bulk1 and 20 MPa for bulk2, and $A,\ C,\ D$ are the parameters relating to the testing results. The high cycle fatigue (HCF) testing was interrupted above 10^7 cycles if the specimen did not fail and if the unbroken specimen was denoted as a "run-out".

For the group method, the selected stress level for bulk1 is in a range from 300 MPa to 540 MPa, with a spacing of 30 MPa. Two specimens were tested in each stress step in consideration of the limited bulk material. The selected stress level for bulk2 is in a range from 480 MPa to 600 MPa, with a spacing of 30 MPa, and the sample size of each stress step is 6–7. Finally, the S-N curve of the two bulks were depicted according to the results of the group method. The fracture surface of the failed specimens was cut by electric discharging machining (EDM) and was cleaned in anhydrous alcohol ultrasonically. Subsequently, a stereo microscope and a scanning electron microscope (ZEISS Gemini 300, Oberkochen, Germany) were used to observe the fracture morphology. The equivalent diameter of the defect and the distance from the free surface were measured under scanning electron microscope (SEM) for subsequent analysis.

3. Results and Discussion

3.1. Microstructure Characterization and Tensile Testing

The oxygen and nitrogen content of the two batches of powder were detected using the ON 3000 type analyzer (NCS TESTING TECHNOLOGY Co., Ltd., Beijing, China), and the results are shown in Table 3. It can be observed that the contaminated powder contains a considerable amount of oxygen and nitrogen elements, which validates the effect of contamination.

Table 3. Oxygen and nitrogen content of the contaminated powder and original powder.

Element	Contaminated Powder	Original Powder
Oxygen	0.10464%	0.02421%
Nitrogen	0.00785%	0.00188%

Figure 4a shows the typical microstructure of the HP-DED fabricated Ti6Al4V alloy in the XOZ plane. It can be observed that the prior β columnar grains grow epitaxially to the top area, which is the result of directional heat conduction and dissipation. The width and length of β grains can reach hundreds of microns and 4–5 mm, respectively. Meanwhile, the width and length of the prior-β grains in the vertical direction exhibit a heterogeneous distribution, and the morphology of prior-β grains is primarily governed by the thermal conditions during solidification [3]. Interestingly, the longitudinal axis of β grains is not strictly parallel to the deposition direction, but exhibits a certain angle with a maximum value of 20 degrees. This phenomenon has been extensively reported by previous investigations in selective laser melting. Prior β columnar grains with different colors can be ascribed to the different crystallographic orientations within the β columnar grains. In addition, the layer band can also be identified in the panoramic view, which is indicated by the red dash line in Figure 4a. According to Brandl [7], the layer band is the result of alternate reheating and cooling processes during the AM process. Furthermore, the spacing of the layer band does not correspond to the Z axis increment (0.9 mm) during the AM process, since it is related to the thermal history of the material.

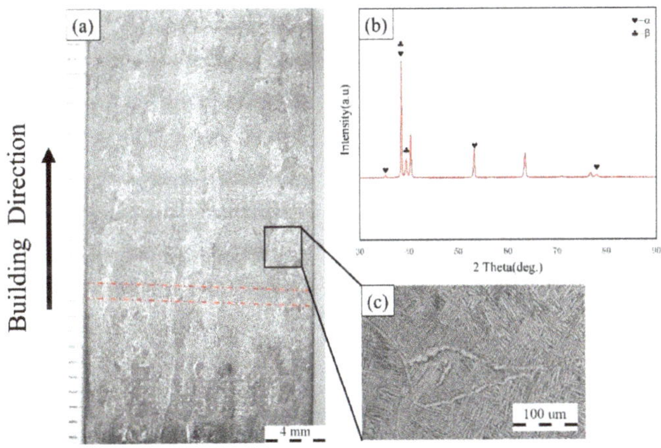

Figure 4. (**a**) Macroscopic morphology of high-power DED bulk; (**b**) XRD result of the black box area in (**a**); (**c**) enlarged view of the microstructure.

Figure 4b shows the XRD result of the microstructure in the middle area of the height direction. The phases within the HP-DED fabricated Ti6Al4V part mainly consist of a hexagonal close-packed α phase and a body-centered cubic β phase. Due to the repeated melting and the rapid cooling of layers, the basket-weave microstructure can be clearly identified and exhibits extreme fine characteristics, as shown in Figure 4c. Meanwhile, a small number of Widmanstätten α-laths could also be observed in the microstructure. Table 4 lists the results of tensile testing, and it can be concluded that the bulk1 demonstrates a higher ultimate tensile strength (UTS) and yield strength (YS) than bulk2, but owns a lower elongation after fracture. The higher UTS of bulk1 may be ascribed to the higher content of oxygen, which can intensify the effect of solution strengthening.

Table 4. The results of tensile testing for HP-DED fabricated Ti6Al4V.

State	Ultimate Tensile Strength (MPa)	Yield Strength (MPa)	Elongation (%)
Bulk1	1033.33	951.67	6.74
Bulk2	933.76	825.67	9.99

3.2. High Cycle Fatigue Testing

3.2.1. The Effect of Porosity Population

Figure 5 shows the S-N curve according to the staircase method and group method. Table 5 lists the fatigue limit strength and its standard deviation. The formula in Figure 5 shows the relationship between applied maximum stress and fatigue life, and the horizontal asymptote indicates that the fatigue life is infinite. It can be clearly observed that the HCF performance of bulk2 is superior than bulk1. The fatigue limit of bulk2 (358.57 MPa) is much higher than that of bulk1 (252.5 MPa) according to Table 5. In addition, the dispersion of the fatigue strength of bulk1 is larger than that of bulk2, suggesting that the fatigue fracture of bulk1 is more stochastic. The large scatter of HCF testing results for bulk2 can be ascribed to the presence of numerous pores, which can be validated by the subsequent fracture analysis. However, the HCF performance of both bulks is still far away when compared with its wrought counterparts [21].

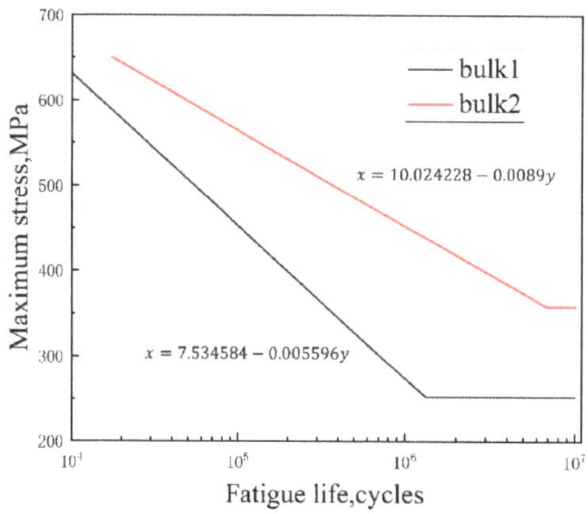

Figure 5. S-N curve for bulk1 and bulk2.

Table 5. Results of the HCF testing data.

	Mean Value of Fatigue Strength	Standard Deviation
Bulk1	252.5 MPa	64.2573 MPa
Bulk2	358.57 MPa	8.87 MPa

Figure 6 shows the typical fracture morphology of the failed specimens. It can be clearly seen that the fracture surface of bulk1 contains a large number of pores of different sizes, which is visible to the naked eye under SEM throughout the whole fracture surface. In the meantime, there is no obvious fatigue source region on the fracture surface of bulk1, which can be attributed to the innumerable pores evenly spread over the whole fracture surface. In this case, fatigue microcracks may initiate from multiple defects. Afterwards, the fatigue microcracks propagate and converge to form the main crack, which finally lead to a premature failure. Therefore, there are no typical radical streaks in the fracture surface of bulk1. In addition, a large population of pores reduced the effective bearing area of the specimen, thus meaning that the fatigue strength of bulk1 is much lower than the wrought Ti6Al4V. This detrimental effect will make the engineering design become unfeasible and excessively conservative due to the high dispersion of fatigue life.

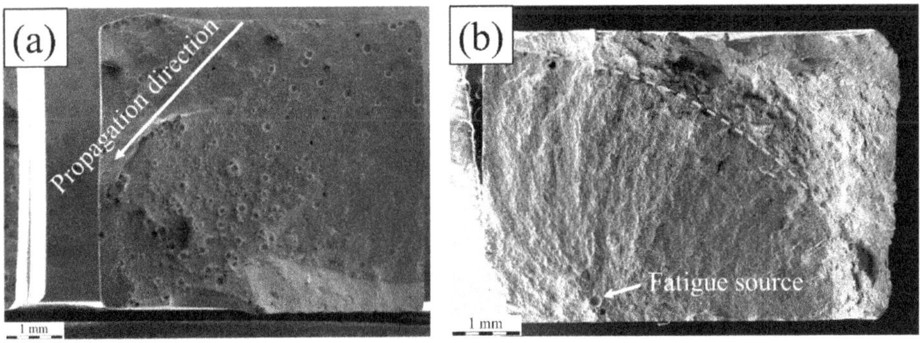

Figure 6. Typical fracture morphology of failed specimen cut from (**a**) bulk1; (**b**) bulk2.

In contrast, however, the fracture morphology of bulk2 exhibits a typical fatigue fracture feature. The crack initiated from the pore in the sub-surface and propagated to the peripheral region, as shown in Figure 6b. The boundary between the crack propagation zone and the final fracture zone is distinct, which is indicated by the yellow dash line.

In summary, the contaminated feedstock materials result in a higher porosity within the as-fabricated part, which causes severe damage to the component integrity. Considering the poor fatigue performance of bulk1, the subsequent study will focus on bulk2.

3.2.2. The Effect of Lack of Fusion

In general, there are three common defects within the AM fabricated part: a lack of fusion (LOF), spherical pores and un-melted particles. In the present work, the LOF defects have also been detected on the fracture surface, which is shown in Figure 7 (white dash circles). Meanwhile, some un-melted particles can also be found inside the LOF defects, indicating that the laser energy density is insufficient in local areas. The length of LOF defects and the corresponding fatigue life are listed in Table 6. The LOF defects have a more detrimental effect on fatigue life than spherical pores, since the defect with an irregular shape can act as a stress raiser. The LOF1 specimen has the largest LOF defect, which consequently leads to premature failure, with a fatigue life of 3800 cycles.

Figure 7. Lack of fusion defects in the fracture morphology of HP-DED.

Table 6. Statistical data of fatigue specimens with a lack of fusion.

Specimen No.	Maximum Defect Length (mm)	Stress Level (MPa)	Fatigue Life (N)
LOF1	1.83 mm	380	3800
LOF2	0.6 mm	510	37,100
LOF3	0.4 mm	480	100,200

3.2.3. The Effect of Spherical Pores

Another porosity defect within the HP-DED fabricated part is the spherical pores, which is commonly reported in other AM technologies [22,23]. Generally, pores within an AM fabricated part can be classified into three categories: surface pores, embedded pores and clustered pores. Pores that are semi-spherical in shape and are on the surface are classified into "surface pores", whereas pores with a circular morphology are classified into "embedded pores". The surface semi-spherical pores were usually formed during the subsequent machining process. In addition, clustered pores were also defect sites that caused fatigue crack initiation in the present study.

Figure 8 presents the conventional S–N format, including the information of the applied maximum stress and the corresponding fatigue life. Meanwhile, the unique symbols are made to represent the crack initiating type and defect size, respectively. The equivalent diameter of the pores was calculated under the scanning electron microscope according to the measured areas of defects. Figure 9 shows the statistical data of the equivalent diameter for different pores that cause fatigue crack initiation.

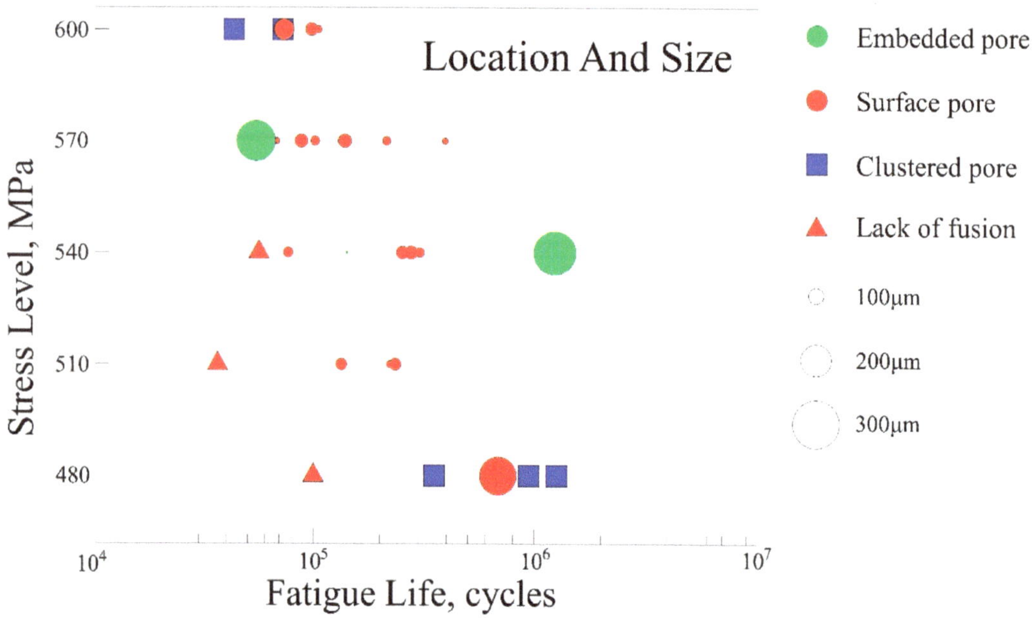

Figure 8. HCF test results with a stress ratio of 0.1, tested in air and ambient temperature. The circles are scaled to the size of the pore diameter (μm).

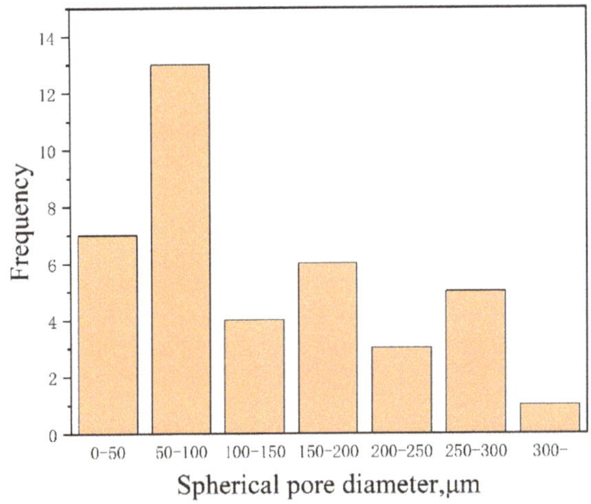

Figure 9. The statistical distribution of the equivalent diameter for pores.

The specimens with LOF defects have the lowest fatigue life compared with other specimens under the same maximum stress, which can be observed in Figure 8 (indicated by triangle symbols). Except for the specimens with LOF defects, the fatigue cracks within other specimens are all initiated from pores or clustered pores, which is marked with circle or box icons in Figure 8. Interestingly, it can be observed from Figure 8 that the average fatigue life under each stress level did not present a linear relationship with the reduction of maximum stress. For instance, the largest fatigue life under 510 MPa is 241,300 cycles, which is lower than that of 540 MPa and 570 MPa. This can be attributed to the stochastic characteristics of the pore distribution and the intrinsic fatigue nature, which is also reported in many other literatures [24].

Under a lower stress level (480 MPa), it can be observed that the fatigue crack of three specimens is initiated from clustered pores. However, a fatigue crack initiated from the clustered pores did not mean that the specimen presented a lower fatigue life than specimens with isolated pores. Other factors, such as pore size and pore location, can also have a significant influence on fatigue life. Figure 10 shows the fracture morphology of different specimens under a 480 MPa stress level, with their fatigue life (cycles) and the diameter of pores (μm) also included in the figure. It can be clearly observed that the 480-2 specimen exhibits a fracture morphology with both surface pores and clustered pores, which contributes to the lowest fatigue life. However, although the 480-5 specimen has clustered pores, it has the largest fatigue life, since the pores exhibit a smaller size and a larger distance from the free surface.

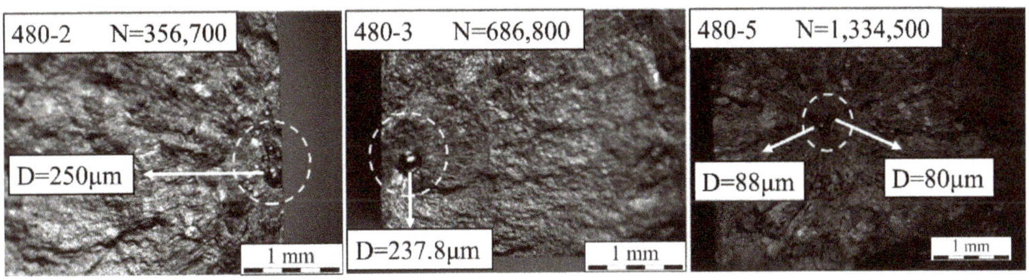

Figure 10. Fracture morphology of different specimens under a 480 MPa stress level.

It is notable that the majority of the pores which contribute to the fatigue crack initiation can be classified into surface pores, which are indicated by the red solid circles in Figure 8. The larger the porosity the as-fabricated part owns, the larger the probability of the formation of surface pores during the post-processing machining. Therefore, it can be concluded that the porosity of the bulk2 is relatively high compared with the Ti6Al4V part fabricated by the selective laser melting (SLM) method, which can realize a porosity lower than 0.5% or even a near-full dense part after a hot isostatic pressing (HIP) post treatment was applied [25]. In order to characterize the porosity of the HP-DED fabricated part quantitatively, an Archimedes drainage method was employed and the result shows that the porosity of bulk2 is 2.2%. Therefore, feasible measures for porosity inhibition should be proposed in future work.

When attention is concentrated on the fatigue failure, which is initiated from a surface pore, it can be found in Figure 8 that the larger pore usually results in a lower fatigue life, which is consistent with the fracture mechanics theory. In addition, a dispersion of fatigue life up to one order of magnitude can be observed, which has also been reported by many other researchers [11,26].

3.2.4. Discussion

As we have discussed in the previous section, a lack of fusion (LOF) has a detrimental effect on the component integrity of AM fabricated parts. The reason for the formation of LOF can be attributed to the process instability and improper processing parameters. During the HP-DED process, the actual laser output power was monitored by accessories within the laser generator and the results can be observed on the computer screen. The actual laser power during the HP-DED process fluctuated near the preset value (i.e., 4 kW) rather than being maintained at a stable value. The unstable laser power may lead to an unstable molten pool. Therefore, the unstable overlapping behavior and interlayer fusion process consequently resulted in the incomplete fusion of Ti6Al4V particles. Although the subsequent processing reheated the precious layer, the energy input was insufficient to eliminate LOF defects.

When attention was concentrated on the surface pore, it can be observed from Figure 8 that some specimens with a larger pore present a longer fatigue life than specimens with smaller pores. This can be ascribed to the classification method in the present study. According to ref. [27], if the distance between the pore and free surface is less than one diameter of the pore, the pore should also be classified as "surface pores" as well. Therefore, there are two kinds of surface pore in Figure 8 (i.e., surface pore with a non-spherical shape and sub-surface pore with a spherical shape), which is schematically shown in Figure 11. It should be mentioned that the pore in Figure 11 is not a genuine pore, but just a sketch map for illustration.

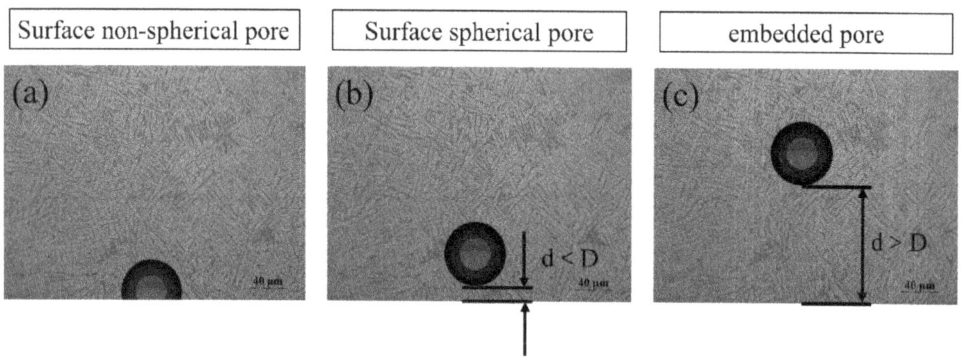

Figure 11. Schematic of pore classification. (**a**) the surface incomplete pore; (**b**)the surface spherical pore; (**c**) the embedded pore.

In previous literatures [19,28], it is well acknowledged that the surface incomplete pores (Figure 11a) have a more detrimental effect on fatigue performance than surface spherical pores (Figure 11b). Therefore, some pores have a relatively smaller size but this is located in the outmost surface, resulting in a decreased fatigue life compared with the specimens with larger ones. Under the same circumstances, pores at the surface can lead to a severer stress concentration, facilitating the initiation and propagation of fatigue cracks. In the same way, this explanation is also suitable for the HP-DED fabricated part, as smaller surface pores have been shown to be more detrimental than larger embedded pores.

In general, there are three main factors that lead to the formation of pores: feedstock material, ambient gas entrapment and the key-hole effect. A key-hole will form in the molten pool during the HP-DED process due to an elevated laser energy density, which is the same as the laser welding process. The stability of the key-hole has a significant influence on the defect formation and microstructure evolution of the HP-DED process. Heat accumulation is a typical phenomenon that occurs during the HP-DED process for titanium alloys. It is well known that the titanium alloys have a relatively low thermal conductivity compared with other metal materials. Therefore, the heat dissipation is insufficient in the interface between the base plate and the deposited part, which leads to an elevated temperature of the deposited part. Subsequently, the molten pool will become larger and deeper under the same laser power due to the pre-heated effect. Consequently, the predefined scanning strategy will become inapplicable and the key-hole will become more unstable. Figure 12 shows the schematic of the pore formation mechanism, the periodic enlargement and closure of the key-hole will lead to gas entrapment, thus increasing the porosity of the as-fabricated part. In future research, the coaxial on-line monitoring and closed-loop feedback system may be a feasible approach for solving this issue.

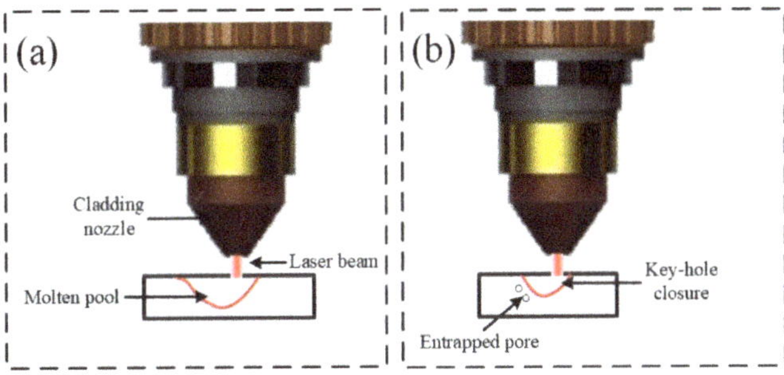

Figure 12. The formation of pores due to the unstable key-hole. (**a**) the enlargement of the key-hole; (**b**) the closure of the key-hole.

4. Conclusions

In the present work, two bulks with different porosities were fabricated by high-power direct energy deposition (HP-DED). Afterwards, the microstructure feature and high cycle fatigue (HCF) performance of the as-build part were characterized and tested. The effect of processing-induced porosity on the HCF performance was systematically discussed, and the following conclusions can be drawn:

(1) The microstructure feature of HP-DED fabricated Ti6Al4V is similar to the other AM technologies, which mainly consist of basket-weave microstructure and an α colony within elongated prior β columnar grains.

(2) The porosity of the as-build part using contaminated powder is extremely large, which consequently leads to an inferior HCF performance compared with parts fabricated with original powder. The HCF performance of both bulks is far from its wrought counterpart.

(3) The fatigue strength of the Ti6Al4V part fabricated with original powder is approximately 358.57 MPa. Lack of fusion, spherical pores and un-melted particles are the main porosity defects in the HP-DED fabricated Ti6Al4V part.
(4) The high-power laser will increase the instability of the molten pool and key-hole, which further leads to a higher porosity in the as-build part. Future research will concentrate on the processing monitoring and closed-loop controlling of the HP-DED process. In addition, feasible measures should be proposed to reduce the porosity in the as-build part of the HP-DED process.

Author Contributions: Conceptualization, H.L., Z.Z. and Y.L.; Data curation, J.L. and Y.C.; Formal analysis, Z.Z. and Y.C.; Funding acquisition, H.H.; Investigation, H.L. and J.L.; Project administration, H.H. and J.C.; Resources, H.H.; Software, J.C.; Supervision, Y.L. and H.C.; Validation, H.C.; Visualization, H.L. and Z.Z.; Writing—original draft, H.L.; Writing—review and editing, Z.Z. and Y.L. All authors have read and agreed to the published version of the manuscript.

Funding: This research was funded by the Sichuan Science and Technology Program (2021YJ0004).

Institutional Review Board Statement: Not applicable.

Informed Consent Statement: Not applicable.

Data Availability Statement: The raw/processed data required to reproduce these findings cannot be shared at this time as the data also forms part of ongoing research.

Acknowledgments: The authors would like to thank Bin Li and Tao Yi from AVIC Chengdu Aircraft Industrial (Group) Co., Ltd. (Chengdu, China) for the device support.

Conflicts of Interest: The authors declare no conflict of interest.

References

1. Xu, R.D.; Yu, H.C. Effects of Building Orientation on Fatigue Behavior of Ti-6Al-4V Alloy Produced by Selective Laser Melting. *Key Eng. Mater.* **2019**, *795*, 208–214. [CrossRef]
2. Azarniya, A.; Colera, X.G.; Mirzaali, M.J.; Sovizi, S.; Bartolomeu, F.; Weglowski, M.K.S.; Wits, W.W.; Yap, C.Y.; Ahn, J.; Miranda, G.; et al. Additive manufacturing of Ti–6Al–4V parts through laser metal deposition (LMD): Process, microstructure, and mechanical properties. *J. Alloy. Comp.* **2019**, *804*, 163–191. [CrossRef]
3. Ren, Y.; Lin, X.; Jian, Z.; Peng, H.; Huang, W. Long fatigue crack growth behavior of Ti–6Al–4V produced via high-power laser directed energy deposition. *Mater. Sci. Eng. A* **2021**, *819*, 141392. [CrossRef]
4. Ren, Y.M.; Lin, X.; Guo, P.F.; Yang, H.O.; Tan, H.; Chen, J.; Li, J.; Zhang, Y.Y.; Huang, W.D. Low cycle fatigue properties of Ti-6Al-4V alloy fabricated by high-power laser directed energy deposition: Experimental and prediction. *Int. J. Fatigue* **2019**, *127*, 58–73. [CrossRef]
5. Ren, Y.M.; Lin, X.; Yang, H.O.; Tan, H.; Chen, J.; Jian, Z.Y.; Li, J.Q.; Huang, W.D. Microstructural features of Ti-6Al-4V manufactured via high power laser directed energy deposition under low-cycle fatigue. *J. Mater. Sci. Technol.* **2021**, *83*, 18–33. [CrossRef]
6. Wang, Y.; Chen, R.; Cheng, X.; Zhu, Y.; Zhang, J.; Wang, H. Effects of microstructure on fatigue crack propagation behavior in a bi-modal TC11 titanium alloy fabricated via laser additive manufacturing. *J. Mater. Sci. Technol.* **2019**, *35*, 403–408. [CrossRef]
7. Brandl, E.; Schoberth, A.; Leyens, C. Morphology, microstructure, and hardness of titanium (Ti-6Al-4V) blocks deposited by wire-feed additive layer manufacturing (ALM). *Mater. Sci. Eng. A* **2012**, *532*, 295–307. [CrossRef]
8. Brandl, E.; Palm, F.; Michailov, V.; Viehweger, B.; Leyens, C. Mechanical properties of additive manufactured titanium (Ti–6Al–4V) blocks deposited by a solid-state laser and wire. *Mater. Des.* **2011**, *32*, 4665–4675. [CrossRef]
9. Tshephe, T.S.; Akinwamide, S.O.; Olevsky, E.; Olubambi, P.A. Additive manufacturing of titanium-based alloys-A review of methods, properties, challenges, and prospects. *Heliyon* **2022**, *8*, e09041. [CrossRef]
10. Li, E.; Zhou, Z.; Wang, L.; Shen, H.; Zou, R.; Yu, A. Particle scale modelling of melt pool dynamics and pore formation in selective laser melting additive manufacturing. *Powder Technol.* **2022**, *397*, 117012. [CrossRef]
11. Akgun, E.; Zhang, X.; Biswal, R.; Zhang, Y.; Doré, M. Fatigue of wire+arc additive manufactured Ti-6Al-4V in presence of process-induced porosity defects. *Int. J. Fatigue* **2021**, *150*, 106315. [CrossRef]
12. Biswal, R.; Zhang, X.; Syed, A.K.; Awd, M.; Ding, J.; Walther, F.; Williams, S. Criticality of porosity defects on the fatigue performance of wire + arc additive manufactured titanium alloy. *Int. J. Fatigue* **2019**, *122*, 208–217. [CrossRef]
13. Hu, Y.N.; Wu, S.C.; Wu, Z.K.; Zhong, X.L.; Ahmed, S.; Karabal, S.; Xiao, X.H.; Zhang, H.O.; Withers, P.J. A new approach to correlate the defect population with the fatigue life of selective laser melted Ti-6Al-4V alloy. *Int. J. Fatigue* **2020**, *136*, 105584. [CrossRef]
14. Li, W.B.; Pang, J.C.; Zhang, H.; Li, S.X.; Zhang, Z.F. The high-cycle fatigue properties of selective laser melted Inconel 718at room and elevated temperatures. *Mater. Sci. Eng. A* **2022**, *836*, 142716. [CrossRef]

15. Liu, F.; He, C.; Chen, Y.; Zhang, H.; Wang, Q.; Liu, Y. Effects of defects on tensile and fatigue behaviors of selective laser melted titanium alloy in very high cycle regime. *Int. J. Fatigue* **2020**, *140*, 105795. [CrossRef]
16. Deng, K.; Wei, H.; Liu, W.; Zhang, M.; Zhao, P.; Zhang, Y. Probabilistic-based random maximum defect estimation and defect-related fatigue life prediction for laser direct deposited 316L parts. *J. Mater. Process. Technol.* **2022**, *299*, 117389. [CrossRef]
17. Tang, D.; He, X.; Wu, B.; Wang, X.; Wang, T.; Li, Y. The effect of porosity defects on the mid-cycle fatigue behavior of directed energy deposited Ti-6Al-4V. *Theor. Appl. Fract. Mech.* **2022**, *119*, 103322. [CrossRef]
18. Biswal, R.; Syed, A.K.; Zhang, X. Assessment of the effect of isolated porosity defects on the fatigue performance of additive manufactured titanium alloy. *Addit. Manuf.* **2018**, *23*, 433–442. [CrossRef]
19. Luo, Y.W.; Zhang, B.; Feng, X.; Song, Z.M.; Qi, X.B.; Li, C.P.; Chen, G.F.; Zhang, G.P. Pore-affected fatigue life scattering and prediction of additively manufactured Inconel 718: An investigation based on miniature specimen testing and machine learning approach. *Mater. Sci. Eng. A* **2021**, *802*, 140693. [CrossRef]
20. Zhan, Z.; Hu, W.; Meng, Q. Data-driven fatigue life prediction in additive manufactured titanium alloy: A damage mechanics based machine learning framework. *Eng. Fract. Mech.* **2021**, *252*, 107850. [CrossRef]
21. Gerhard, W.; Boyer, R.R.; Collings, E.W. *Materials Properties Handbook: Titanium Alloys*; ASM International: Almere, The Netherlands, 1994.
22. Wang, X.; Meng, Q.; Hu, W. Continuum damage mechanics-based model for the fatigue analysis of welded joints considering the effects of size and position of inner pores. *Int. J. Fatigue* **2020**, *139*, 105749. [CrossRef]
23. Wu, Z.; Kou, H.; Chen, N.; Zhang, Z.; Qiang, F.; Fan, J.; Tang, B.; Li, J. Microstructural influences on the high cycle fatigue life dispersion and damage mechanism in a metastable β titanium alloy. *J. Mater. Sci. Technol.* **2021**, *70*, 12–23. [CrossRef]
24. Zhan, Z.; Ao, N.; Hu, Y.; Liu, C. Defect-induced fatigue scattering and assessment of additively manufactured 300M-AerMet100 steel: An investigation based on experiments and machine learning. *Eng. Fract. Mech.* **2022**, *264*, 108352. [CrossRef]
25. Rans, C.; Michielssen, J.; Walker, M.; Wang, W.; Hoen-Velterop, L.T. Beyond the orthogonal: On the influence of build orientation on fatigue crack growth in SLM Ti-6Al-4V. *Int. J. Fatigue* **2018**, *116*, 344–354. [CrossRef]
26. Ferro, P.; Fabrizi, A.; Berto, F.; Savio, G.; Meneghello, R.; Rosso, S. Defects as a root cause of fatigue weakening of additively manufactured AlSi10Mg components. *Theor. Appl. Fract. Mech.* **2020**, *108*, 102611. [CrossRef]
27. Pineau, A.; Antolovich, S.D. Probabilistic approaches to fatigue with special emphasis on initiation from inclusions. *Int. J. Fatigue* **2016**, *93*, 422–434. [CrossRef]
28. Wang, J.; Zhang, M.; Wang, B.; Tan, X.; Wu, W.J.; Liu, Y.; Bi, G.J.; Tor, S.B.; Liu, E. Influence of surface porosity on fatigue life of additively manufactured ASTM A131 EH36 steel. *Int. J. Fatigue* **2021**, *142*, 105894. [CrossRef]

coatings

MDPI

Article

Crack Behavior of Ni60A Coating Prepared by Laser Cladding on a Tilted Substrate

Bowen Shi [1,*], Xiaokai Mu [1,*], Huan Zhan [1], Linhui Deng [1], Tao Li [1] and Hongchao Zhang [2]

[1] School of Mechanical Engineering, Dalian University of Technology, Dalian 116024, China; zhan__huan@163.com (H.Z.); denglinhui321@163.com (L.D.); litao@dlut.edu.cn (T.L.)
[2] Department of Industrial, Manufacturing, and Systems Engineering, Texas Tech University, Lubbock, TX 79409, USA; hongchao18@163.com
* Correspondence: shibowen850@163.com (B.S.); muxiaokai@dlut.edu.cn (X.M.)

Abstract: Almost all of the research on cracks in laser cladding is based, at present, on a horizontal substrate, which cannot be directly applied to prepare high performance coatings, especially high hardness coatings, on tilted substrate. In this work, the influence of the substrate's tilt angle on the crack behavior of high hardness Ni60A coating is studied, based on the laser intensity distribution and energy attenuation models on the tilted substrate. Results show that the cracking rate (the crack number in the unit's cladding length) of the coating increases with the increasing substrate tilt angle, but the tilt angle has no significant influence on the crack mechanism. The different lap direction has a certain influence on the crack, and the coating prepared by downward lap cladding has a larger cracking rate due to the greater laser energy loss. Furthermore, with the increasing substrate tilt angle, the residual stress increases due to the decreased plastic flow, and the fracture strength decreases due to the decreased dilution rate, which results in the increase in the cracking rate of the Ni60A coating. This work will broaden the application of laser cladding technology on repairing complex parts such as gear and blades.

Keywords: laser cladding; substrate tilt angle; cracking rate; lap direction; fracture strength

1. Introduction

Laser cladding is a surface process utilizing a high-power laser beam to deposit a metal layer on the substrate, which has the advantages of low dilution rate, minimal deformation, good metallurgical combination with the substrate and almost unrestricted powder material. Thus, it is used in many branches of the industry, including aerospace, automobile, shipbuilding, oil and gas, transport, power engineering, et al. [1–4]. However, laser cladding is a non-equilibrium metallurgical process, which increases the inhomogeneity of the local temperature field and results in a great residual stress in the specimen. Especially, the residual tensile stress significantly reduces the static load strength and fatigue strength of the specimens, and even leads to cracking [5–7]. The residual stress and crack behavior are always the focus of laser cladding.

The fracture surface topography is a common method to study crack behavior, which is very effective in revealing the mechanism and process of cracking. By studying the fracture surface morphology of tensile specimens, Buet et al. [8] obtained the mechanical behavior of specific 2.5D SiC/SiC composites reinforced with hi-nicalon S fibers and the influence of interphase texture and thickness of pyrocarbon coatings. Macek et al. [9] studied the fracture morphology subject to bending–torsion fatigue loadings, discussing the correlation between the fractal dimension and standard surface topography parameters and loading conditions. It predicted the fracture behavior and cracking process of the material. When high hardness coatings are prepared by laser cladding, cracks are almost impossible to avoid, but there are relatively few methods to study the crack's behavior due to the high temperature process and rapid solidification. Additionally, the stress and microstructure evolution during laser cladding can be understood by analyzing fracture surface morphology. Niu et al. [10] studied the fracture morphology of Al_2O_3 and found that the crack fracture showed typical characteristics of transgranular and intergranular brittle fracture, and the liquid film formed by the eutectic structure with a low melting point was the main reason for crack initiation and propagation. Zhou et al. [11] found that the fracture surface appears to contain small dimples when a Al–Zn–Mg–Cu alloy coating with co-incorporation of Si and TiB2 is prepared by selective laser melting (SLM), which indicates ductile fracture behavior and the reason for cracking elimination. Compared with other laser cladding schemes, coaxial feeding powder laser cladding has been in widespread use because it is prone to automation and big area cladding. A variety of material powders are suitable for this scheme, such as Ni60A self-fusing alloy powder. It has been widely applied in many special purpose applications due to its outstanding wear resistance, corrosion resistance and relatively low cost. However, because of a large number of nonuniformly distributed Cr-rich precipitates and eutectic structures, Ni60A coating prepared by laser cladding has a high cracking susceptibility [12,13]. Yu et al. [14] found that thermal stress was the main source of residual stress, and the low plasticity caused by the high hard brittle phase and the excessive residual tensile stress were the main reasons for the cracking of Ni60A coating. Based on the temperature gradient in laser cladding, Fu et al. [6] established a thermal stress analytical model and analyzed the factors affecting cracking, including laser power, scanning speed, laser diameter and the properties of the material. By summarizing the existing research, Song and Lin et al. [15,16] proposed a variety of crack control methods for Ni60A coating, such as preheating the substrate, optimizing process parameters, adding pure nickel powder or rare earth elements, introducing ultrasonic vibration or an electric–magnetic compound field, et al.

Most of the research on coaxial laser cladding is, at present, based on the horizontal substrate, which is to say that the laser nozzle is always kept vertical and perpendicular to the horizontal substrate [17,18]. Actually, most surfaces of damaged parts are not flat due to deformation, and the area to be repaired for some inconvenient mobile parts is generally not in a horizontal attitude; therefore, the research based on horizontal substrates has some limitations [19]. In order to further broaden the application of coaxial laser cladding, some research had been conducted on non-horizontal substrates. Lin et al. [20] studied the influence of the substrate tilt angle on laser energy and powder concentration distribution in the vertical status of the laser nozzle by using a Gaussian model, and obtained the influence rules of the substrate tilt angle on the section morphology of cladding layers by experiments and theoretical analyses. Zhu et al. [21] investigated the influence of a large tilt angle (more than 90°) of the substrate on the morphology and microstructure of the cladding layer, in the condition that the laser head and the substrate are always perpendicular to each other, and analyzed the internal forces of the molten pool at different tilt angles. Paul et al. [22] established a theoretical model concerning the cladding layer's dimension and displacement offset relative to the center position of the molten pool, considering the influence of gravity, and obtained the optimal process parameters of prototyping. By numerical simulation, Yu et al. [23] obtained the temperature and residual stress field of the two scanning modes, i.e., along the horizontal direction and the tilt direction of the tilted substrate, respectively. Results showed that the two scanning modes had no obvious influence on the residual stress distribution of the cladding layer, and the residual stress of the latter was slightly larger.

Due to the change in laser cladding conditions, the crack behavior of the cladding layer on the tilted substrate is different from that on the horizontal substrate. However, most of the existing research has been focused on the influence of the tilted substrate on the morphology and size of the cladding layer, and there has been little research on the residual stress and crack behavior. In this work, based on the laser intensity distribution and energy attenuation models on the tilted substrate, the influence of the substrate tilt angle on the crack behavior of a Ni60A coating is studied. Furthermore, the reasons why the cracking rate changes with the substrate tilt angle are studied from the two aspects of residual stress and fracture strength of the coating. This work lays a foundation to prepare a high-performance coating, especially a high hardness coating, on a non-horizontal reference plane.

2. Materials and Methods

All laser cladding experiments in this work are carried out on a laser direct manufacturing system, which includes a YAG laser generator with a 4-kW maximum output power, a laser cladding head, a six-axis industrial robot, a negative pressure airborne powder feeding system and auxiliary equipment (water cooler, air compressor, et al.). The focusing lens of the cladding head has a focal length of 150 mm, and the spot size is 3 mm. The cladding material and the substrate used in the experiment are Ni60A self-fusing alloy powder and 45# steel, respectively. Their chemical compositions are listed in Table 1. The size of the spherical powder particle is 70–126 μm and the size of the substrate is 40 × 30 × 8 mm^3. Before laser cladding, the powder is dried in a drying chamber at 120 °C for 2 h to remove moisture in the powder. The oxide layer of the substrate is ground mechanically and dried after rinsing with acetone. The substrate is placed on special supports with different tilt angles (0°–60°) for experiments. The special supports are made of 316L stainless steel. One side of the support is horizontal and the other side is designed at a specific tilt angle to place the substrate for laser cladding. The schematic diagram of the laser cladding process on a tilted substrate is shown in Figure 1. The processing parameters are adopted as follows: laser power—1400 W; scanning speed—3 mm/s; powder feed rate—0.133 g/s; overlap rate—40%. The multi-tracks' cladding test has 10 tracks, and the length of a single track is 30 mm. Pure argon with 99.99% purity is used as the carrier gas and shielding gas, with a flow rate of 600 L/h. An infrared thermal imager with 100 Hz sampling frequency is aimed at a fixed point in the laser scanning track to detect the thermal cycle of the molten pool. The infrared thermal imager is fixed with a tripod and adopts a paraxial measurement method to measure the molten pool temperature. The schematic diagram is shown in Figure 1. The infrared thermal imager is recalibrated every time after adjusting the base substrate tilt angle.

Table 1. Chemical compositions of Ni60A powder and 45# steel substrate (wt.%).

Materials/Elements	C	B	Cr	Si	Mn	Cu	P, S	Fe	Ni
Ni60A	0.8–1.0	3.1–3.8	15.5–17.8	3.8–4.5	-	-	-	3.5~4.5	Bal.
45# steel	0.42–0.5	-	≤0.25	0.17–0.37	0.5–0.8	≤0.25	≤0.045	Bal.	≤0.25

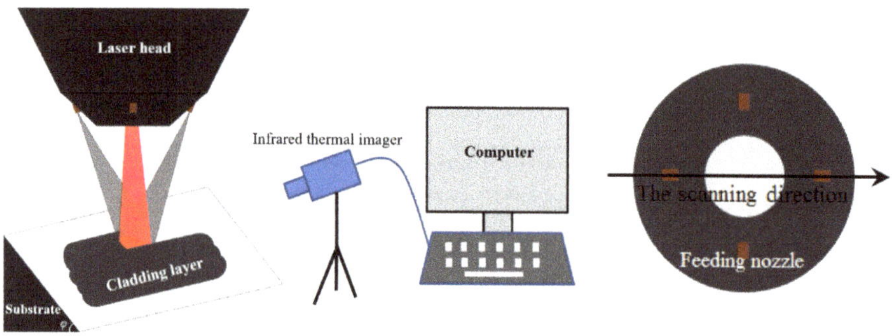

Figure 1. Schematic diagram of the laser cladding process on tilted substrate.

After laser cladding, the surface cracks of Ni60A coating are characterized by dye penetration test. The cracking rate R_c (N/mm) is assessed via the crack number (N) in unit cladding length (mm), i.e., the ratio of the total crack number to the total length of cladding tracks [24]. To reveal the microstructure of the Ni60A coating, the cross-section of specimens is chemically etched with a mixed acid consisting of 75 vol.% HCl and 25 vol.% HNO_3, and the etched time is 15 s. The microstructure is observed and analyzed using an optical microscope (OM) (ECLIPSE MA100, Nikon, Tokyo, Japan) and a scanning electronic microscope (SEM) (SUPRA 55, Carl Zeiss, Oberkochen, Germany). The main software used in this paper is MATLAB, SmartSEM Origin, Visio, ImageJ, et al.

3. Results

3.1. Laser Intensity Distribution and Energy Attenuation on Tilted Substrate

When the substrate is in a horizontal position for laser cladding, a circular laser spot with a diameter of *bc* is formed on the surface of the substrate due to the divergence angle 2β. Laser energy intensity of the spot has a Gaussian distribution, which can be expressed as [20]

$$I(x,y) = \frac{2P}{\pi R_0^2} \exp\left[-\frac{2x^2 + y^2}{R_0^2}\right] \tag{1}$$

where P is the laser power and R_0 is the characteristic radius of the laser spot on horizontal substrate. According to Equation (1), the energy intensity is the largest at the spot's center, and the energy intensity decreases with the increasing distance from the spot's center.

Assume that the influence of the concentration distribution of the powder stream on laser energy is ignored. When the substrate is tilted, the laser spot on the substrate surface becomes elliptical, as shown in Figure 2. The characteristic radius in the X direction (S direction) changes from R_0 to R, while the characteristic radius in the Y direction remains unchanged. In this case, laser energy intensity is still in a Gaussian distribution in the elliptical region, and the center of laser energy intensity on the tilted substrate shifts l compared with that on the horizontal substrate. Thus, the short axis of the elliptical spot $bc = R_0$, and the long axis

$$ef = R_0\left[\frac{\cos\beta}{\cos(\varphi+\beta)} + \frac{\cos\beta}{\cos(\varphi-\beta)}\right] \tag{2}$$

where β is one-half of the divergence angle of the laser beam and φ is the tilt angle of the substrate. The characteristic radius of the laser spot in S direction $R = \frac{1}{2}ef$. Since the scanning path always coincides with the short axis direction of the ellipse, this work only focuses on the laser intensity distribution of the tilted substrate in a direction perpendicular to the scanning path. Therefore, laser intensity distribution $I(y,s)$ on the tilted substrate is

$$I(y,s) = \frac{2P}{\pi R_0 R} \exp\left[\frac{-2s^2}{R^2} + \frac{-2y^2}{R_0^2}\right] \tag{3}$$

where s is the one-dimensional vector of the coordinate axis on the tilted substrate's surface. Thus, the relationship between the spot area on the substrate surface and the tilt angle is

$$S_\varphi = \frac{\pi R_0^2}{2}\left[\frac{\cos\beta}{\cos(\varphi+\beta)} + \frac{\cos\beta}{\cos(\varphi-\beta)}\right] \tag{4}$$

It can be seen that the spot area increases with the increasing substrate tilt angle, which is in agreement with the experimental observation. Therefore, the average laser energy intensity, i.e., the ratio of laser energy to spot area (P/S_φ), decreases with the increasing tilt angle.

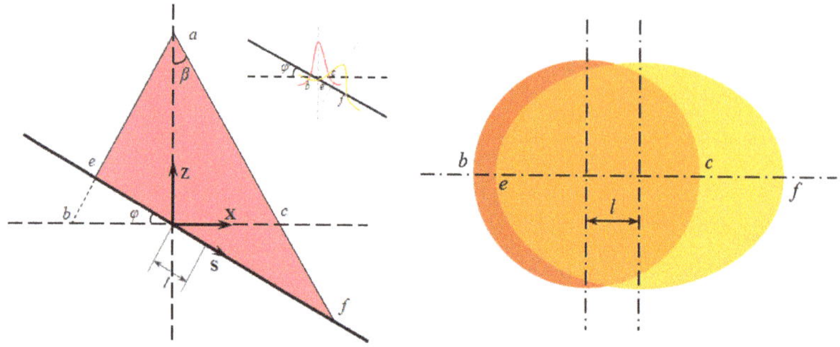

Figure 2. Gaussian distribution model diagram of laser beam.

In the laser cladding process, the laser energy will be attenuated because of the scattering and absorption of the laser beam by the powder stream. Since the laser wavelength (1.06 μm) is smaller than the size of the powder particle (70–126 μm), Mie scattering will occur [25]. Assume that the influence of the substrate's tilt angle on the reflection of laser energy is ignored. The energy attenuation rate of the laser beam is

$$A = 1 - e^{-nS_p\mu_e l_0} \tag{5}$$

where n is the powder concentration, S_p is the projection area of the powder particle, μ_e is the extinction coefficient and l_0 is the distance of the laser through the powder stream. According to Figure 2, the general tendency of l_0 increases with the increasing tilt angle.

In this work, the powder is fed by carrier gas synchronously. Assume that powder particles have the same size and uniform distribution, and the influence of gravity and the interaction between particles are ignored. Therefore, the powder particle concentration is [26]

$$n = \frac{3V_f}{4\pi r_p^3 \rho_p v_p s_n} \tag{6}$$

where V_f is the powder feed rate, r_p is the radius of the powder particle, ρ_p is the density of the powder particle, v_p is the velocity of the powder particle and s_n is the cross-sectional area of the laser nozzle.

According to the theory of projection, the projection area S_p of powder particles per unit time on the substrate is

$$S_p = \frac{n\pi r_p^2}{\cos\varphi} \tag{7}$$

Combined with Equations (5)–(7), the energy attenuation rate can be expressed as

$$A = 1 - e^{-\frac{3\mu_e V_f l_0}{4r_p\rho_p v_p s_n \cos\varphi}} \tag{8}$$

It can be seen that the energy attenuation rate increases with the increasing powder feed rate V_f and substrate tilt angle φ. Therefore, the increased tilt angle will partly reduce the utilization rate of laser power.

3.2. Influence of Tilt Angle on Cracking Rate

Unlike the multi-track one-way laser cladding on the horizontal substrate, there are two lap directions on the tilted substrate, i.e., downward lap (the latter cladding track below the previous one) and upward lap (the latter cladding track above the previous one), as shown in Figure 3. The different lap direction has a certain influence on the coating.

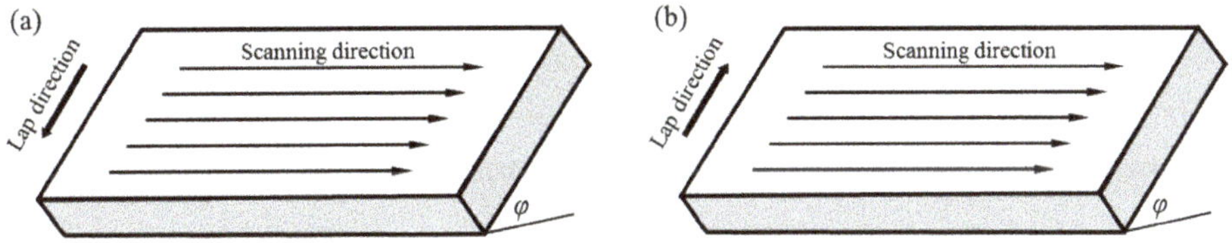

Figure 3. Lap directions: (**a**) downward; (**b**) upward.

As shown in Figure 4, the cracking rate of Ni60A coating increases with the increasing substrate tilt angle. As the tilt angle increases from 0° to 60°, the cracking rate of the coating prepared by upward lap cladding increases 3.5 times, and that of the coating prepared by downward lap cladding increases 4 times. There are two main reasons for this result. Firstly, the average laser energy intensity decreases with the increasing substrate tilt angle, which reduces the gained laser energy of the unit area coating and increases the cooling rate of the molten pool. A large cooling rate not only decreases the rise time of slag and bubble in the molten pool but also reduces the influence of high temperature plastic flow on stress relaxation [27,28], which can increase defects and residual stress of the coating, making the coating more prone to cracking. Secondly, the energy attenuation rate increases with the increasing substrate tilt angle, which reduces the dilution rate (as shown in Figure 5) due to less energy being obtained by the substrate. A low dilution rate can hardly affect the hard precipitates of chromium borides and chromium carbides [29], which reduces the plasticity of the coating and increases the susceptibility to cracking. Therefore, the cracking rate increases with the increase in the substrate tilt angle, with the same process parameters.

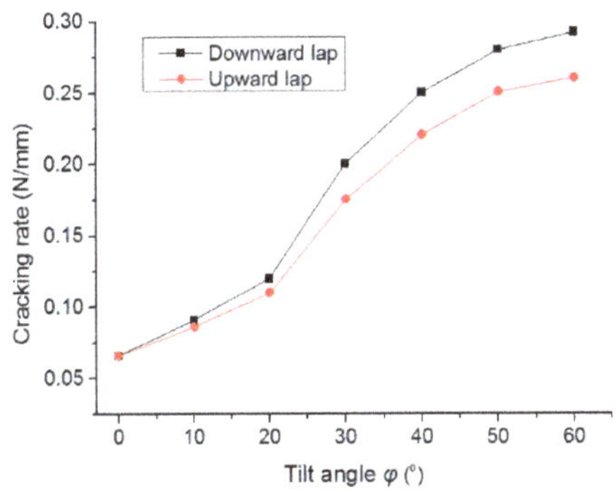

Figure 4. Influence of substrate tilt angle on cracking rate.

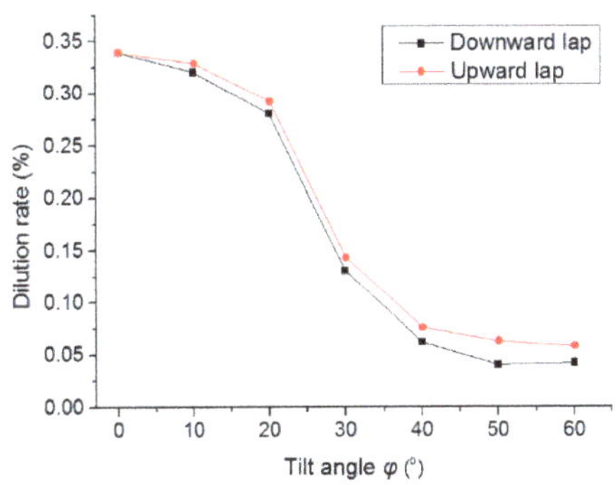

Figure 5. Influence of substrate tilt angle on dilution rate.

When the substrate tilt angle is large (>40°), the cracking rate increases slowly, as shown in Figure 4. In this case, more powder particles skim over the surface of the molten pool rather than fall into it [30]. This results in a rapid decrease in the utilization rate of powder η, which is greater than that of the average laser energy intensity, as shown in Figure 6. Thus, the gained energy of a unit mass powder particle injected into the molten pool increases. In this work, the effective mass energy, i.e., the ratio of the average laser energy intensity P/S_{φ} to effective powder feed rate V_{fe}, is used to represent this trend, where $V_{fe} = \eta V_f$. As shown in Figure 6, the effective mass energy increases significantly when the tilt angle is large, which partly counteracts the influence of the increasing substrate tilt angle and causes the cracking rate to increase slowly. While the effective mass energy is small, without an obvious change when the tilt angle is small, so the cracking rate increases rapidly at this stage.

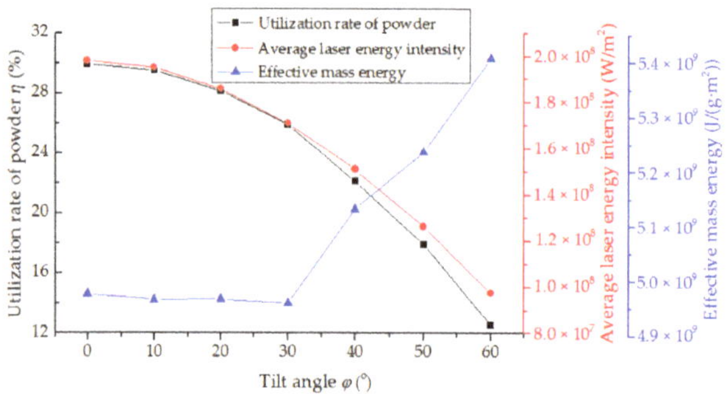

Figure 6. Influence of substrate tilt angle on three parameters.

Some SEM fracture morphology images of different angles and different lap directions are shown in Figure 7. Figure 7a shows the fracture morphology when the substrate tilt angle is 0°. For the downward lap cladding, Figure 7b,d shows the fracture morphologies when the substrate tilt angle is 20° and 50°, respectively. For the upward lap cladding, Figure 7c shows the fracture morphology when the substrate tilt angle is 30°. It can be seen that there are some obvious characteristics of non-cleavage (as at the red arrows) at all fractures. This is mainly because hard Cr-rich precipitates and eutectic and amorphous phases are the main hard phases in Ni60A coatings, and the crack in the hard precipitate continues to propagate to break the eutectic and amorphous phases under the residual tensile stress, until a complete crack is formed [31]. Therefore, the substrate tilt angle has no influence on the cracking mechanism of Ni60A coating, and the fracture mode is quasi-dissociation.

As shown in Figure 8, the EDS patterns (a)(b)(c) of the crack fracture correspond to labels (I)(II)(III) in Figure 7c, respectively. Combined with the chemical composition analysis in Table 2, labels (I)(II)(III) are hard Cr-rich precipitates, γ'(Ni) and eutectic and amorphous phases, respectively. These high hardness Cr-rich precipitates and eutectic phases provide easy routes for crack initiation and propagation.

Table 2. Chemical compositions corresponding to Figure 8 (wt.%).

Elements	Ni	Cr	Fe	Si	B	C
Figure 8a	2.55	86.91	2.17	—	6.96	1.40
Figure 8b	80.61	7.69	2.50	6.04	0.14	3.03
Figure 8c	72.14	12.92	5.63	3.08	0.42	2.88

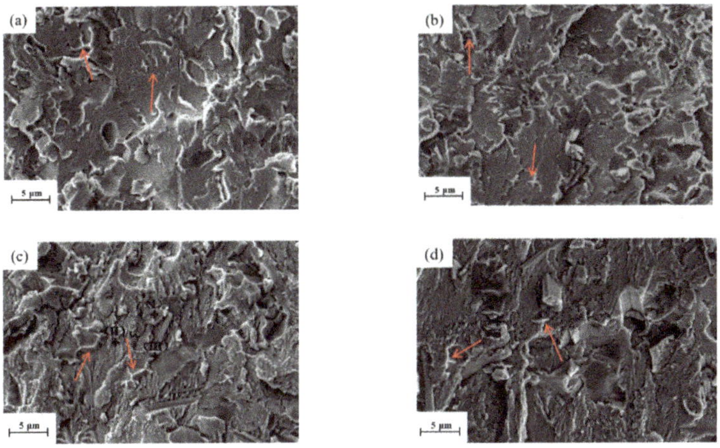

Figure 7. SEM fracture morphology: (**a**) 0°; (**b**) 20°, downward; (**c**) 30°, upward; (**d**) 50°, downward.

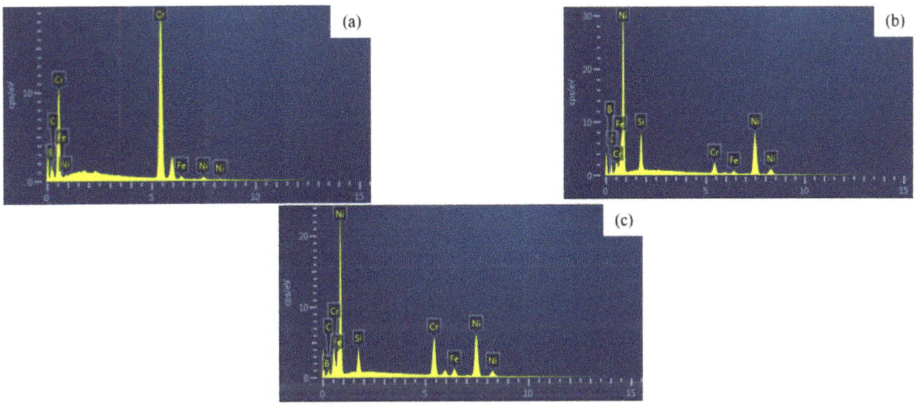

Figure 8. EDS patterns of the crack fracture in Figure 7c: (**a**) hard Cr-rich precipitates; (**b**) γ'(Ni) (**c**) eutectic and amorphous phases.

As shown in Figure 4, the cracking rate of the coating prepared by downward lap cladding is larger than that prepared by upward lap cladding with the same tilt angle, and the difference between them increases with the increasing tilt angle. The main reason is that the laser beam is blocked and reflected by part of the previous track, and the energy loss increases with the increasing size of the previous track in the laser beam range. Compared with the upward lap cladding, there is a larger size of the previous track in the laser beam range in the case of downward lap cladding, as shown in Figure 9. Hence, the downward lap cladding obviously reduces laser energy obtained by the latter track and leads to a larger cracking rate. There is a larger substrate tilt angle, a greater difference in blocking and reflection to laser energy, and a greater difference in the cracking rate between them. When the tilt angle is 60°, the cracking rate of the coating prepared by downward lap cladding is 15% higher than that of the coating prepared by upward lap cladding. In addition, in downward lap cladding, it is difficult to obtain enough powder and energy at the toe of the previous track (label the blue circle in Figure 8) due to its own obstruction of the powder stream and laser beam, which leads to defects such as holes, causing cracks [32]. This is also one of the reasons why the coating prepared by downward lap cladding has a larger cracking rate.

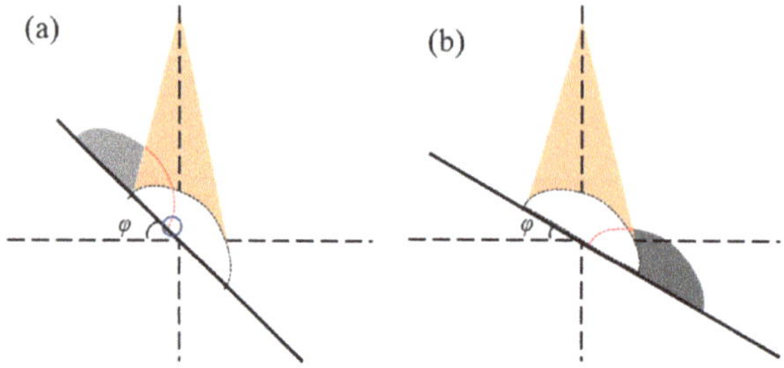

Figure 9. Laser beam range of different lap directions: (**a**) downward; (**b**) upward.

4. Discussion

The main reasons for cracks in a Ni60A coating prepared by laser cladding are excessive residual tensile stress and nonuniformly distributed Cr-rich precipitates. When the residual tensile stress exceeds the fracture strength of the coating, cracks are formed [6]. Due to the difference in thermophysical properties between the substrate and the coating, thermal stress is the main source of residual tensile stress. Compared with the width of the track (about 3.5 mm), the length of the track (30 mm) is much larger. Due to the constraint of the substrate, the stress along the cladding direction is the largest, leading to cracking of the coating.

4.1. Residual Stress of the Coating

The generation of thermal stress in the coating is mainly because of the stage at which the molten pool is cooled from the solidus curve to room temperature. Without considering the plastic deformation, as the thermal expansion coefficient of the coating is greater than that of the substrate, the thermal tensile stress along the cladding direction σ_x can be expressed as [13]

$$\sigma_x = \frac{E_c E_s h_s (\alpha_c - \alpha_s)\Delta T}{(1-\mu)(E_s h_s + E_c h_c)} \tag{9}$$

where μ is the Poisson ratio of the coating, α, E, and h are defined as the coefficient of thermal expansion (CTE), elastic modulus and thickness, respectively, the subscripts c and s represent the coating and the substrate, respectively, and ΔT is the difference between the solidus curve and the room temperature.

In the cooling process, the thermal stress and yield strength increase with the decreasing temperature. When the stress exceeds its yield strength, plastic flow will occur, which releases residual stress inside the coating. Since the coating is with good plastic at elevated temperature, the increased dwell time of the coating at the high temperature can greatly benefit the thermal plastic flow and stress relaxation of the coating. t_{8-5}, the dwell time in the temperature range of 800 °C–500 °C, is an important concept in welding, largely affecting its residual stress. Since the laser cladding is similar to the welding, it is appropriate to study the residual stress in the coating by t_{8-5}, as well [28,33,34]. In this work, t_{8-5} is applied to represent the dwell time at high temperature to assess the influence of plastic flow on stress relaxation. This influence can be expressed as $f(t_{8-5})$. As t_{8-5} is positively correlated with stress relaxation, the residual stress of the coating is

$$\sigma_{res} = \sigma_x - f(t_{8-5}) \tag{10}$$

Thus, the residual stress is negatively correlated with t_{8-5}. The value of t_{8-5} can be obtained by empirical formula or experiment. The empirical formula of t_{8-5} is [35]

$$t_{8-5} = \frac{\mu_2 P}{2\pi\lambda V_s}\left(\frac{1}{500 - T_0} - \frac{1}{800 - T_0}\right) \tag{11}$$

where μ_2 is the utilization rate of laser power, λ is the thermal conductivity and T_0 is the ambient temperature. As the energy attenuation and substrate reflection energy increase with the increasing substrate tilt angle, the utilization rate of laser power μ_2 decreases, which results in t_{8-5} decrease, according to Equation (11). The experimental data of t_{8-5} are obtained by analyzing the molten pool data gained by the infrared thermal imager. Moreover, the same trend can be found in the experimental data, as shown in Figure 10. Fitted data, the relationship between t_{8-5} and tilt angle φ can be approximatively expressed as

$$t_{8-5,\varphi} = t_{8-5,0} \times \cos\varphi \tag{12}$$

where $t_{8-5,\varphi}$ is the t_{8-5} value when the substrate tilt angle is φ and $t_{8-5,0}$ is the t_{8-5} value when the substrate is horizontal. In this work, $t_{8-5,0}$ = 0.667 s. When the substrate tilt angle increases to 60°, the value of t_{8-5} decreases 55%.

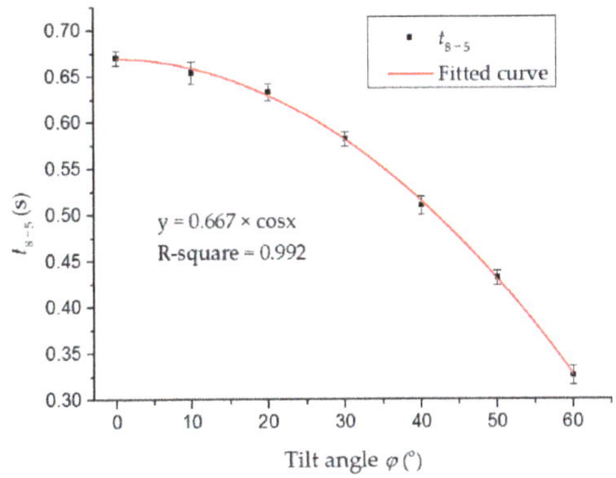

Figure 10. Influence of substrate tilt angle on t_{8-5}.

Smaller t_{8-5} will weaken the influence of plastic flow on stress relaxation, which causes the residual stress of the coating to increase with the increasing tilt angle. Therefore, the coating tends to crack more easily when the substrate tilt angle increases.

4.2. Fracture Strength of the Coating

Laser cladding is a cycle process in which powder and substrate are heated from room temperature to melting point and then cooled to room temperature in a very short time. The microscopic residual stress brought on by rapid solidification can reach the yield strength and even the tensile limit of the material, and the local stress concentration can easily cause microscopic defects or even microcracks. These damages determine the properties of the coating, including its fracture strength. According to thermodynamics and the energy balance of classical mechanics, Irwin and Orowan modified the Griffith fracture strength model in relation to material properties and damage, taking the influence of plastic work into account [36].

$$\sigma_f = \sqrt{\frac{E(2\gamma + \gamma_p)}{\pi C_z}} \tag{13}$$

where E is the elastic modulus, γ is the surface energy of the material, γ_p is the plastic work required for the microcrack or microdefect expansion unit length and C_z is the size of microcrack and microdefect. The fracture surface energy γ is $a_dE/100$, and a_d is the lattice distance of the crystal. The plastic work γ_p is 2~3 magnitude larger than γ. In the microstructure of Ni60A coating, the hard Cr-rich precipitates of Cr_7C_3, $Cr_{23}C_6$ and Cr_5B_3 are interstice compounds with complex close-packed crystal structures, and the lattice distance varies with different crystal systems. Thus, the average lattice distance value is selected for calculation [37,38]. Due to high brittleness and rapid formation, hard Cr-rich precipitates themselves can be regarded as a microscopic defect, so the maximum size of the hard precipitate is the maximum size of the microdefect in the coating.

In addition, the fracture strength of the coating is mainly affected by the dilution rate that determines the Fe content in Ni60A coating. XRD results showed that the Fe element can form a ductile phase, such as Fe_3Ni with Ni, and inhibit the hard precipitations such as chromium borides and chromium carbides, reducing the brittleness of the coating [29]. When the Fe content exceeds 40%, the precipitation of primary chromium boride is completely inhibited, and the size of the chromium carbides is greatly reduced. In this case, the hardness of the coating decreases by 40%–50%, and the main hard phase is a eutectic structure with high Fe. As shown in Figure 5, the dilution rate decreases with the increasing tilt angle, resulting in the decrease in Fe content to weaken the inhibition on hard precipitates. The size of hard Cr-rich precipitates is determined by the dilution rate of the coating, which effects the fracture strength. Any cross-section of the coating was selected, and SEM and ImageJ were used to observe and measure the maximum size of the hard precipitate. Since the dilution rates of the coatings prepared by two lap directions' cladding have the same changing trend and are close in value, the observation results of the two groups are combined in Figure 11, considering the randomness of cross-section selection. As shown in Figure 11, the maximum size of the hard precipitate (i.e., C_z) increases significantly with the increasing substrate tilt angle. When the tilt angle is 60°, C_z increases about 10 times. According to Equation (13), the fracture strength of the coating is reduced to 1/3. Combined with Sections 4.1 and 4.2, it can be found that the residual stress increases and the fracture strength of the coating decreases as the substrate tilt angle increases, which leads to the increase in the cracking rate of the coating. Therefore, it is helpful to obtain a crack-free coating by selecting the smallest substrate tilt angle possible when preparing a high hardness coating on a tilted substrate.

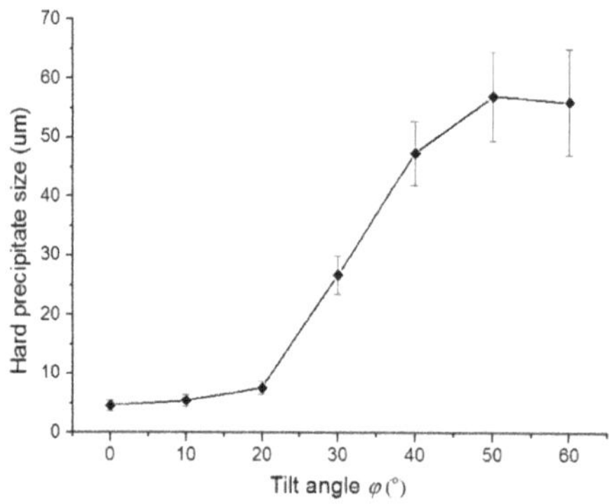

Figure 11. Influence of substrate tilt angle on hard precipitate size.

There are still some limitations in the present method. When studying the influence of the substrate tilt angle on residual stress and fracture strength, there is a lack of corresponding research method and sufficient quantitative research. In the study of residual stress, only the changing rule of residual stress can be obtained, but no specific value and distribution are obtained. Furthermore, this work does not consider the stress and structure of different parts of the cladding layer, which results in differences between study results and the actual cladding. These limitations and deficiencies will be improved on in future studies.

5. Conclusions

In this work, Ni60A coating was prepared by coaxial laser cladding on tilted 45# steel substrates, and the influence of the substrate tilt angles on the cracking rate of Ni60A coating was studied. Some of conclusions can be summarized as follows:

(1) Since the average laser energy intensity decreases and the energy attenuation rate increases with the increasing substrate tilt angle, the energy obtained by the coating decreases, resulting in the increase in cracking rate. As the tilt angle increases from 0° to 60°, the cracking rate of the coating prepared by upward lap cladding increases 3.5 times, and that of the coating prepared by downward lap cladding increases 4 times. The effective mass energy increases significantly when the tilt angle is large, which causes the cracking rate to increase slowly, while the substrate tilt angle has no significant influence on the crack mechanism of Ni60A coating, and the fracture mode is quasi-dissociation.

(2) As the laser beam is blocked and reflected by the larger size of the previous track, less laser energy is obtained by the latter track in downward lap cladding, which leads to a larger cracking rate than that in the condition of upward lap cladding. The larger the substrate tilt angle, the greater the difference in blocking and reflection of laser energy, and the greater the difference in the cracking rate between them. When the tilt angle is 60°, the cracking rate of the coating prepared by downward lap cladding is 15% higher than that of the coating prepared by upward lap cladding.

(3) When the substrate tilt angle increases to 60°, the value of t_{8-5} decreases 55%, which can reduce the influence of plastic flow on stress relaxation and increase the residual stress of the coating. Meanwhile, the maximum size of hard precipitates increases by approximatively 10 times due to the decreased dilution rate, which results in the decreased fracture strength of the coating. These are the main reasons for the increase in the cracking rate of Ni60A coating.

Author Contributions: Conceptualization, B.S. and H.Z. (Hongchao Zhang); Data curation, B.S.; Formal analysis, B.S. and L.D.; Funding acquisition, T.L. and H.Z. (Hongchao Zhang); Investigation, H.Z. (Huan Zhan) and T.L.; Methodology, X.M. and T.L.; Project administration, B.S. and T.L.; Resources, X.M. and T.L.; Supervision, T.L. and H.Z. (Hongchao Zhang); Validation, B.S. and H.Z. (Huan Zhan); Writing—original draft, B.S. and X.M.; Writing—review and editing, B.S. and L.D. All authors have read and agreed to the published version of the manuscript.

Funding: This research was funded by the National Natural Science Foundation of China, grant number 52005081, 51975099.

Institutional Review Board Statement: Not applicable.

Informed Consent Statement: Not applicable.

Data Availability Statement: Not applicable.

Conflicts of Interest: The authors declare no conflict of interest.

Nomenclature

α (1/°C)	Coefficient of thermal expansion (CTE)
a_d (nm)	The lattice distance of the crystal
β (°)	One-half of the divergence angle
C_z (mm)	The size of microcrack and microdefect
μ	Poisson ratio of the coating
μ_e (L/(mol·cm))	Extinction coefficient
E (GPa)	Elastic modulus
γ (J/m^2)	The surface energy
γ_p (J/m^2)	The plastic work required for microcrack or microdefect expansion unit length
v_p (m/s)	The velocity of powder particle

h (mm)	The thickness
ρ_p (kg/m^3)	The density of powder particle
l_0 (mm)	The distance of the laser through the powder stream
n (kg/m^3)	The powder concentration
P (W)	Laser power
R (mm)	The characteristic radius of the laser spot on horizontal substrate
R_0 (mm)	The characteristic radius of the laser spot in S direction
R_c (N/mm)	Cracking rate
r_p (mm)	The radius of powder particle
ΔT (°C)	The difference between the solidus curve and the room temperature
S_p (mm^2)	The projection area of the powder particle
S_φ (mm^2)	Laser spot area
s_n (mm^2)	The cross-sectional area of laser nozzle
t_{8-5} (s)	The dwell time in the temperature range of 800 °C–500 °C
$t_{8-5,\varphi}$ (s)	The t_{8-5} value when the substrate tilt angle is φ
$t_{8-5,0}$ (s)	The t_{8-5} value when the substrate is horizontal
η	The utilization rate of powder
V_f (g/s)	Powder feed rate
V_{fe} (g/s)	Effective powder feed rate
φ (°)	The tilt angle of the substrate

References

1. Chen, Z.H.; Li, R.F.; Gu, J.Y.; Zhang, Z.Y.; Tao, Y.W.; Tian, Y.T. Laser cladding of Ni60+17-4PH composite for a cracking-free and corrision resistive coating. *Int. J. Mod. Phys. B* **2020**, *34*, 2040042. [CrossRef]
2. Sexton, L.; Lavin, S.; Byrne, G.; Kennedy, A. Laser cladding of aerospace materials. *J. Mater. Process. Technol.* **2002**, *122*, 63–68. [CrossRef]
3. Zhang, P.R.; Liu, Z.Q. Machinability investigations on turning of Cr-Ni-based stainless steel cladding formed by laser cladding process. *Int. J. Adv. Manuf. Technol.* **2016**, *82*, 1707–1714. [CrossRef]
4. Kusinski, J.; Kac, S.; Kopia, A.; Radziszewska, A.; Rozmus-Górnikowska, M.; Major, B.; Major, L.; Marczak, J.; Lisiecki, A. Laser modification of the materials surface layer—A review paper. *Bull. Pol. Acad. Sci. Tech.* **2012**, *60*, 711–728. [CrossRef]
5. Suárez, A.; Amado, J.M.; Tobar, M.J.; Yáñez, A.; Fraga, E.; Peel, M.J. Study of residual stresses generated inside laser cladded plates using FEM and diffraction of synchrotron radiation. *Surf. Coat. Technol.* **2010**, *204*, 1983–1988. [CrossRef]
6. Fu, F.X.; Zhang, Y.L.; Chang, G.G.; Dai, J. Analysis on the physical mechanism of laser cladding crack and its influence factors. *Optik* **2016**, *127*, 200–202. [CrossRef]
7. de Oliveira, U.; Ocelík, V.; De Hosson, J.T.M. Analysis of coaxial laser cladding processing conditions. *Surf. Coat. Technol.* **2005**, *197*, 127–136. [CrossRef]
8. Buet, E.; Braun, J.; Sauder, C. Influence of texture and thickness of pyrocarbon coatings as interphase on the mechanical behavior of specific 2.5D SiC/SiC composites reinforced with hi-nicalon S fibers. *Coatings* **2022**, *12*, 573. [CrossRef]
9. Macek, W. Correlation between fractal dimension and areal surface parameters for fracture analysis after bending-torsion fatigue. *Metals* **2021**, *11*, 1790. [CrossRef]
10. Niu, F.Y.; Wu, D.J.; Yan, S.; Ma, G.Y.; Zhang, B. Process optimization for suppressing cracks in laser engineered net shaping of Al_2O_3 ceramics. *JOM* **2017**, *69*, 557–562. [CrossRef]
11. Zhou, S.Y.; Su, Y.; Wang, H.; Enz, J.; Ebel, T.; Yan, M. Selective laser melting additive manufacturing of 7xxx series Al-Zn-Mg-Cu alloy: Cracking elimination by co-incorporation of Si and TiB2. *Addit. Manuf.* **2020**, *36*, 101458. [CrossRef]
12. Arias-González, F.; del Val, J.; Comesaña, R.; Penide, J.; Lusquiños, F.; Quintero, F.; Riveiro, A.; Boutinguiza, M.; Pou, J. Fiber laser cladding of nickel-based alloy on cast iron. *Appl. Surf. Sci.* **2016**, *374*, 197–205. [CrossRef]
13. Hemmati, I.; Ocelík, V.; De Hosson, J.T.M. Effects of the alloy composition on phase constitution and properties of laser deposited Ni-Cr-B-Si coatings. *Phys. Procedia* **2013**, *41*, 302–311. [CrossRef]
14. Yu, T.; Deng, Q.L.; Zhang, W.; Dong, G.; Yang, J.G. Study on Cracking Mechanism of Laser Clad NiCrBSi Coating. *J. Shanghai Jiaotong Univ.* **2012**, *46*, 1043–1048. [CrossRef]
15. Song, J.L.; Deng, Q.L.; Ge, Z.J.; Chen, C.Y.; Hu, D.J. The cracking control technology of laser rapid forming nickel-based alloys. *J. Shanghai Jiaotong Univ.* **2006**, *40*, 548–552. [CrossRef]
16. Lin, Y.H.; Yuan, Y.; Wang, L.; Hu, Y.; Zhang, Q.L.; Yao, J.H. Effect of electric-magnetic compound field on the microstructure and crack in solidified Ni60 alloy. *Acta Metall. Sin.* **2018**, *54*, 1442–1450. [CrossRef]
17. Zhou, S.F.; Zeng, X.Y.; Hu, Q.W.; Huang, Y.J. Analysis of crack behavior for Ni-based WC composite coatings by laser cladding and crack-free realization. *Appl. Surf. Sci.* **2008**, *255*, 1646–1653. [CrossRef]
18. Marzban, J.; Ghaseminejad, P.; Ahmadzadeh, M.H.; Teimouri, R. Experimental investigation and statistical optimization of laser surface cladding parameters. *Int. J. Adv. Manuf. Technol.* **2015**, *76*, 1163–1172. [CrossRef]

19. Tan, H.; Fan, W.; Qian, Y.H.; Chen, Y.G.; Liu, S.Q.; Lin, X. Influence of inclined substrate on process characteristics of directed energy deposition. *Opt. Laser Technol.* **2020**, *129*, 106288. [CrossRef]
20. Lin, J.M.; Hwang, B.C. Coaxial laser cladding on an inclined substrate. *Opt. Laser Technol.* **1999**, *31*, 571–578. [CrossRef]
21. Zhu, G.X.; Shi, S.H.; Fu, G.Y.; Shi, J.J.; Yang, S.; Meng, W.D.; Jiang, F.B. The influence of the substrate-inclined angle on the section size of laser cladding layers based on robot with the inside-beam powder feeding. *Int. J. Adv. Manuf. Technol.* **2017**, *88*, 2163–2168. [CrossRef]
22. Paul, C.P.; Mishra, S.K.; Kumar, A.; Kukreja, L.M. Laser rapid manufacturing on vertical surfaces: Analytical and experimental studies. *Surf. Coat. Technol.* **2013**, *224*, 18–28. [CrossRef]
23. Yu, T.B.; Qiao, R.Z.; Han, J.B.; Chen, Y.D.; Ren, H. Residual stress numerical simulation of laser cladding on inclined substrate. *Hot Work. Technol.* **2020**, *49*, 75–79. [CrossRef]
24. Huang, Y.J.; Zeng, X.Y. Investigation on cracking behavior of Ni-based coating by laser-induction hybrid cladding. *Appl. Surf. Sci.* **2010**, *256*, 5985–5992. [CrossRef]
25. Ren, Z.B.; Lu, Z.W.; Zhu, H.D.; Sun, Q. Study on microsphere light scattering. *Infrared Laser Eng.* **2004**, *33*, 401–404. [CrossRef]
26. Li, H.B.; Li, T.; Wang, X.L.; Xu, H.Y.; Zhang, H.C. Study on the influence of the inclined substrate to the energy distribution of laser cladding. *Appl. Laser* **2017**, *37*, 333–339. [CrossRef]
27. Hahn, T.A. Thermal stress relaxation due to plastic flow in the fiber coating of a continuous fiber reinforced composite. *J. Compos. Mater.* **1993**, *27*, 1545–1577. [CrossRef]
28. Brückner, F.; Lepski, D.; Beyer, E. Modeling the influence of process parameters and additional heat sources on residual stresses in laser cladding. *J. Therm. Spray Technol.* **2007**, *16*, 355–373. [CrossRef]
29. Hemmati, I.; Ocelík, V.; De Hosson, J.T.M. Dilution effects in laser cladding of Ni-Cr-B-Si-C hardfacing alloys. *Mater. Lett.* **2012**, *84*, 69–72. [CrossRef]
30. Yin, Y.J.; Yang, W.; Liu, J.Y.; Zeng, D.X. The effects of substrate tilt and coaxial nozzle deflection on laser cladding morphology and powder utilization rate. *Appl. Laser* **2020**, *40*, 800–805. [CrossRef]
31. Hemmati, I.; Ocelík, V.; De Hosson, J.T.M. Advances in laser surface engineering: Tackling the cracking problem in laser-deposited Ni-Cr-B-Si-C alloys. *J. Min. Met. Mat. S.* **2013**, *65*, 741–748. [CrossRef]
32. Zhang, P.; Ma, L.; Yuan, J.P.; Cai, Z.H. Analysis of stress and strain fields of laser cladding process on ring circular orbit. *J. Shanghai Jiaotong Univ.* **2011**, *16*, 296–301. [CrossRef]
33. Wang, D.Z.; Hu, Q.W.; Zeng, X.Y. Residual stress and cracking behaviors of $Cr_{13}Ni_5Si_2$ based composite coatings prepared by laser-induction hybrid cladding. *Surf. Coat. Technol.* **2015**, *274*, 51–59. [CrossRef]
34. Dats, E.P.; Tkacheva, A.V. Technological thermal stresses in the shrink fitting of cylindrical bodies with consideration of plastic flows. *J. Appl. Mech. Tech. Phy.* **2016**, *57*, 569–576. [CrossRef]
35. Lazić, V.N.; Sedmak, A.S.; Živković, M.M.; Aleksandrović, S.M.; Čukić, R.D.; Jovičić, R.D.; Ivanović, I.B. Theoretical-experimental determining of cooling time ($t_{8/5}$) in hard facing of steels for forging dies. *Therm. Sci.* **2010**, *14*, 235–246. [CrossRef]
36. Weertman, J. Fracture mechanics: A unified view for Griffith-Irwin-Orowan cracks. *Acta Metall.* **1978**, *26*, 1731–1738. [CrossRef]
37. Gong, X.G.; Cui, C.; Yu, Q.; Wang, W.; Xu, W.W.; Chen, L.J. First-principles study of phase stability and temperature-dependent mechanical properties of (Cr, M)$_{23}C_6$ (M = Fe, Mo) phases. *J. Alloys Compd.* **2020**, *824*, 153948. [CrossRef]
38. Liu, Y.Z.; Jiang, Y.H.; Zhou, R. First-principles study on stability and mechanical properties of Cr_7C_3. *Rare Met. Mat. Eng.* **2014**, *43*, 2903–2907. [CrossRef]

coatings

MDPI

Article

Investigations on Microstructures and Properties of (Fe, Cr, W)$_7$C$_3$ Carbides by First Principles and Experiments

Chao Chen [1], Junfa Wang [1,*], Yiyuan Ge [2] and Lili Ma [2]

1 School of Materials Science and Engineering, Jiamusi University, Jiamusi 154007, China
2 School of Mechanical Engineering, Jiamusi University, Jiamusi 154007, China
* Correspondence: wangjunf2934@sina.com

Highlights:

- The structural and elastic properties of (Fe, Cr, W)$_7$C$_3$ were investigated by first principles.
- Tungsten doping can improve the ductility and the indentation modulus of (Fe, Cr)$_7$C$_3$ carbides.
- The elastic anisotropy of M$_7$C$_3$ became weaker after tungsten doping.
- A new carbide (Fe$_{3.27}$Cr$_{2.99}$W$_{0.74}$) C$_3$ was found to be a combination of mechanical properties.

Abstract: Tungsten doping is critical for the wear resistance and application of High-Chromium Cast Iron (HCCI). A series of investigations of (Fe, Cr, W)$_7$C$_3$ carbides were performed by first principles calculations and experimental analysis. The calculated results showed that with the increase in tungsten content in M$_7$C$_3$, the equilibrium cell volumes and the density gradually increased, and the formation energy of M$_7$C$_3$ carbides gradually decreased. The TEM results showed that the (Fe, Cr, W)$_7$C$_3$ carbides were (Fe$_{3.27}$Cr$_{2.99}$W$_{0.74}$) C$_3$ with a hexagonal structure after adding 2.13 wt % tungsten into laser cladding coatings of High-Chromium Cast Iron with a composition of Fe-26.8 wt % Cr-3.62 wt % C. These results from calculations and in situ nanoindentation show that tungsten doping could improve the ductility and indentation modulus of (Fe, Cr)$_7$C$_3$ carbides, and the composition of (Fe, Cr, W)$_7$C$_3$ was expected to be a high hardness and softness material. The wear test results showed that the wear resistance of tungsten-bearing HCCI was better than ordinary HCCI.

Keywords: (Fe, Cr, W)$_7$C$_3$; HCCI; first principles calculations; laser cladding coatings; nanoindentation

1. Introduction

Hypereutectic High-Chromium Cast Iron (HCCI) is an important wear resistance material, widely used in the mining and crushing field due to its easy production, low cost and excellent wear resistance [1–3]. The microstructure of HCCI primarily contains hard M$_7$C$_3$-type carbides (with a high hardness of 1300–1800 HV) and martensitic matrix [4]. There are three types of HCCI: hypoeutectic HCCI, eutectic HCCI and hypereutectic HCCI [5]. For hypereutectic HCCI, the M$_7$C$_3$ eutectic carbides solidify firstly, and could improve the hardness and the wear resistance of HCCI. HCCI could be prepared by casting [6], an Electrothermal Exial Plasma Accelerator (EAPA) [7] and so on. Heydari et al. [6] prepared HCCI with 22% Cr, 10%–12% tungsten by casting. The results showed that the coarse chromium carbides are distributed in the matrix when the carbon content was low (2.3 wt %). With the increase in carbon content, the carbides will become finer. Efremenko et al. [7] studied the effect of layered morphology and heat treatment on the microstructure and hardness after the pulsed plasma deposition of Fe-C-Cr-W coating on HCCI. Post-deposition heat treatment resulted in the precipitation of M$_7$C$_3$ carbides, the carbide precipitation led to a substantial increase in the coating hardness to 1240–1445 HV$_{0.05}$. Thus, M$_7$C$_3$ carbides are important for HCCI.

M$_7$C$_3$ (M = Fe, Cr or other element) carbides are the main hardening phases in HCCI, showing a high hardness, a high strength, a high elastic moduli and a ductile property [8]. Furthermore, M_7C_3 carbides enhance the mechanical properties and oxidation resistance of HCCI at high temperatures [9]. The effect mechanism of alloy elements on the structure and mechanical properties of M_7C_3 has been comprehensively studied by first principles calculations [10–14]. Experiments from Coronado [15] showed that M_7C_3 carbides in HCCI were rod-like single crystals with strong anisotropy. The abrasion resistance of the M_7C_3 carbides in the transverse section ([0001] direction) was higher than in the longitudinal section (non-[0001] direction) [15]. Moreover, the wear resistance of ZTA_P/HCCI composites with a honeycomb structure was higher than High-Chromium Cast Iron [16,17]. Therefore, the extensive application of M_7C_3 multi-component carbides is of great significance to future wear-resistant materials.

One method to improve the hardness and the wear resistance would be to add alloying elements in HCCI, such as vanadium, tungsten, titanium and niobium [18]. As a strong carbide-forming element, tungsten could improve the wear resistance of HCCI [19]. The extensive experiments on the effect of tungsten on microstructure and properties of HCCI have been published over the past few decades [18–21]. Cortés-Carrillo et al. [18] analyzed the effects of tungsten on the microstructure, hardness, microhardness and abrasive wear of High-Chromium Cast Iron with 17 wt % Cr. The results showed that when tungsten content was 4 wt %, the hardness of the alloys increased due to the dispersion of tungsten into the matrix and the M_7C_3 carbides. When the added amount of tungsten exceeded 4 wt %, M_2C and M_6C carbides appear in the microstructure of the alloy. The work from Lv et al. [19] found that tungsten considerably improved the wear resistance of HCCI, and the wear resistance of HCCI with 1.03 wt % W increased by 205% compared to HCCI without tungsten. Results from Anijdan et al. [20] also showed that the wear resistance of High-Chromium Cast Iron increased after adding the tungsten. The experimental results from Guerra et al. [21] verified that tungsten partially distributed in the different phases, increasing the microhardness and refining the eutectic carbides. However, the effects of tungsten on the structure and mechanical properties of M_7C_3 are not yet clear and should be further explored to reveal the mechanism.

First principles calculations are an effective way to reveal the mechanism. Zhang et al. [8] investigated the optimization of mechanical properties of $Fe_{7-x}Cr_xC_3$ carbides by first principles calculations, including Fe_6CrC_3, $Fe_4Cr_3C_3$, $Fe_3Cr_4C_3$ and $FeCr_6C_3$, but they ignored M_7C_3 carbides. Chong et al. [22] designed the anisotropic mechanical properties of M_7X_3 (M = Fe, Cr, W, Mo. X = C, B) by multi-alloying. The results showed that the ductility could be increased by doping of W + B and W + Mo without sacrificing the mechanical modulus of $Cr_4Fe_3C_3$, and the hardness of $Cr_4Fe_3C_3$ could be improved by doping of Mo + B and Mo + W + B with a finite decrease in ductility. The anisotropy of M_7C_3 carbides is important in the performance of High-Chromium Cast Iron [23]. However, the effect of tungsten on the anisotropy has not been researched and the mechanism is not clear to date. In addition, unfortunately, the stability, properties and electronic structure of M_7C_3 (M = Fe, Cr, W) carbides are seldom comprehensively investigated in the literature.

Therefore, the electronic structures, stability, chemical bonds and existing form of M_7C_3 (M = Fe, Cr and W) in Fe—26.8 wt % Cr—3.62 wt % C—2.13 wt % W High-Chromium Cast Iron were thoroughly investigated by first principles calculation experiments, which are helpful to improve the whole performance of HCCI.

2. Experimental Details

2.1. Calculation Details

First principles calculations are an effective way of investigating the electronic structures, stability and chemical bonds of M_7C_3 carbides in HCCI. The Density Functional Theory (DFT) calculation based on the pseudopotential plane-wave within the Generalized Gradient Approximation (GGA), as implemented in the Cambridge Serial Total Energy Package (CASTEP), was performed in the present work. The Broyden–Fletcher–Goldfarb–Shanno (BFGS) algorithm [24] was applied in the relaxation process of models to optimize the structures. In the structural optimization process, the maximal displacement was

11.0×10^{-3} Å, the largest force was 0.03 eV/Å and the energy change convergence value was less than 1×10^{-5} eV/atom. After convergence tests, the cutoff energy of 360 eV and k-point of $8 \times 8 \times 8$ were selected for these carbides. Although Fang et al. [25] demonstrated that orthorhombic Fe_7C_3 is more stable than an orthorhombic or hexagonal structure, Chong et al. [22] found that the crystal structure of (Cr, Fe)$_7C_3$ carbides is hexagonal with the space group of $P6_3mc$ (No. 186) by XRD and TEM analysis. Therefore, a hexagonal structure was selected in this work. Zhang et al. [8] investigated the mechanical properties of $Fe_{7-x}Cr_xC_3$ carbides based on first principles calculations, including Fe_6CrC_3, $Fe_4Cr_3C_3$, $Fe_3Cr_4C_3$ and $FeCr_6C_3$, but they neglected to calculate the other type, M_7C_3 carbides. Therefore, the properties of $Fe_5Cr_2C_3$, $Fe_2Cr_5C_3$ and $Fe_3Cr_4C_3$ (the crystal structure shown in Figure 1) were calculated by first principles, and the results were compared to the calculated results from references [8,22]. To study the influencing mechanism of tungsten on the structure and mechanical properties of M_7C_3, different amounts of tungsten were doped into M_7C_3 carbides. The M_7C_3 structures consist of nonequivalent Fe (Cr) atoms, and the lattice parameters, elastic moduli, etc., depend strongly on the substitution sites. We calculated all the formation energy of various sites of doped atoms in the unit cell, and then the unit cell obtained by the minimum formation energy was selected as the final result. Some unit cells of M_7C_3 carbides with crystal structures were built, including of $Fe_3Cr_{3.5}W_{0.5}C_3$, $Fe_3Cr_3W_1C_3$, $Fe_3Cr_{2.5}W_{1.5}C_3$ and $Fe_3Cr_2W_2C_3$, as shown in Figure 1.

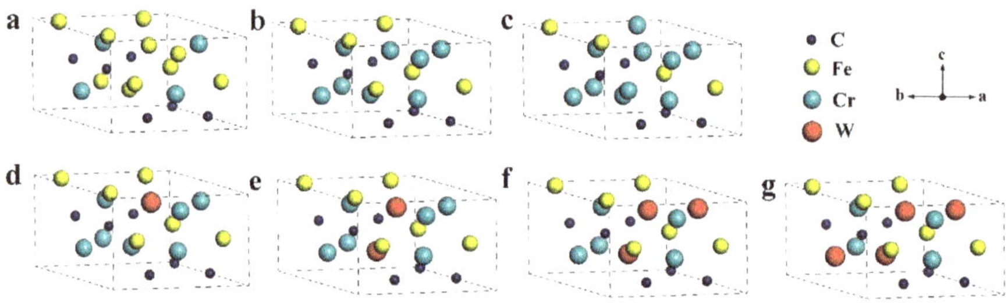

Figure 1. Unit cells of M_7C_3 carbides with crystal structures: (**a**) $Fe_5Cr_2C_3$; (**b**) $Fe_3Cr_4C_3$; (**c**) $Fe_2Cr_5C_3$; (**d**)$Fe_3Cr_{3.5}W_{0.5}C_3$; (**e**) $Fe_3Cr_3W_1C_3$; (**f**) $Fe_3Cr_{2.5}W_{1.5}C_3$; (**g**) $Fe_3Cr_2W_2C_3$.

2.2. Experimental Data

In this work, Q235 carbon steel with a chemical composition of Fe-0.18% wt % C-0.22% wt % Si-0.45% wt % Mn-0.02% wt % P-0.02% wt % S was used as the substrate, and the cladding materials were High-Chromium Cast Iron powders and tungsten powders (70 wt % W and 29 wt % Fe). The diameter of the above powders was 75–105 μm. The cladding layer was made by an IPG fiber laser system (YLS-6000) with a continuous wave, with a laser beam size of 5 mm × 5 mm. The parameters were as follows: powder feeding rate k = 15 g/min, scanning speed v = 4 mm/s, laser power P = 2000 W, flow rate of high-purity argon shielding gas = 15 L/min.

The microstructures were observed using a JEM-2100F Transmission Electron Microscope (TEM, manufacturer, city, country) with an energy-dispersive X-ray (EDX, manufacturer, city, country). TEM samples were prepared by ion milling. The wear resistance of the cladding layer was tested by a MM-200 block-on-ring wear testing machine (manufacturer, city, country, the working principle of machine can be found in reference [26]). In situ nanoindentation experiments were performed using the NanoFlip InForce 50 (manufacturer, city, country) to investigate the hardness and modulus of M_7C_3 carbides.

For the wear resistance of HCCI, the reported value of weight loss is the average of five results. All data graphs were drawn by Origin 8.0 software, and show the formation energy, mechanical properties, hardness and modulus and weight loss.

3. Results and Discussion

3.1. The Equilibrium Lattice Constants and Stability

To ensure the accuracy of the calculation, the optimized lattice constants were calculated, and the calculated results were compared to other calculated results and experimental values, as shown in Table 1. The optimized lattice constants of $Fe_3Cr_4C_3$ were 6.8358, 6.8358 and 4.5471 Å, which are similar to other calculated values [8] and experimental values [9] (error less than 1.2%). Considering the differences in experimental equipment and calculation methods, it was proved that the parameters adopted were reliable. With the increase in Cr content, the equilibrium lattice parameter *a* increased and *c* decreased, meaning that *a* decreases when *c* increases, which is consistent with the conclusion of reference [22]. With the increase in tungsten content, the equilibrium lattice parameters *a* and *b* gradually increased. The calculated result of equilibrium cell volume (185.61 Å) for $Fe_3Cr_4C_3$ was similar to the calculated values (181.09 Å) from Zhang et al. [8] (error less than 3%). With the increase in tungsten content, the equilibrium cell volumes and the density gradually increased.

Table 1. The lattice parameters of different M_7C_3 carbides.

Species	Space Group	a (Å)	b (Å)	c (Å)	Volume (Å^3)	Density (g/cm^3)
$Fe_5Cr_2C_3$	$P6_3mc$ (186)	7.0054	7.0054	4.3188	183.00	7.61
$Fe_2Cr_5C_3$	$P6_3mc$ (186)	6.7795	6.7795	4.6500	187.21	7.23
$Fe_3Cr_4C_3$	$P6_3mc$ (186)	6.8358	6.8358	4.5471	185.61	7.37
$Fe_3Cr_4C_3$ [11]	$P6_3mc$ (186)	6.8258	6.8258	4.4948	180.0	7.48
$Fe_3Cr_4C_3$ [8]	$P6_3mc$ (186)	–	–	–	181.09	–
$Fe_3Cr_4C_{3Exp.}$ [11]	$P6_3mc$ (186)	6.9	–	4.52	–	–
$Fe_3Cr_{3.5}W_{0.5}C_3$	$P6_3mc$ (186)	6.7556	6.7558	4.8070	190.12	8.34
$Fe_3Cr_3W_1C_3$	$P6_3mc$ (186)	6.8609	6.8609	4.8098	195.45	9.23
$Fe_3Cr_{2.5}W_{1.5}C_3$	$P6_3mc$ (186)	6.8998	6.9721	4.7962	199.98	10.12
$Fe_3Cr_2W_2C_3$	$P6_3mc$ (186)	6.9818	6.9818	4.8382	203.78	11.00

In order to predict whether M_7C_3 (M = Fe, Cr and W) carbides were easy to compound, the formation energy was calculated by the following equation [8]:

$$\Delta E_{M7C3\,(M = Fe,\,Cr,\,W)} = E_{tot}(M_7C_3) - xE_{tot}(Fe) - yE_{tot}(Cr) - zE_{tot}(W) - 3E_{tot}(C) \quad (1)$$

where $E_{tot}(M_7C_3)$, $E_{tot}(Fe)$, $E_{tot}(Cr)$, $E_{tot}(W)$ and $E_{tot}(C)$ are the total energies of M_7C_3 carbides, Fe, Cr, W and C systems, respectively. $\Delta E_{M7C3\,(M = Fe,\,Cr,\,W)}$ is the formation energy of M_7C_3 carbides. The formation energy can describe the relative stability of these carbides. If $\Delta E_{M7C3} > 0$, the crystal is unstable or metastable; if $\Delta E_{M7C3} < 0$, the crystal can exist stably [27]. Additionally, if the stability of carbides is better, the carbides may be synthesized easily; if the M_7C_3 carbides are not stable, it may not be easy to compound them in the experimental stage [28]. According to Equation (1), the formation energies of M_7C_3 carbides were calculated, as shown in Figure 2. Except for Fe_7C_3, the formation energies of other carbides are less than zero, indicating that they are stable crystal structures. With the increase in W and Cr content, the formation energy was increased gradually, indicating that the stability of carbides increased with W and Cr content.

3.2. Mechanical Properties

M_7C_3 carbides are the main hard phases in wear-resistant material, so the elastic of M_7C_3 compounds has an important role in the application of wear-resistant material. The elastic constants C_{ij} depend mainly on the response of the crystal to external forces, and can be calculated by the bulk, shear and Young's modulus, Poisson's ratio, etc. The bulk modulus and the shear modulus can be calculated by the following method [29]:

$$B_V = \left(\frac{1}{9}\right)[2(C_{11} + C_{12}) + 4C_{13} + C_{33}] \quad (2)$$

$$G_V = (\frac{1}{30})\,[(C_{11} + C_{12} + 2C_{33} - 4C_{13} + 12C_{44} + 12C_{66}) \quad (3)$$

$$B_R = [(C_{11} + C_{12})\,C_{33} - 2C_{13}^2]/(C_{11} + C_{12} + 2C_{33} - 4C_{13}) \quad (4)$$

$$G_R = (\frac{5}{2})\,[(C_{11} + C_{12})\,C_{33} - 2C_{13}^2]\,C_{44}C_{66}/\{3B_VC_{44}C_{66} + [C_{11} + C_{12})\,C_{33} - 2C_{13}^2]\,(C_{44} + C_{66})\} \quad (5)$$

$$B = (B_V + B_R)/2 \quad (6)$$

$$G = (G_V + G_R)/2 \quad (7)$$

$$E = 9BG/(3B + G) \quad (8)$$

$$\sigma = (3B - 2G)/(6B + 2G) \quad (9)$$

where B and G are the bulk modulus and the shear modulus, respectively. E and σ are Young's modulus and Poisson's ratio, respectively. C_{11}, C_{22} and C_{33} represent the uniaxial deformation along the $[\bar{1}2\bar{1}0]$, $[2\bar{1}\bar{1}0]$ and [0001] directions, respectively. C_{44}, C_{55} and C_{66} represent the pure shear deformation on $(\bar{1}2\bar{1}0)$, $(2\bar{1}\bar{1}0)$ and (0001) crystal planes, respectively. C_{12} is the shear deformation on the $(\bar{1}100)$ crystal plane along the $[1\bar{1}00]$ direction.

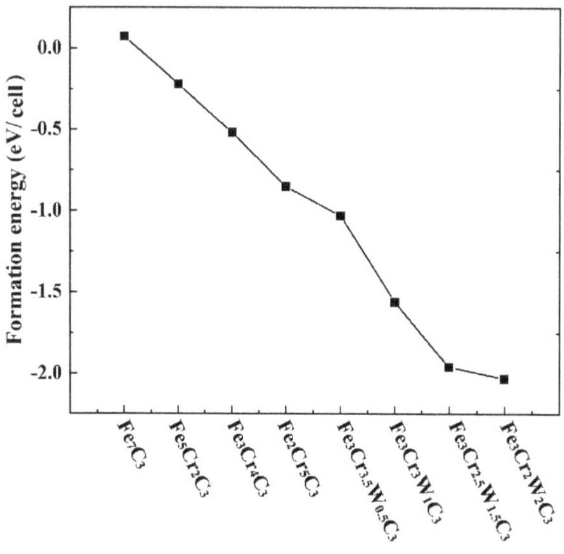

Figure 2. Calculated formation energy of M_7C_3 carbides (ΔE for, eV/cell).

For the hexagonal phase, the mechanical stability criteria are given by [29]:

$$C_{44} > 0,\ C_{11} > |C_{12}|,\ (C_{11} + 2C_{12})C_{33} > 2C_{13} \quad (10)$$

The calculated elastic constants of M_7C_3 carbides satisfied the above formula, indicating that these M_7C_3 carbides were stable structures. Table 2 lists the calculated elastic constants (C_{ij}) of different M_7C_3 carbides, which showed good agreement with the data of other researchers [22]. The largest C_{11}, C_{12} and C_{13} appeared on $Fe_3Cr_2W_2C_3$, but the largest C_{66} and C_{44} appeared on $Fe_2Cr_5C_3$. Because the hardness of M_7C_3 carbides is large, the relatively simple semi-empirical equation of hardness can be used, which is [30]:

$$H_V = 1.92\, K^{1.137} G^{0.708},\ K = G/B \quad (11)$$

Figure 3 shows the mechanical properties of M_7C_3 carbides. The shear modulus and Young's modulus of $Fe_2Cr_5C_3$ are the largest. With the addition of tungsten, the shear modulus and Young's modulus began to decrease, indicating that the alloying tungsten decreases the strength of M_7C_3. Poisson's ratio can reflect the flexibility; the larger Poisson's ratio is, the softer the material is. Poisson's ratio of $Fe_3Cr_{2.5}W_{1.5}C_3$ is the largest, indicating that the $Fe_3Cr_{2.5}W_{1.5}C_3$ compound is the softest. Pugh's modulus ratio B/G and Cauchy pressure (C_{11}–C_{44}) can explain the ductile/brittle properties. When G/B is smaller than 0.571, the M_7C_3 has good ductile property [8]. With the addition of tungsten, G/B is reduced, indicating that tungsten can improve the toughness of M_7C_3. Cauchy pressure of M_7C_3 carbides increased with tungsten content, as shown in Figure 3f, which indicates that tungsten doping can improve the ductility of (Fe, Cr)$_7C_3$ carbides. The composition of (Fe, Cr, W)$_7C_3$ is expected to be a high hardness and softness material; thus, the synthesis of this carbide would be of great interest.

Table 2. The elastic constants (C_{ij}, GPa) of different M_7C_3 carbides.

Species	C_{11}	C_{33}	C_{44}	C_{12}	C_{13}	C_{66}
$Fe_5Cr_2C_3$	534.6	560.8	123.8	193.5	255.4	152.3
$Fe_2Cr_5C_3$	562.1	559.2	157.9	159.2	258.3	193.0
$Fe_3Cr_4C_3$	549.2	531.6	122.5	179.6	248.1	179.6
$Fe_3Cr_4C_3$ [9]	550.7	532.8	110.6	185.2	229.0	182.7
$Fe_3Cr_{3.5}W_{0.5}C_3$	546.1	523.6	100.9	251.5	246.0	166.5
$Fe_3Cr_3WC_3$	543.2	504.9	99.1	236.8	243.6	155.8
$Fe_3Cr_3WC_3$ [9]	565.5	415.4	87.8	252.0	249.2	156.8
$Fe_3Cr_{2.5}W_{1.5}C_3$	578.9	403.4	92.3	276.5	279.5	149.1
$Fe_3Cr_2W_2C_3$	581.5	377.5	90.9	300.1	288.1	166.9

Anisotropy of M_7C_3 effects the wear resistance of HCCI [8]. Therefore, the analysis of anisotropy is important for understanding the properties of M_7C_3 carbides. The anisotropy of Young's modulus for hexagonal M_7C_3 carbides along different directions is expounded by the 3D surface contours, following [22]:

$$1/E = (1 - l_1^2)^2\, S_{11} + l_1^4 S_{33} + l_1^2\, (1 - l_1^2)\, (2S_{13} + S_{44}) \quad (12)$$

where S_{ij} is the elastic compliance constant. S_{ij} is the inverse matrix of C_{ij}. $l_3 = \cos\varphi$ is the directional cosine. The results are shown in Figure 4. In Figure 4a,b,d,e, it can be seen that $Fe_5Cr_2C_3$, $Fe_3Cr_4C_3$, $Fe_3Cr_{3.5}W_{0.5}C_3$ and $Fe_3Cr_3W_1C_3$ showed strong anisotropy. As the Cr content increased, the elastic anisotropy of (Fe, Cr)$_7C_3$ became weaker. Similarly, as the tungsten content increased, the elastic anisotropy of (Fe, Cr, W)$_7C_3$ became weaker, as shown in Figure 4f,g. The results indicate that alloying could weaken the elastic anisotropy of M_7C_3 carbides, which is in agreement with other calculated results [22].

3.3. The Electronic Structures

As is known, the properties of carbides are associated with electronic states [31]. In order to gain some insight into the nature of bonding in M_7C_3 carbides, the band structure and the Partial Density of States (PDOS) were calculated. Figure 5 shows the band structure of M_7C_3, and the dashed line indicates the Fermi level. All calculated M_7C_3 carbides in this work exhibited metallic properties. Figure 6 shows the Total Density of States (TDOS) and the Partial Density of States (PDOS). PDOS can analyze the electronic hybridization states quantitatively with chemical bonding. $Fe_5Cr_2C_3$, $Fe_3Cr_4C_3$, Fe_2Cr_5C and $Fe_3Cr_3W_1C_3$ carbides show large shifts, because the up and down spin channels are not symmetric. However, $Fe_3Cr_{3.5}W_{0.5}C_3$, $Fe_3Cr_{2.5}W_{1.5}C_3$ and $Fe_3Cr_2W_2C_3$ are symmetric, which may indicate non-magnetic characteristics of these carbides. In Figure 6, the DOS on both sides of the Fermi level were determined mainly by the Fe-d and Cr-d. From −15 to −10 eV, TDOS mainly consists of C-s orbit, but from −7.5 to 10 eV, TDOS of M_7C_3 mainly consists of W-d, Fe-d, Cr-d and C-p orbit, as shown in Figure 6d–g. From −7.5 to −2.5 eV, TDOS mainly consists of Fe-d, W-d and Cr-d orbit, and their peak shape and peak intensity are similar, indicating that there is orbital hybridization. The d orbit of Cr, Fe, W and the p orbit of carbon have strong hybridization, suggesting a covalent bond between the Cr, Fe, W atom and carbon atom. The total electron density distribution is shown in Figure 7. For an ideal single crystal, the magnitude of the mechanical modulus is related to the chemical bond strength. In Figure 7a, Fe-C-Cr and Fe-C-Fe covalent chains can be observed in $Fe_3Cr_4C_3$ carbides, which is in agreement with other calculated results [8]. In Figure 7b, Cr-W-C, Fe-C-Cr and Fe-C-W covalent chains can be observed in $Fe_3Cr_2W_2C_3$ carbides, explaining the decrease in formation energy after adding tungsten.

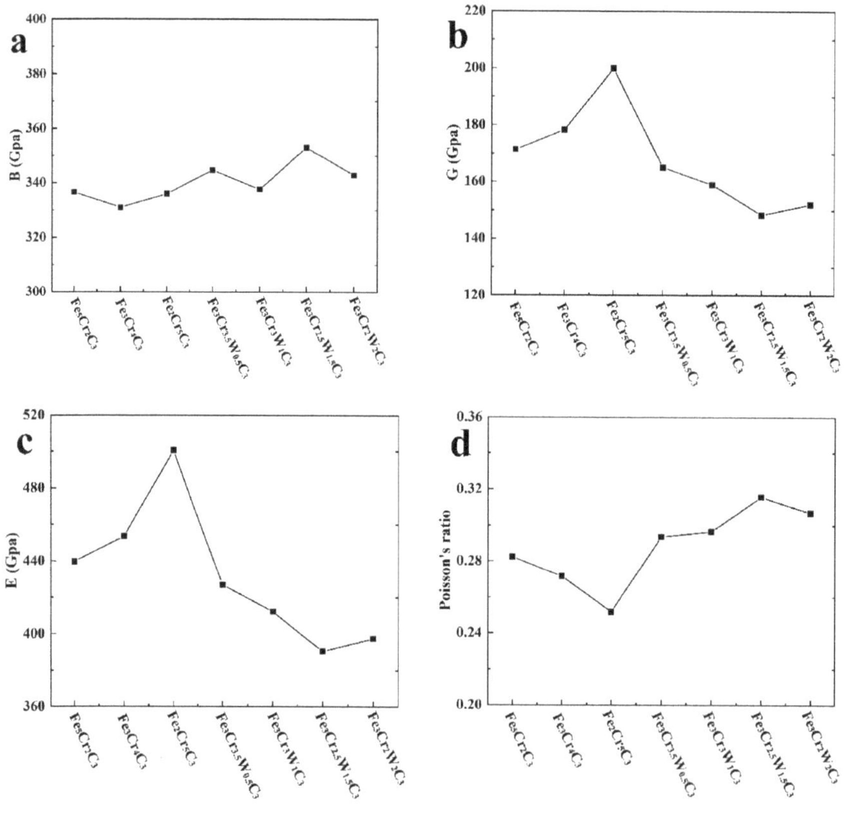

Figure 3. *Cont.*

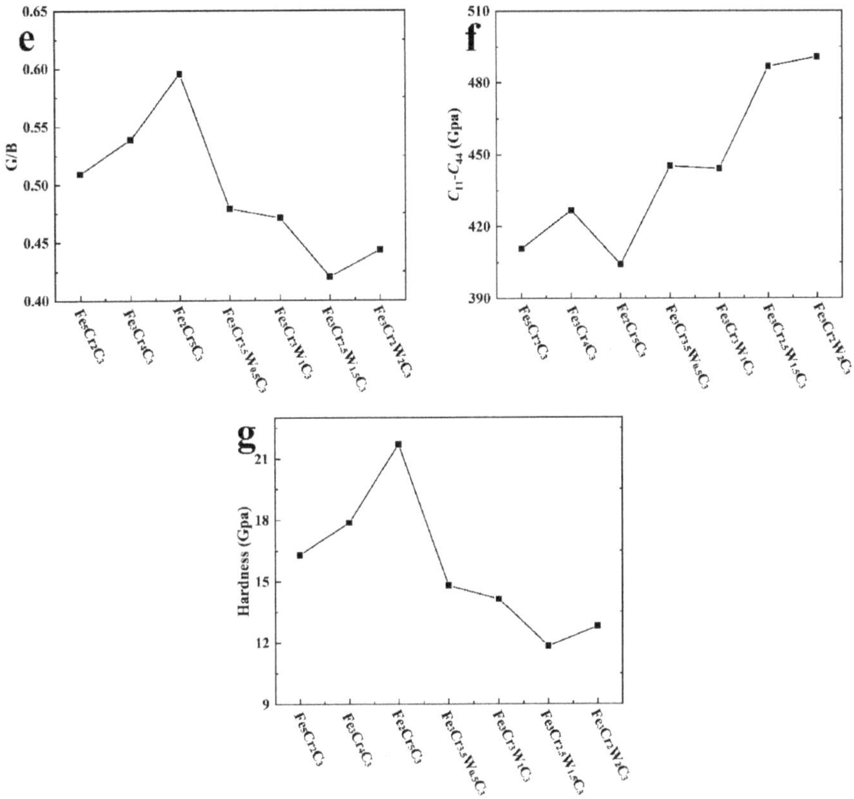

Figure 3. The mechanical properties of M_7C_3: (**a**) the bulk modulus (B); (**b**) the shear modulus (G); (**c**) Young's modulus (E); (**d**) Poisson's ratio (σ); (**e**) Pugh's modulus ratio G/B; (**f**) Cauchy pressure C_{12}–C_{44}; (**g**) calculated hardness.

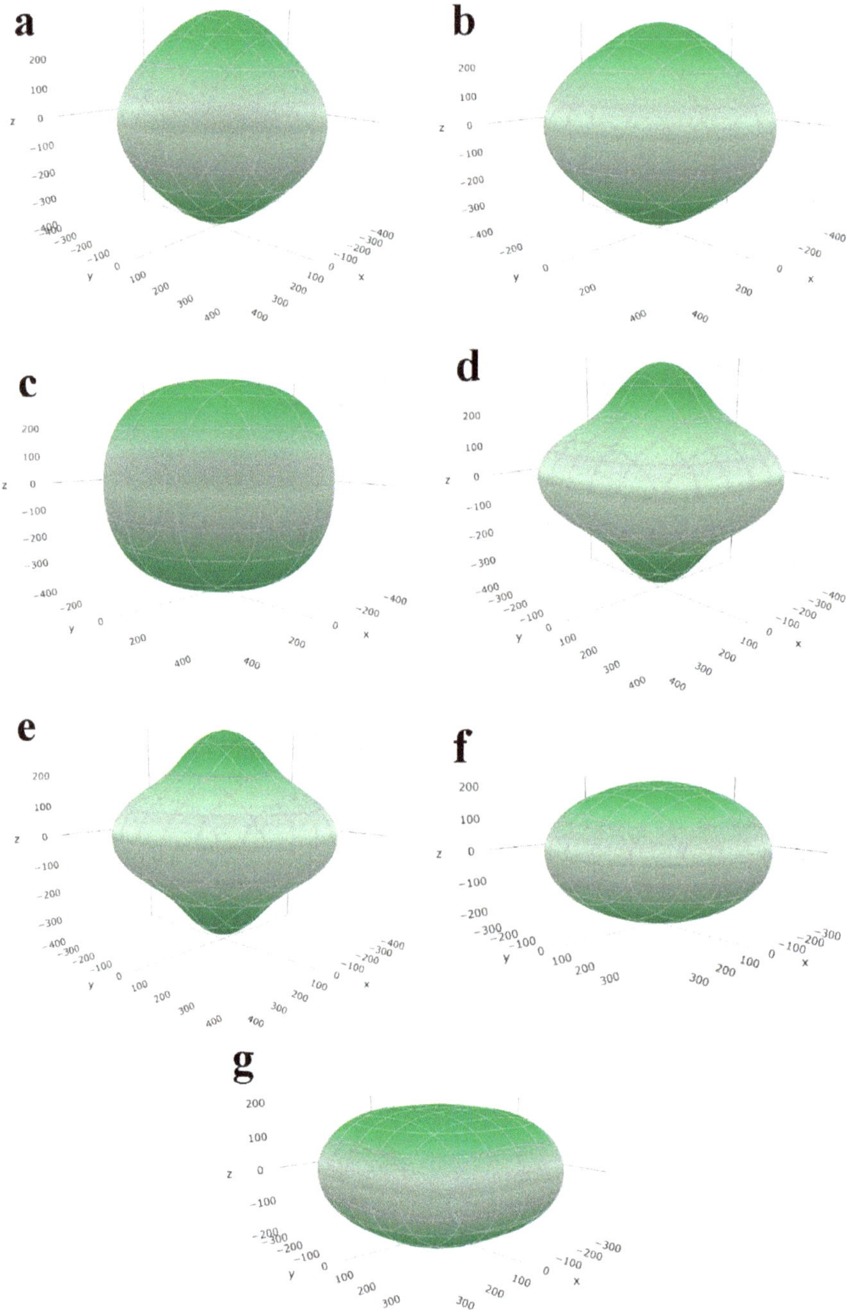

Figure 4. Surface constructions of Young's modulus of M_7C_3 carbides: (**a**) $Fe_5Cr_2C_3$; (**b**) $Fe_3Cr_4C_3$; (**c**) $Fe_2Cr_5C_3$; (**d**)$Fe_3Cr_{3.5}W_{0.5}C_3$; (**e**) $Fe_3Cr_3W_1C_3$; (**f**) $Fe_3Cr_{2.5}W_{1.5}C_3$; (**g**) $Fe_3Cr_2W_2C_3$.

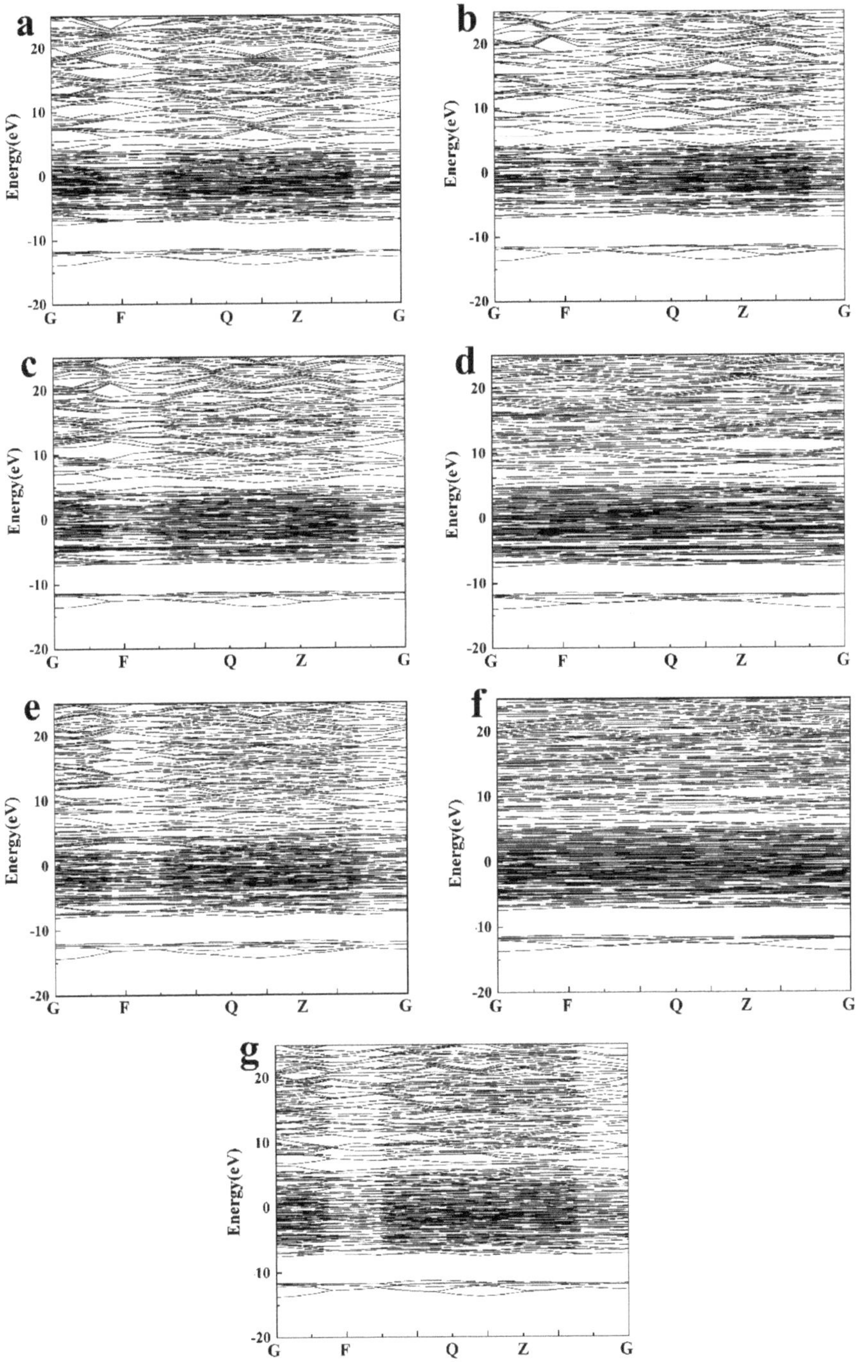

Figure 5. Calculated band structure: (**a**) $Fe_5Cr_2C_3$; (**b**) $Fe_3Cr_4C_3$; (**c**) $Fe_2Cr_5C_3$; (**d**) $Fe_3Cr_{3.5}W_{0.5}C_3$; (**e**) $Fe_3Cr_3W_1C_3$; (**f**) $Fe_3Cr_{2.5}W_{1.5}C_3$; (**g**) $Fe_3Cr_2W_2C_3$.

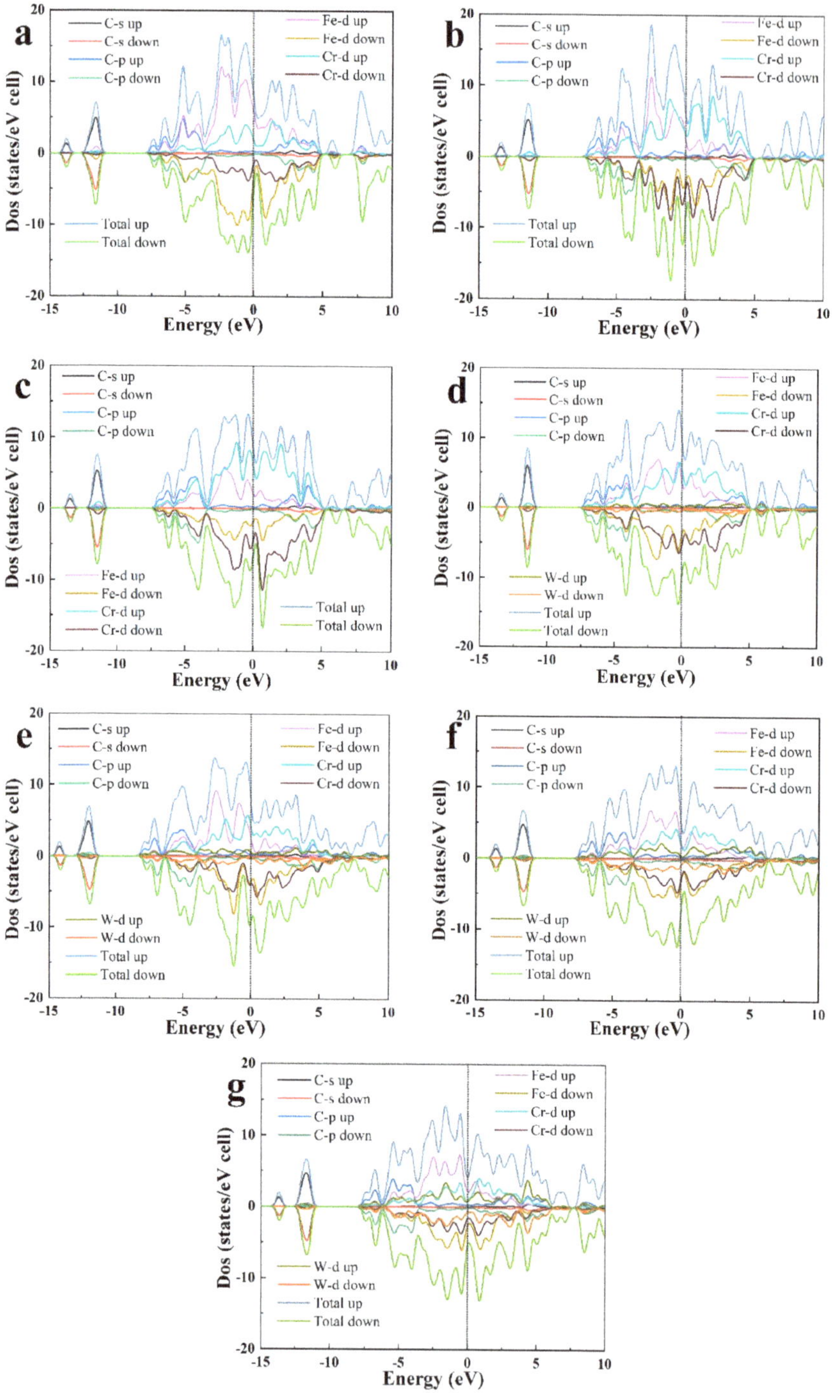

Figure 6. Calculated PDOS: (**a**) $Fe_5Cr_2C_3$; (**b**) $Fe_3Cr_4C_3$; (**c**) $Fe_2Cr_5C_3$; (**d**)$Fe_3Cr_{3.5}W_{0.5}C_3$; (**e**) $Fe_3Cr_3W_1C_3$; (**f**) $Fe_3Cr_{2.5}W_{1.5}C_3$; (**g**) $Fe_3Cr_2W_2C_3$.

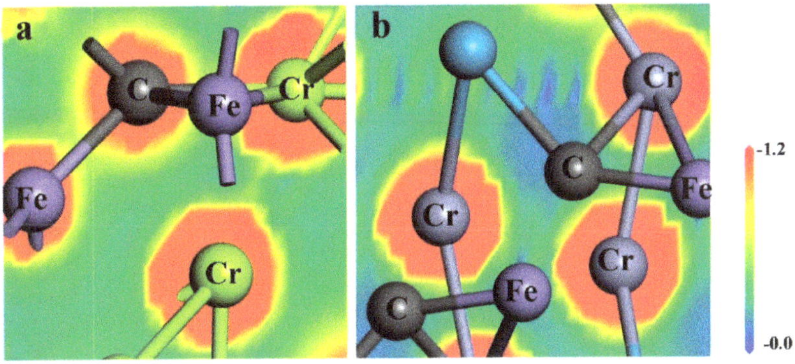

Figure 7. The valence electron density for (**a**) $Fe_3Cr_4C_3$ and (**b**) $Fe_3Cr_2W_2C_3$.

3.4. TEM Analysis

To study the existence form of carbides after adding tungsten to HCCl, the microstructures of High-Chromium Cast Iron with a composition of 3.44C-26.7Cr-1.25Mn-2.3 wt % W was characterized by TEM. Figure 8a shows the bright-field TEM micrographs, and Figure 8b shows the Selected Area Diffraction Pattern (SADP) of M_7C_3. The results show that the carbides are M_7C_3 phase with a hexagonal structure, and the space group is $P6_3mc$ (No. 186). According to the accurate measurement using EDX at 10 different areas of M_7C_3, the calculated analysis suggests that (Fe, Cr, W)$_7C_3$ has a stoichiometry of $(Fe_{3.27}Cr_{2.99}W_{0.74})$ C_3. Moreover, the lattice constant of $(Fe_{3.27}Cr_{2.99}W_{0.74})$ C_3 carbides is a = 0.6833 nm, b = 0.6833 nm, c = 0.4796 nm.

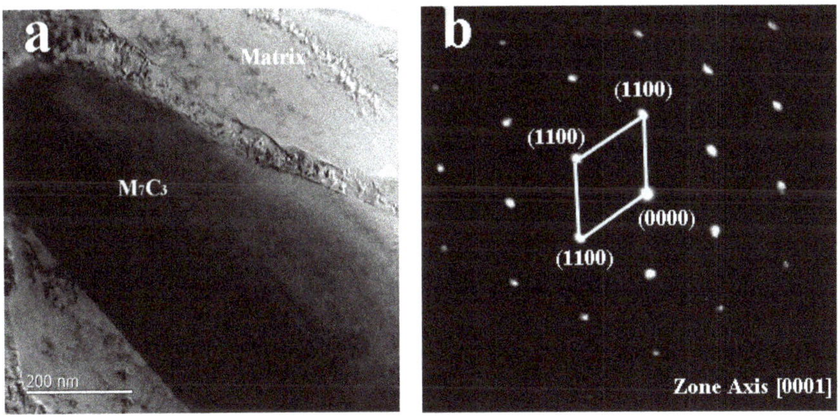

Figure 8. Bright-field TEM micrographs and corresponding Selected Area Diffraction Patterns (SADPs): (**a**) bright-field TEM micrographs; (**b**) corresponding SADPs of M_7C_3.

3.5. Nanoindentation Experiments

The hardness and modulus of carbides were investigated by a NanoFlip InForce 50. Figure 9a,b shows the indentation hardness HIT and the indentation modulus EIT of M_7C_3 carbides, respectively. With indentation depths larger than 100 nm, EIT and HIT of M_7C_3 phase reached a constant level, which indicated that the intrinsic material properties of the hard phases were measured in this experiment. Furthermore, the crack formation was not observed at an indentation depth of 200 nm, and the triangular indentations could be observed on M_7C_3 carbides, indicating that the credibility of the data is high. The indentation hardness values of (Fe, Cr)$_7C_3$ and (Fe, Cr, W)$_7C_3$ were 17.55 and 17.39 GPa, respectively. The indentation modulus values of (Fe, Cr)$_7$C$_3$ and (Fe, Cr, W)$_7$C$_3$ were 367.87 and 385.48 GPa, respectively.

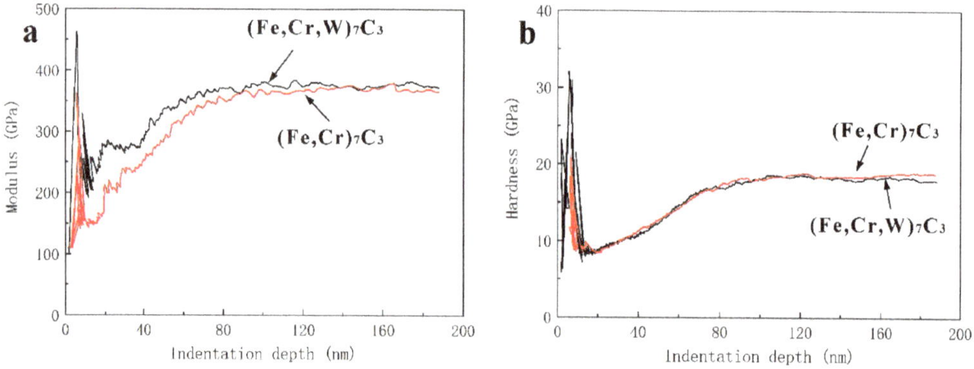

Figure 9. Indentation hardness and indentation modulus of M$_7$C$_3$ phase: (**a**) indentation hardness; (**b**) indentation modulus.

3.6. Wear Resistance

The wear resistance of HCCI was tested by a MM-200 block-on-ring wear testing machine. It was apparent that the wear resistance of HCCI after adding tungsten exceeds that of HCCI without tungsten. With the increase in wear load, the wear loss increased, as shown in Figure 10. The wear resistance of HCCI was closely related to the hardness of carbides, but also to the hardness of the matrix. Therefore, the abrasion resistance slightly increased after tungsten was added.

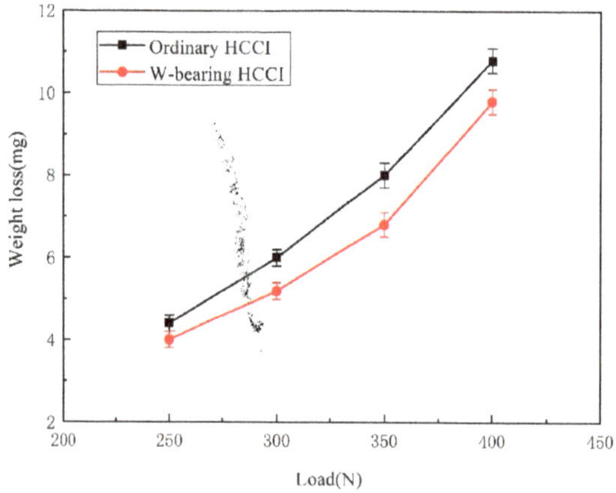

Figure 10. The wear resistance of HCCI.

To better reflect the wear resistance of HCCI after the addition of tungsten, the worn surface was characterized by a JSM-6510 Scanning Electron Microscope (SEM) and a VK-9710 color 3D laser scanning microscope. Figure 11a,b is the SEM images of ordinary HCCI and HCCI with tungsten, respectively. Figure 11c,d is the 3D laser morphologies of the worn surfaces of ordinary HCCI and HCCI with tungsten, respectively. Some obvious scratches were found on the specimen surface, and the wear surfaces of both HCCIs were consistent [32]. However, the groove scratches and fine wear of ordinary HCCI were obviously deep, indicating that the wear resistance of HCCI is better after adding tungsten.

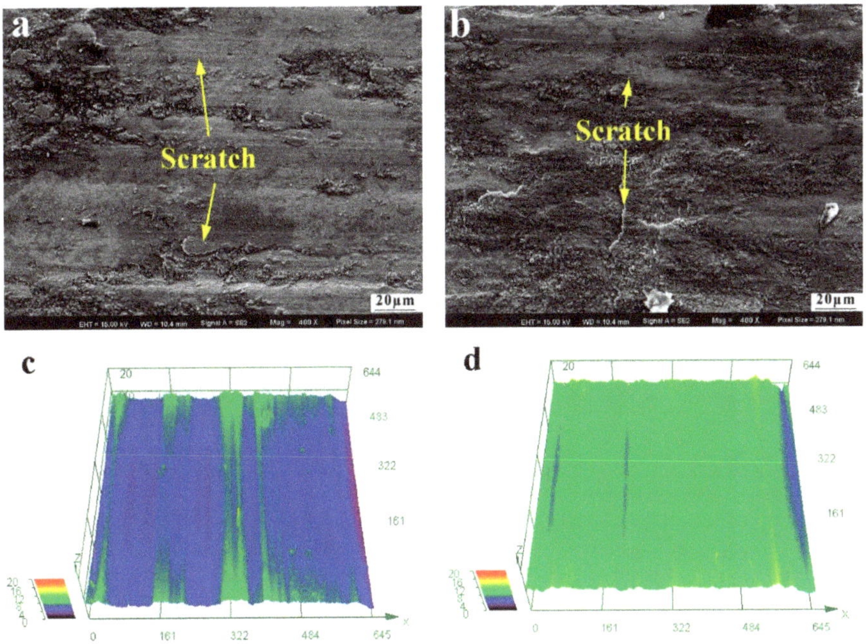

Figure 11. The worn surfaces of HCCI: (**a**,**c**) ordinary HCCI; (**b**,**d**) tungsten-bearing HCCI.

4. Conclusions

In this work, we added tungsten to High-Chromium Cast Iron to investigate the microstructures and properties of M_7C_3 carbides by first principles and experiments.

(a) With the increase in W and Cr content in M_7C_3 carbides, the formation energy of M_7C_3 carbides gradually decreased. Tungsten doping can improve the ductility of (Fe, Cr)$_7C_3$ carbides, and the composition of (Fe, Cr, W)$_7C_3$ is expected to be a high hardness and softness material.

(b) TEM results showed that the (Fe, Cr, W)$_7C_3$ carbides are ($Fe_{3.27}Cr_{2.99}W_{0.74}$) C_3 with a hexagonal structure after adding 2.13 wt % W into Fe—26.8 wt % Cr—3.62 wt % C High-Chromium Cast Iron.

(c) Wear test results showed that the wear resistance could be improved after adding tungsten to HCCI.

Author Contributions: C.C.: Investigating, Methodology, Formal analysis, Writing, Review, Editing. J.W.: Supervision, Funding acquisition, Writing, Review, Editing. Y.G.: Investigation, Experiments, Data curation. L.M.: Software, Writing, Review, Editing. All authors have read and agreed to the published version of the manuscript.

Funding: This research was supported by the National Natural Science Foundation of China (grant 51875252), the Natural Science Foundation of Heilongjiang Province (LH2020E026) and the Basic Scientific Research Business Fee Project of Heilongjiang Provincial Department of Education (2020-KYYWF-0266).

Institutional Review Board Statement: Not applicable.

Informed Consent Statement: Not applicable.

Data Availability Statement: The datasets generated during and/or analyzed during the current study are available from the corresponding author on reasonable request.

Conflicts of Interest: The authors declare no conflict of interest.

References

1. Giourntas, L.; Brownlie, F.; Hodgkiess, T.; Galloway, A.M. Influence of metallic matrix on erosion-corrosion behaviour of high chromium cast irons under slurry impingement conditions. *Wear* **2021**, *477*, 203834. [CrossRef]
2. Li, Y.; Shen, D.; Xu, N.; Tong, W. Cladding high chromium cast irons on low carbon steel: Microstructure and mechanical properties. *J. Mater. Res. Technol.* **2020**, *9*, 3856. [CrossRef]
3. Wiengmoon, A.; Pearce, J.T.H.; Nusen, S.; Chairuangsri, T. Electron microscopy study of carbides precipitated during destabilization and tempering heat treatments of 25 wt.%Cr-0.7 wt.%Mo high chromium cast irons. *Micron* **2021**, *143*, 103025. [CrossRef] [PubMed]
4. Nayak, U.P.; Guitar, M.A.; Mücklich, F.A. A Comparative Study on the Influence of Chromium on the Phase Fraction and Elemental Distribution in As-Cast High Chromium Cast Irons: Simulation vs. Experimentation. *Metals* **2020**, *10*, 30. [CrossRef]
5. Bedolla-Jacuinde, A.; Guerra, F.V.; Guerrero-Pastran, A.J.; Sierra-Cetina, M.A.; Valdez-Medina, S. Microstructural effect and wear performance of high chromium white cast iron modified with high boron contents. *Wear* **2021**, *476*, 203675. [CrossRef]
6. Heydari, D.; Skandani, A.A.; Haik, M.A. Effect of carbon content on carbide morphology and mechanical properties of A.R. white cast iron with 10-12% tungsten. *Mater. Sci. Eng. A.* **2012**, *542*, 113–126. [CrossRef]
7. Efremenko, V.G.; Chabak, Y.G.; Lekatou, A.; Karantzalis, A.E.; Efremenko, A.V. Pulsed plasma deposition of Fe-C-Cr-W coating on high-Cr-cast iron: Effect of layered morphology and heat treatment on the microstructure and hardness. *Surf. Coat. Technol.* **2016**, *304*, 293–305. [CrossRef]
8. Zhang, P.; Zhou, Y.; Yang, J.; Li, D.; Ren, X.; Yang, Y.; Yang, Q. Optimization on mechanical properties of Fe7-xCrxC3 carbides by first-principles investigation. *J. Alloys Compd.* **2013**, *560*, 49–53. [CrossRef]
9. Zhao, Z.; Song, R.; Zhang, Y.; Yu, P.; Pei, Y. Co-orientation relationship between secondary carbides and adjacent ferrite after quenching and tempering in high chromium cast iron. *Vacuum* **2021**, *184*, 109911. [CrossRef]
10. Xiao, B.; Xing, J.D.; Feng, J.; Zhou, C.T.; Cheng, Y.H. A comparative study of Cr_7C_3, Fe_3C and Fe_2B in cast iron both from ab initio calculations and experiments. *J. Phys. D* **2009**, *42*, 115415. [CrossRef]
11. Xiao, B.; Feng, J.; Zhou, C.T.; Xing, J.D.; Xie, X.J.; Chen, Y.H. First principles study on the electronic structures and stability of Cr_7C_3 type multi-component carbides. *Chem. Phys. Lett.* **2008**, *459*, 129–132. [CrossRef]
12. Konyaeva, M.A.; Medvedeva, N.I. Electronic structure, magnetic properties, and stability of the binary and ternary carbides (Fe, $Cr)_3C$ and (Fe, $Cr)_7C_3$. *Phys. Solid State* **2009**, *51*, 2084. [CrossRef]
13. Music, D.; Kreissig, U.; Mertens, R.; Schneider, J.M. Electronic structure and mechanical properties of Cr_7C_3. *Phys. Lett. A* **2004**, *326*, 473. [CrossRef]
14. Sun, L.; Ji, X.; Zhao, L.; Zhai, W.; Xu, L.; Dong, H.; Liu, Y.; Peng, J. First principles investigation of binary chromium carbides Cr_7C_3, Cr_3C_2 And $Cr_{23}C_6$: Electronic structures, mechanical properties and thermodynamic properties under pressure. *Materials* **2022**, *15*, 558. [CrossRef] [PubMed]
15. Coronado, J.J. Effect of (Fe, $Cr)_7C_3$ carbide orientation on abrasion wear resistance and fracture toughness. *Wear* **2011**, *270*, 287–293. [CrossRef]
16. Zhou, M.; Sui, Y.; Chong, X.; Jiang, Y.H. Wear resistance mechanism of ZTA_P/HCCI composites with a honeycomb structure. *Metals* **2018**, *8*, 588. [CrossRef]
17. El-Aziz, K.A.; Zohdy, K.; Saber, D.; Sallam, H.E.M. Wear and corrosion behavior of high-Cr white cast iron alloys in different corrosive media. *J. Bio- Tribo. Corros.* **2015**, *1*, 25. [CrossRef]
18. Cortés-Carrillo, E.; Bedolla-Jacuinde, A.; Mejía, I.; Zepeda, C.M.; Guerra-Lopez, F.V. Effects of tungsten on the microstructure and on the abrasive wear behavior of a high chromium white iron. *Wear* **2017**, *376-377*, 77–85. [CrossRef]
19. Lv, Y.; Sun, Y.; Zhao, J.; Yu, G.; Shen, J.; Hu, S. Effect of tungsten on microstructure and properties of high chromium cast iron. *Mater. Des.* **2012**, *39*, 303–308. [CrossRef]
20. Anijdan, H.S.M.; Bahrami, A.; Varahram, N.; Davami, P. Effects of tungsten on erosion–corrosion behavior of high chromium white cast iron. *Mater. Sci. Eng. A* **2007**, *454*, 623–628. [CrossRef]
21. Guerra, F.V.; Bedolla-Jacuinde, A.; Zuno-Silva, J.; Mejia, I.; Cardoso-Legorreta, E.; Arenas-Flores, A. Effect of the simultaneous Ti and W addition on the microstructure and wear behavior of a high chromium white cast iron. *Met. Res. Technol.* **2019**, *116*, 602. [CrossRef]
22. Chong, X.Y.; Hu, M.Y.; Wu, P.; Shan, Q.; Jiang, Y.H.; Li, Z.L.; Feng, J. Tailoring the anisotropic mechanical properties of hexagonal M_7X_3 (M=Fe, Cr, W, Mo; X=C, B) by multialloying. *Acta. Mater.* **2019**, *169*, 193–208. [CrossRef]
23. Xiao, B.; Feng, J.; Zhou, C.T.; Jiang, Y.H.; Zhou, R. Mechanical properties and chemical bonding characteristics of Cr_7C_3 type multicomponent carbides. *J. Appl. Phys.* **2011**, *109*, 1–9. [CrossRef]
24. Fischer, T.H.; Almlof, J. General methods for geometry and wave function optimization. *J. Phys. Chem.* **1992**, *96*, 9768. [CrossRef]
25. Fang, C.M.; van Huis, M.A.; Zandbergen, H.W. Structural, electronic, and magnetic properties of iron carbide Fe_7C_3 phases from first-principles theory. *Phys. Rev. B* **2009**, *80*, 224108. [CrossRef]
26. Yang, P.; Fu, H.; Zhao, X.; Lin, J.; Lei, Y. Wear behavior of CADI obtained at different austenitizing temperatures. *Tribol. Int.* **2019**, *140*, 105876.
27. Lin, C.M.; Chang, C.M.; Chen, J.H.; Wu, W. The effects of additive elements on the microstructure characteristics and mechanical properties of Cr–Fe–C hard-facing alloys. *J. Alloys Comp.* **2010**, *498*, 30–36. [CrossRef]

28. Sun, S.; Fu, H.; Lin, J.; Guo, G.; Lei, Y.; Wang, R. The stability, mechanical properties, electronic structures and thermodynamic properties of (Ti, Nb)C compounds by first principles calculations. *J. Mater. Res.* **2018**, *4*, 495–506. [CrossRef]
29. Wu, Z.; Zhao, E.; Xiang, H.; Hao, X.; Liu, X.; Meng, J. Crystal structures and elastic properties of superhard IrN_2 and IrN_3 from first principles. *Phys. Rev. B.* **2007**, *76*, 1–15.
30. Tian, Y.; Xu, B.; Zhao, Z. Microscopic theory of hardness and design of novel superhard crystals. *Int. J. Refract. Met. Hard. Mater.* **2012**, *33*, 93–106. [CrossRef]
31. Ganguly, A.; Murthy, V.; Kannoorpatti, K. Structural and electronic properties of chromium carbides and Fe-substituted chromium carbides. *Mater. Res. Express* **2020**, *7*, 056508. [CrossRef]
32. Ţălu, Ş. Micro and nanoscale characterization of three dimensional surfaces. In *Basics and Applications*; Napoca Star Publishing House: Cluj-Napoca, Romania, 2015; pp. 21–27.

coatings

MDPI

Article

Effect of Cr_3C_2 Content on the Microstructure and Wear Resistance of Fe_3Al/Cr_3C_2 Composites

Yingkai Feng [1,2], Jiguo Shan [1,*], Lin Zhao [3,*], Zhenbo Liu [3], Karin Gong [4] and Changhai Li [4]

1 Department of Mechanical Engineering, Tsinghua University, Beijing 100084, China
2 Beijing Technology Research Branch of Tian Di Science & Technology Co., Ltd., Chinese Institute of Coal Science, Beijing 100013, China
3 Welding Institute, Central Iron and Steel Research Institute, Beijing 100081, China
4 Department of Industrial and Materials Science, Chalmers University of Technology, SE-41296 Gothenburg, Sweden
* Correspondence: shanjg@tsinghua.edu.cn (J.S.); hhnds@aliyun.com (L.Z.)

Abstract: In this paper, an engine piston ring coating comprising composite material of Fe_3Al and Cr_3C_2 mixed powder was prepared by laser cladding onto carbon structural steel. The microstructure and tribological properties of the cladding materials were investigated through X-ray diffraction (XRD), scanning electron microscopy (SEM), energy dispersive spectroscopy (EDS), transmission electron microscopy (TEM), and wear tests. The influence mechanism of the Cr_3C_2 content in cladding powder was studied. During the process of wear, the soft Fe_3Al/Fe_2AlCr matrix is first ground off, and the hard Cr_7C_3 phase initially supports the abrasive surface before being worn away into hard particles, resulting in abrasive wear. With the increase in Cr_3C_2 content, the hardness of the cladding layer increases, the proportion of the Cr_7C_3 phase increases, and the morphology changes from a sparse network to a dense floccule. Of the cladding layers with different Cr_3C_2 content, the 15 wt.% Cr_3C_2 cladding layer had the lowest friction coefficient, and the 25 wt.% Cr_3C_2 cladding layer had the lowest wear rate. The low wear rate of the 25 wt.% Cr_3C_2 cladding layer can be attributed to the fact that adhesive wear does not easily occur and the fine microstructure of the strengthening phase, which facilitates better separation in the grinding surfaces.

Keywords: carbide behavior; Fe_3Al/Cr_3C_2 composites; laser cladding; wear resistance

1. Introduction

With the development of high-temperature, wear-resistant structural materials, the research into intermetallic compounds and their composites has been gradually increasing. TheFe_3Al alloy and its composite materials have good resistance against wear, high temperatures, and oxidation [1–4]. Compared with other intermetallic compounds, Fe and Al are relatively cheap, representing a large raw material cost advantage. Thus, alloys comprising Fe and Al have great application potential in ships, automobiles, aviation, and other fields.

Due to the importance of developing a new generation of wear-resistant piston ring materials, Chen et al. [5] utilized a TiC-reinforced Fe_3Al base composite material for coating via laser cladding onto the stainless steel base material. This type of coating has good wear resistance, and its performance is gradually enhanced due to the increase in the strengthening phase. As the load is increased, the wear type changes from abrasive to adhesive. Zhang et al. [6] prepared a nano-Fe_3Al/Al_2O_3 gradient coating using the laser-cladding method. Their experimental results showed that the nanocoating had significantly improved wear resistance, hardness, and corrosion resistance, and the microstructure and properties of the gradient material changed significantly at different locations.

Cr_3C_2 has a high thermal hardness, good corrosion resistance, and high oxidation resistance, which are compatible with the thermal expansion coefficient of most alloys. Therefore, the addition of Cr_3C_2 to the matrix as a hard strengthening phase will greatly improve the wear resistance of a material. Zhao et al. [7] obtained a high-temperature, wear-resistant material coating with excellent comprehensive performance by preparing an Ni_3Al base material composite reinforced with Cr_3C_2 particles. They found that the laser-clad Ni_3Al alloy and Ni_3Al/Cr_3C_2 composite coatings differed in terms of their microstructures, mechanical properties, and friction and wear properties. Chen et al. [8] prepared wear-resistant cladding materials by laser cladding onto a 304 stainless steel base material, using either pure Ni_3Al or Ni_3Al/Cr_3C_2 mixed powder. By analyzing its structure, the cladding layer that was prepared from mixing powder was found to be mainly composed of an Ni_3Al matrix phase and in situ, self-generated M_7C_3. The wear resistance of the cladding layer was also characterized.

Laser cladding refers to the use of a laser as a heat source to deposit the required material onto a substrate. The dilution rate of laser cladding is very low due to the higher concentration of energy. Due to the rapid heating and cooling in this process, the heat-affected zone of the material is small. In particular, when used to melt different materials, the characteristics of laser cladding are vastly superior to other heating methods [9]. After cladding, the material has good grain structure and properties similar or even superior to the intrinsic material.

The research on the Fe_3Al alloy and Fe_3Al matrix composite materials has mainly focused on Fe-Al/Al_2O_3, Fe-Al/WC, and other composite materials, and in the research on Fe-Al/Cr_3C_2 composite materials, only thermal spraying coating technology has been largely considered. Systematic investigations on the wear resistance of the Fe_3Al/Cr_3C_2 cladding layer, the influence of Cr_3C_2 on the properties of the Fe_3Al/Cr_3C_2 composite, and the friction and wear mechanism of the Fe_3Al/Cr_3C_2 composite laser cladding are still lacking. Compared with previous studies, the main novelty of this work is that we studied the effect of Cr_3C_2 content on the microstructure and wear resistance of Fe_3Al/Cr_3C_2 composites. The influence mechanism of carbide behavior on the wear resistance of Fe_3Al/Cr_3C_2 composites was also investigated.

2. Experimental Materials and Method

2.1. Composite Powders

Carbon structural steel was used as the base material. The cladding powder was a mixture of Fe_3Al and Cr_3C_2 powder of particle size 75–125 μm. The Cr_3C_2 powder was a commercial pure Cr_3C_2 prepared by crushing a large Cr_3C_2 block. The chemical composition of Fe_3Al powder is shown in Table 1.

Table 1. Chemical composition of Fe_3Al powder (wt.%).

Al	B	Cr	Fe	Mn	Ni	Zr
15.23	0.11	5.78	76.82	0.052	0.54	0.26

Aluminum blocks and nickel powder (CISRI, Beijing, China), applied as raw materials for powder atomization, were melted in a vacuum furnace at 1900~2000 °C for 1 h and then atomized into powder. In this paper, cladding layers with 0, 5, 15, 25, and 35 wt.% Cr_3C_2 content were prepared and investigated.

2.2. Preparation of the Cladding Layer

A YLS-6000 fiber laser-cladding system composed of a YLS-6000 fiber laser (IPG Photonic, Oxford, MA, USA), KUKA mechanical arm, and powder feeder was utilized for the laser-cladding experiment, and images and details of the setup are shown in Figure 1. The laser wavelength was 1.07 μm. A Gaussian distribution focusing lens was selected for use on the laser head to generate a rectangular laser spot of size 2 mm × 5 mm. The powder feeding type was coaxial powder feeding; the shielding gas and powder-carrying gas were high-purity argon (99.99%), and the flow of shielding gas was 15 L/min. Before cladding, the surface of the base material was polished with 320# SiC sandpaper to reduce its brightness and remove the surface oxide layer, which was followed by cleaning and wiping of the polished base material with acetone.

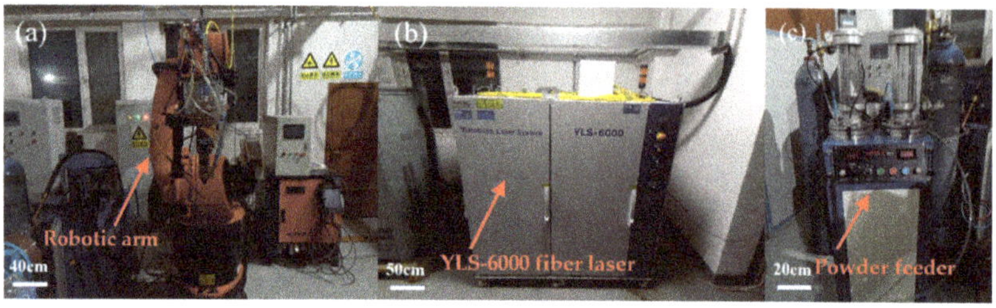

Figure 1. Laser cladding system: (**a**) KUKA robotic arm, (**b**) YLS-6000 fiber laser, and (**c**) powder feeder.

Through the preliminary experiment, process parameters that led to fewer cracks and better forming were determined, as were the following parameters: laser power of 2.2 kW, laser scanning speed of 0.002 m/s, and powder feed rate of 0.644 kg/h. These process parameters were used to prepare cladding layers with five different levels of Cr_3C_2 content.

2.3. Microstructure Characterization

Sections of the cladding layer along the direction perpendicular to the cladding layer were obtained by wire-electrode cutting. The sample sizes were 10 mm × 10 mm and the full thickness of the section. We sanded the sections with 320#, 600#, and 1000# sandpaper in succession, and the samples were then polished with a 5 μm diamond-polishing agent. After polishing, the samples were cleaned and dried. Phase analysis of the alloyed powder and cladding layer was carried out using a Bruker D8 ADVANCE X-ray diffractometer (XRD, BRUKER, Karlsruhe, Germany) with a scanning speed of 2°/min. In order to observe the microstructure of the cladding layer, Zeiss high-resolution field emission scanning electron microscopy (SEM, ZEISS, Jena, Germany) was used to observe the cross section of the cladding layer. An Auriga focused ion beam Zeiss FIB was used to prepare TEM samples. The selected observation area was coated and marked using a vapor deposition system (GIS), and the samples were then sheared to an appropriate thickness using an ion beam. The prepared samples were observed by high-resolution transmission electron microscopy (JEM-2010, JEOL, Tokyo, Japan). The hardness distribution from the top of the cladding layer to the substrate material was measured using an FM300 microhardness tester (F-T, Tokyo, Japan).

2.4. Wear Test

A multifunctional friction and wear testing machine (UMT, BRUKER, Madison, MA, USA) was used for the wear test. As shown in Figure 2 below, the friction type was pin disk surface contact cyclic friction. The pins were Ø3 mm in size and made of the Fe_3Al/Cr_3C_2 cladding layer and common wear-resistant vermicular cast iron. The discs were made of gray cast iron and were cut from the inner wall of the engine to a size of Φ 24 mm × 7.88 mm. The friction and wear test conditions were as follows: dry friction and wear, load of 30 N, rotation speed of 200 r/min, rotation radius of 7 mm, testing time of 60 min, and room temperature. In this investigation, two friction indices are mainly considered: the friction coefficient and the wear rate. There are many influencing factors of friction and wear, among which the surface roughness of the contact pin and disc has a great influence on the friction coefficient. Therefore, before the experiment, we adopted finish grinding to adjust the surface roughness to Ra0.4. To more accurately measure the wear rate, the grinding pins and discs were ultrasonically cleaned with acetone for 10 min. Following this, the cleaned grinding pin and disc were weighed, and the measured masses were recorded as M_1 and M_2, respectively. After the experiment, the grinding pin and disc were again ultrasonically cleaned with acetone for 10 min to remove residual wear debris and determine the mass of the grinding pin and disc, M_3 and M_4, by weighing. We calculated M_1-M_3 and M_2-M_4, which represent the wear loss, for use in the subsequent wear rate calculation. The friction coefficient was obtained in real time during the friction experiment. Thereafter, the pins were scanned using a 3D white light interferometer. The morphology of the worn surface and section were observed by high-resolution field emission scanning electron microscopy (SEM, ZEISS, Jena, Germany).

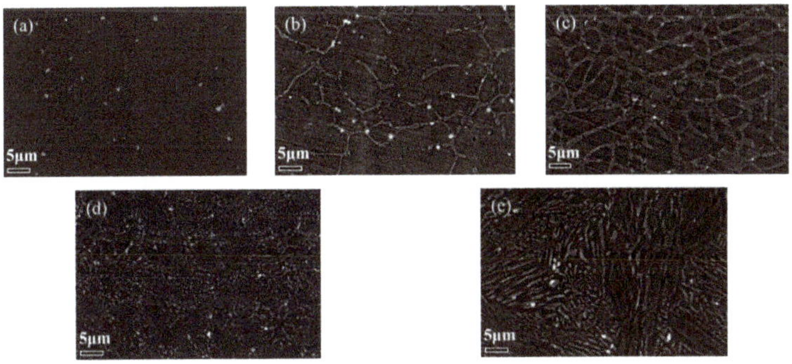

Figure 2. BSE morphology of cladding layers prepared using the same process and different Cr_3C_2 content: (**a**) 0 wt.%, (**b**) 5 wt.%, (**c**) 15 wt.%, (**d**) 25 wt.%, and (**e**) 35 wt.%.

3. Results

3.1. Microstructure and Phase Composition of the Cladding Layer

A large number of studies [10–12] have shown that the size and distribution of the strengthening phase in a wear-resistant coating have a large influence on its performance. In order to find the reasons for different wear rate results, the microstructure morphology was first analyzed. From Figure 2, we can see that the Cr_3C_2 content has a notable influence on the microstructure of the cladding layer. Figure 2a shows that no apparent strengthening phase can be observed in the 0 wt% Cr_3C_2 cladding layer. Figure 3b–d show that at 5, 15, and 25 wt.% Cr_3C_2 content, a strengthening phase appears as a reticular structure in the base material. With the increase in Cr_3C_2 content, the reticular structure becomes denser, and the area of voids decrease. Figure 3e shows that the strengthening phase structure changes from its original reticular structure to a block-shaped structure in the 35 wt% Cr_3C_2 cladding layer.

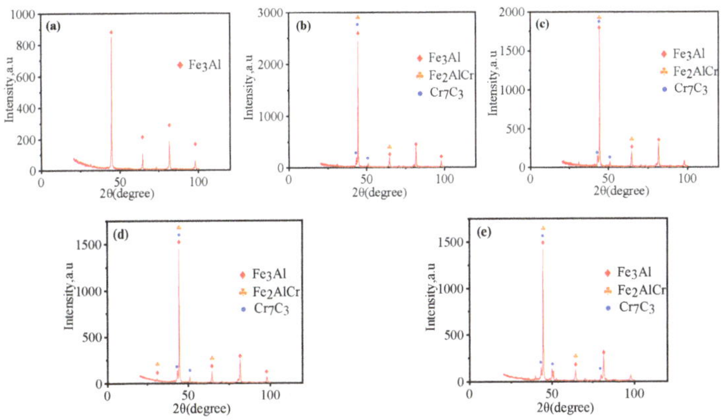

Figure 3. XRD spectra of cladding layers prepared using the same process and different Cr_3C_2 content: (**a**) 0 wt.%, (**b**) 5 wt.%, (**c**) 15 wt.%, (**d**) 25 wt.%, and (**e**) 35 wt.%.

The phase composition of the cladding layer was analyzed. XRD phase analysis of the cladding layers with different Cr_3C_2 content was carried out, and the results are shown in Figure 3. Figure 3a shows the presence of only one phase of Fe_3Al in the 0 wt.% Cr_3C_2 cladding layer. Figure 3b–e show that the 5, 15, 25, and 35 wt% Cr_3C_2 cladding layers contain Fe_3Al, Fe_2AlCr, and Cr_7C_3. Because the peak characteristic of Fe_3Al coincides with that of Fe_2AlCr, the existence of an Fe_2AlCr phase could not be determined from the XRD results alone. As shown in Table 2 below, the number of cracks in the cladding layer gradually decreases and then increases with increasing Cr_3C_2 content. This indicates that the addition of Cr_3C_2 initially improves the plasticity of brittle Fe_3Al [13]. Fe_2AlCr phases are characterized by good plasticity [14], and thus, the appearance of this layer results in reduced cracking of the cladding layer. With the increase in Cr_3C_2 content, the brittle phase of Cr_7C_3 in the cladding layer increases. Although the content of the Fe_2AlCr phase also increases, it is still brittle as a whole, so the cracks increase again.

Table 2. The number of cracks generated in samples of different cladding materials of the same length (15 cm).

Group	Number of Cracks
0 wt.% Cr_3C_2 cladding layer	6
5 wt.% Cr_3C_2 cladding layer	5
15 wt.% Cr_3C_2 cladding layer	0
25 wt.% Cr_3C_2 cladding layer	0
35 wt.% Cr_3C_2 cladding layer	6

Figure 4 shows TEM images and selected electron diffraction patterns of the 15 and 35 wt.% Cr_3C_2 cladding layers. Figure 4a,b show the TEM images of the 15 and 35 wt.% Cr_3C_2 cladding layers, respectively. Figure 4c,d show the electron diffraction pattern of the Fe_3Al/Fe_2AlCr and Cr_7C_3 structures, respectively. The strengthening phase exists in the form of Cr_7C_3. In the 15 wt.% Cr_3C_2 cladding layer, Cr_7C_3 has a strip or punctate structure, whereas in the 35 wt.% Cr_3C_2 cladding layer, it has a polygon structure. In terms of size, the strengthening phase is larger in the 35 wt.% Cr_3C_2 cladding layer.

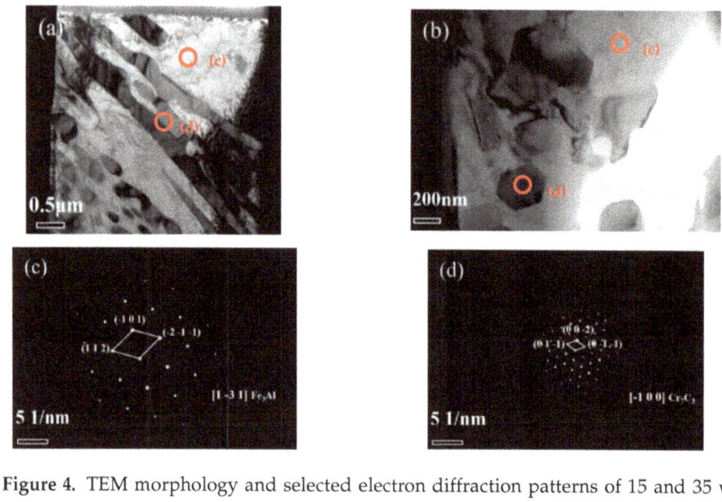

Figure 4. TEM morphology and selected electron diffraction patterns of 15 and 35 wt.% Cr3C2 cladding layers. Microstructure morphology of (**a**) 15 wt.% Cr3C2 cladding layer and (**b**) 35 wt.% Cr3C2 cladding layer. Diffraction patterns of (**c**) Fe3Al/Fe2AlCr in the [1–31] crystal band direction and (**d**) Cr7C3 in the [−100] crystal band direction.

3.2. Friction Coefficient and Wear Loss

The friction coefficient curves of each group are shown in Figure 5. Due to the significant amount of time required for the wear test, in the last stage of the test, most of the friction coefficient curves were relatively stable and could be directly obtained through reading. According to the curves, the vermicular cast iron group had a stable friction coefficient of 0.7. In the cladding layer groups, except the layers with 35 wt.% Cr_3C_2, the values for stable friction coefficients were all below 0.65. The coefficient of the 15 wt.% Cr_3C_2 group was 0.32, which was the lowest. In terms of the fluctuation of the friction coefficient, except the 5 wt.% Cr_3C_2 cladding group fluctuating in the end, the friction coefficient curves of the other groups were relatively stable in the last stage. To demonstrate that this phenomenon of friction coefficient instability was not accidental or the result of a single test, we repeated the tests with the 5 wt.% Cr_3C_2 content group, and the friction coefficient curves were obtained as shown in Figure 6. Table 3 shows the wear loss for all three tests. The differences among the coefficients of each curve were small, and the overall trends of the curves were the same. The error was within the allowable range of the test. We can draw the conclusion that fluctuation is a common phenomenon in specific groups.

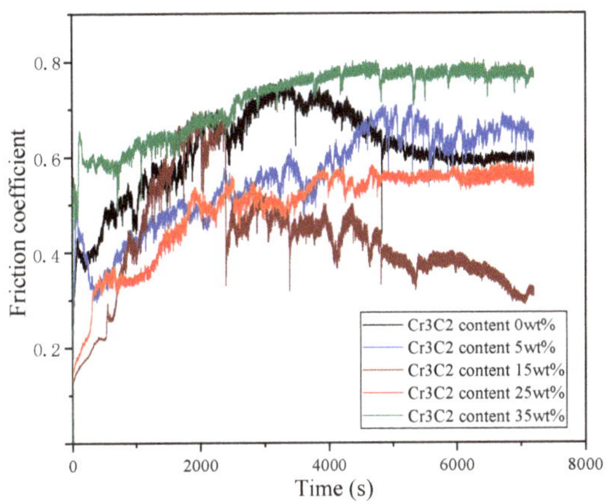

Figure 5. Friction coefficient curves of vermicular graphite cast iron and cladding layers with the same process and different Cr_3C_2 content.

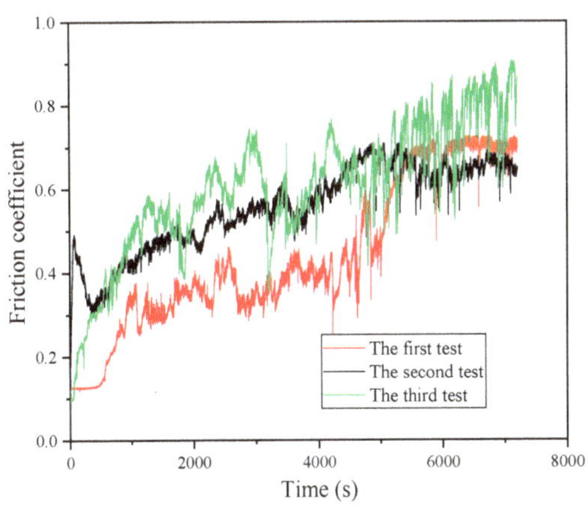

Figure 6. Repeated-test friction coefficient curves of 5 wt.% Cr_3C_2 cladding layer.

Table 3. Wear loss of 5 wt.% Cr_3C_2 cladding layer in repeated tests.

	First Test	Second Test	Third Test
Pin wear rate (1×10^{-5} $mm^3/N{\cdot}m$)	0.910	1.017	1.203
Disc wear loss (g)	0.00227	0.00201	0.00283

Table 4 shows the wear loss in the wear tests. From the results of the pin wear rate, all the groups were below 1×10^{-5} $mm^3/N{\cdot}m$, except the 5 wt.% Cr_3C_2 group. Figure 7 shows that the hardness of the cladding layer increases with the increase in Cr_3C_2 content. In the inner parts of the cladding layer, the hardness of different positions is similar. When the Cr_3C_2 content was 15 and 25 wt.%, the wear loss was quite low and was correlated with hardness. The 5 and 35 wt.% groups demonstrated deviance with large wear rates. In terms of disc wear loss, the results of the 0, 5, and 15 wt.% groups were close, with the lowest determined for 25 wt.% Cr_3C_2 content and the highest for 35 wt.% Cr_3C_2 content.

Table 4. Pin and disc wear loss of different pin material groups.

Pin Material	Disc Wear Loss (g)	Pin Wear Rate (1×10^{-5} $mm^3/N{\cdot}m$)
0 wt.% Cr_3C_2 cladding layer	0.00247	0.514
5 wt.% Cr_3C_2 cladding layer	0.00201	1.017
15 wt.% Cr_3C_2 cladding layer	0.00239	0.396
25 wt.% Cr_3C_2 cladding layer	0.00074	0.285
35 wt.% Cr_3C_2 cladding layer	0.00961	0.964
Vermicular graphite cast iron	0.00227	3.102

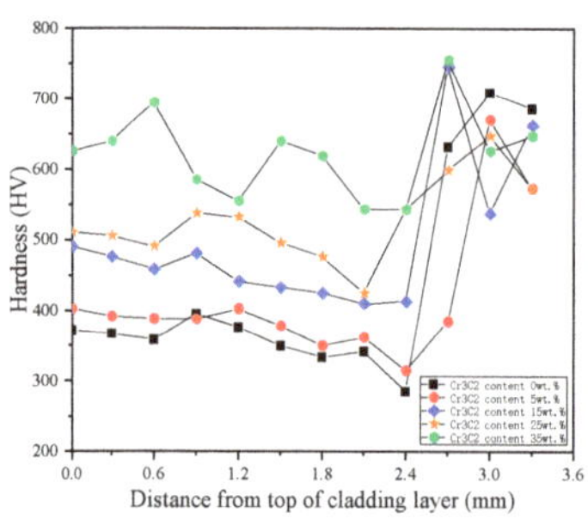

Figure 7. Microhardness of the cladding layers with different Cr_3C_2 content.

3.3. Wear Mechanism of Cladding Layers

A 3D white light interferometer was used to carry out the observation of the worn pins and discs. The results are shown in Figure 8. Figure 8a shows the 3D profiles of the 0 wt.% Cr_3C_2 cladding layer. Figure 8b,f show the 3D profiles of the 5 wt.% Cr_3C_2 cladding layer. Figure 8c shows 3D profiles of the 15 wt.% Cr_3C_2 cladding layer. Figure 8d shows the 3D profiles of the 25 wt.% Cr_3C_2 cladding layer. Figure 8e shows the 3D profiles of the 35 wt.% Cr_3C_2 cladding layer. Very obvious furrows can be observed on the surfaces of the pins and discs, which indicate the obvious occurrence of abrasive wear. Moreover, with the increase in Cr_3C_2 content, the surface-height difference of pins gradually increases, and abrasive wear became increasingly serious. In addition to obvious furrows, deep pits could also be found on the surface of the grinding pin of the 5 wt.% Cr_3C_2 group, which demonstrates that not only abrasive wear, but also serious adhesive wear occurred [15].

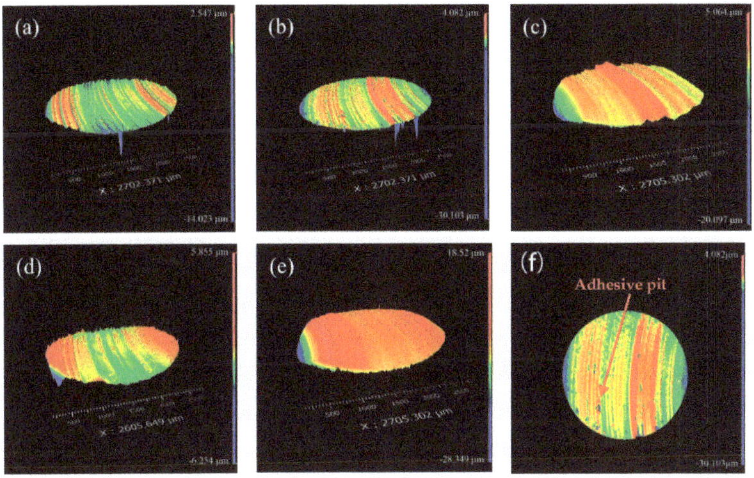

Figure 8. Three-dimensional profiles after wear of grinding pins prepared using the same process and different Cr_3C_2 content: (**a**) 0 wt.%, (**b**,**f**) 5 wt.%, (**c**) 15 wt.%, (**d**) 25 wt.%, and (**e**) 35 wt.%.

Through SEM observation of the worn surfaces, we found varying numbers of black spots on the wear surface of each pin. Through EDS analysis, we found that the oxygen content of the black spots was very high. This indicates that they correspond to an oxide layer generated by oxidative wear on the surface. In the process of friction and wear, the oxide layer gradually falls off and becomes wear debris or adheres to the surface of the pin, as shown in Figure 9a. According to the size of the black spots, we could intuitively judge the severity of oxidative wear in each pin. From Figure 9b–e, we can draw the conclusion that the 5, 15, 25, and 35 wt.% Cr_3C_2 cladding layers had a small degree of oxidative wear, whereas a relatively large degree of oxidative wear occurred on the 0 wt.% Cr_3C_2 cladding layer, for which some large pieces of fallen oxide layer that adhered to the worn surface can be clearly observed.

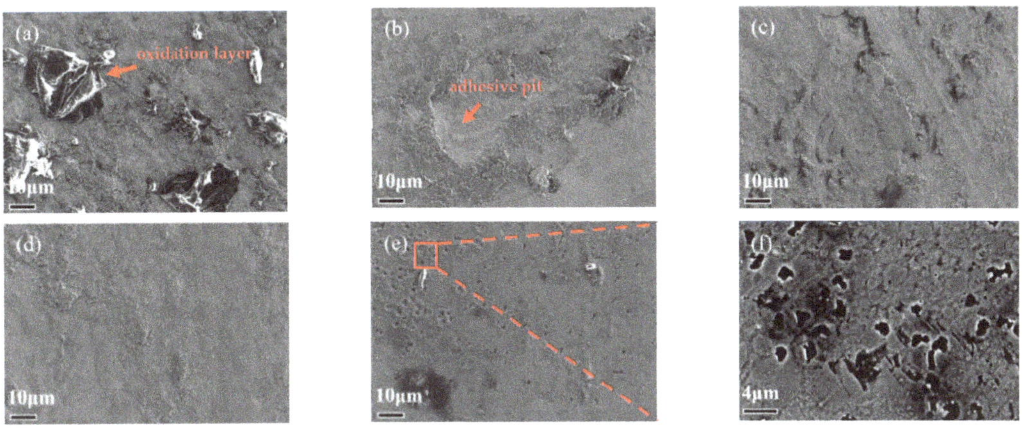

Figure 9. SEM morphology of grinding pin surfaces prepared using the same process and different Cr_3C_2 content: (**a**) 0 wt.%, (**b**) 5 wt.%, (**c**) 15 wt.%, (**d**) 25 wt.%, and (**e**,**f**) 35 wt.%.

By cutting the 0 wt.% Cr_3C_2 grinding pin longitudinally and observing this section (as shown in Figure 10a,b), we found that there was no obvious and continuous oxidation layer on the pin surface, which indicates that oxidative wear is not the main wear mechanism even in cases where the pin shows relatively serious oxidative wear.

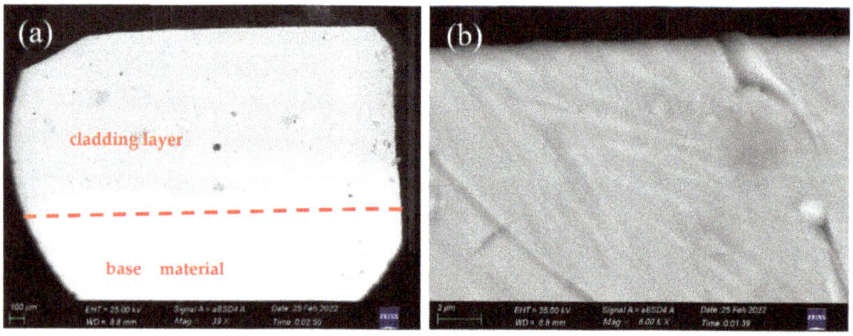

Figure 10. BSE morphology of the longitudinal section of a grinding pin with 0 wt.% Cr_3C_2 content captured at (**a**) low power and (**b**) high power.

4. Discussion

Generally, wear resistance and material hardness are linked; when a material is harder, its wear resistance is often higher [16]. In the wear loss results of our test data, however, this was not the case. Instead, the microstructure characteristics of the cladding layer had a decisive influence on its wear resistance.

Cr_7C_3, as a type of carbide, has a high hardness and also represents the strengthening phase of composite cladding materials by which antiwear ability is enhanced. There is no doubt that Cr_7C_3 has superior wear resistance to Fe_2AlCr and Fe_3Al. Moreover, the composite cladding material in the friction and wear process can be transformed into the following model: the soft part is the base, and the hard, wear-resistant strengthening phase becomes the particles or stripes embedded in the base. With the progress of friction and wear, the soft Fe_3Al and Fe_2AlCr phases are gradually worn away and disappear, resulting in the wear-resistant hard phase becoming exposed outside the base material. The exposed hard phase serves as a fulcrum that contacts the friction pair and will be preferentially consumed in the subsequent process until the exposed hard phase is almost worn out. The soft base of the base material will then again come into contact with the friction pair, and the process is repeated. Liu et al. [17] proposed a similar model for the friction and wear mechanism of Ni_3Al/Cr_3C_2. The strengthening phase of the 35 wt.% Cr_3C_2 cladding layer was found to be block-shaped, whereas in the 5, 15, and 25 wt.% layers, it become reticular. From the wear loss results, we found that, with the exception of the 35 and 5 wt.% Cr_3C_2 groups, the pin wear rate was similar. Based on the above analysis of friction types, we found that the main friction type of these three groups in the friction process is abrasive wear, which is the main reason for the disappearance of the base material. According to the model, some of the abrasive particles come from the grinding disc and some from the grinding pin, and the part from the pin can be subdivided into Cr_7C_3 from the strengthening phase and the softer base material. The hardness of the abrasive particles from the disc and the base material part is low, so in the process of abrasive wear, their contribution to the abrasive wear of the pin is very limited; most of the wear loss caused by abrasive wear can be considered to be due to the strengthening phase. Next, the reason for different wear loss caused by differing Cr_3C_2 content was analyzed.

By comparing the differences between the block-shaped strengthening phase and the reticular strengthening phase, the following two points were found: the size of the block-shaped strengthening phase is much larger than that of the reticular strengthening phase, and from the point of view of carbide content in the whole material, the proportion of the block-shaped strengthening phase is larger than that of the reticular strengthening phase. According to the wear loss data, we found that the wear loss was generally higher for the block-shaped strengthening phase than for the reticular strengthening phase. This indicates that the reticular strengthening phase has advantages in the process of friction and wear because it can also reduce the contact between the grinding surfaces and base material to a certain extent. In addition, the reticular strengthening phase ensures not only higher hardness of the base material, but also a reduced proportion of the hard-phase content and, therefore, generated abrasive particles. From the morphology of the reticular strengthening phase, it is not easy to produce large, hard, abrasive particles.

It can be seen from the SEM morphology of the grinding pin with the 35 wt.% Cr_3C_2 content that there were many small block pits on the grinding surface. Combined with the strengthening phase morphology photos taken earlier, we found that the block-shaped strengthening phase broke into small pieces and peeled off from the large strengthening phase in the process of friction and wear due to its brittleness. The exfoliated strengthening phase was converted into large, hard, abrasive particles, which greatly exacerbated abrasive wear between the pin and disc. We found that the wear rate of the 35 wt.% Cr_3C_2 pin was the most serious of all groups. However, the large, hard strengthening phase, which still existed in the pin, could block the wear of abrasive particles in the base material, and high-content strengthening phases also enhanced the strength of the base material, thus representing a double-edged sword [18,19]. Although there were more large, hard, abrasive particles in the grinding pair, the wear rate of the pin was still not very large, and the ability to resist the wear of abrasive particles was stronger than in other groups.

Next, the cladding layers with 15 and 25 wt.% Cr_3C_2 content were analyzed. Since the strengthening phase shape was a network of fine strips and dots, in the process of friction and wear, the exfoliated strengthening phase did not produce large particles, only small or slender ones. Compared with the 35 wt.% Cr_3C_2 cladding layer, there was less hard phase contained in the 15 and 25 wt.% cladding layers when the same volume material was worn away, and the hard-phase content was lower when introduced to the friction pair. As a result, the cladding layer was less prone to serious abrasive wear, and there was less wear loss through abrasive wear on the cladding layer. This further reduced the content of hard, abrasive particles in the grinding pair, forming a virtuous cycle and reducing the wear rate of the cladding layer. From the BSE morphology (Figure 3), it was found that the cladding layer with 25 wt.% Cr_3C_2 was almost covered with a fine, striated, and dotted strengthening phase. Such a fine, dispersive strengthening phase can cause the matrix to separate from the grinding pair more smoothly, so as to better resist abrasive wear. The 25 wt.% Cr_3C_2 cladding layer also had higher hardness than the 15 wt.% Cr_3C_2 cladding layer, factors that led to its higher wear resistance. The layer produced less hard-phase abrasive particles, reducing the wear loss of the grinding disc.

The wear rate of the 5 wt.% Cr_3C_2 cladding layer was the highest among all Cr_3C_2 content groups. We found that not only abrasive wear, but also adhesive wear occurred. Adhesive wear is more likely to occur in materials with good plasticity than in brittle materials. Although the cladding layers with 15 and 25 wt.% Cr_3C_2 content had better plasticity, there was no obvious adhesive wear because most of the base material was isolated from friction by the carbide strengthening phase exposed after wear. However, in the cladding layer with 5 wt.% Cr_3C_2, the distribution of the strengthening phase was sparse, and the content was small. The base material of the cladding layer could not be well isolated from the grinding disc, and they remained in contact with each other over a large area, resulting in serious adhesive wear. The occurrence of adhesive wear also explains the instability of the friction coefficient in the process of friction and wear (Figure 7).

The vermicular cast iron group also showed serious adhesive wear. Figure 11a,b show the surface morphology after wear, where obvious adhesive pits are observed on the grinding pin surface, and more materials were pulled off, resulting in a significant wear rate of the grinding pin. Compared with the Fe_3Al material, vermicular cast iron has lower hardness and is closer to the grinding disc in composition, resulting in more serious adhesive wear.

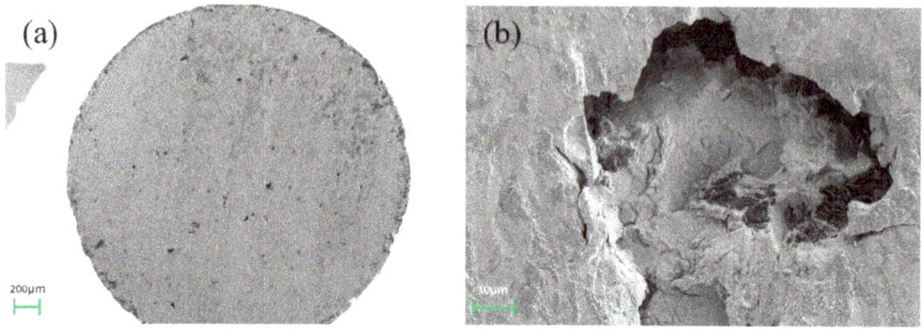

Figure 11. Overall surface morphology of vermicular cast iron after wear, captured at (**a**) low power and (**b**) high power.

The cladding layer with 0 wt.% Cr_3C_2 content did not contain Cr_3C_2, so the isolating friction pair mentioned above was not relevant to the wear reduction mechanism. Therefore, a large area of the Fe_3Al base material was in direct contact with the grinding disc and was also the reason for the relatively severe oxidative wear on the surface. With the wear of the oxide layer, the oxide layer falls off and adheres to the cladding layer.

5. Conclusions

In this paper, through wear tests, the friction coefficient and wear rate of cladding layers with different Cr_3C_2 content were obtained. The friction mechanism was further judged by white light interference, SEM, BSE observation, and EDS analysis. The friction and wear mechanisms were obtained by combining the test results with the microstructure. The microstructure and wear resistance of the cladding layers were affected not only by the factor of Cr_3C_2 content, but also by the cladding process parameters. The influence of process parameters on the microstructure and wear resistance of the cladding layer were subsequently studied and resulted in the following observations and conclusions.

(1) With the increase in Cr_3C_2 content, the number of cracks in the Fe_3Al cladding layer first decreased and then increased, and the 15 and 25 wt.% Cr_3C_2 cladding layers had the fewest number of cracks. In terms of hardness, it also increased gradually with the increase in Cr_3C_2 content in the cladding powder and remained relatively stable inside the cladding layer.

(2) With the increase in Cr_3C_2 content, the morphology of the strengthening phase in the cladding layer changed from a sparse to a dense reticular structure and, finally, to a large block-shaped structure.

(3) For cladding layers prepared using the same process and different Cr_3C_2 content, the group with the smallest friction coefficient was found to be the 15 wt.% Cr_3C_2 cladding layer, with large friction coefficients determined for the remaining groups. In terms of wear rate, the lowest values were found for the 15 and 25 wt.% Cr_3C_2 cladding layers. Compared with cast iron, the Fe_3Al/Cr_3C_2 cladding layer had obvious advantages in various indices.

(4) Judging by the visible furrows on the pin surface, abrasive and oxidative wear appeared in all groups. Adhesive wear only occurred in the pin with 5 wt.% Cr_3C_2 content. The occurrence of adhesive wear or more serious abrasive wear will lead to greater wear.

(5) The friction and wear mechanisms were determined to be as follows: The relatively soft Fe_3Al base material is first worn away, and the harder carbide supports the grinding surface to reduce friction. Then, the carbide gradually disappears from the pin and is transformed into hard-phase particles. The morphology and distribution of the strengthening phase in the cladding layer play an important role in wear resistance. Compared with the block-shaped strengthening phase, the reticular strengthening phase has more advantages; while ensuring higher base material strength, it reduces the proportion of hard particles and size of the hard phase.

Author Contributions: Methodology, J.S. and L.Z.; formal analysis, Y.F.; investigation, Y.F. and Z.L.; resources, L.Z.; writing—original draft preparation, Y.F.; writing—review and editing, K.G. and C.L.; visualization, Y.F. and Z.L.; supervision, J.S. and L.Z.; project administration, L.Z.; funding acquisition, L.Z. All authors have read and agreed to the published version of the manuscript.

Funding: This work was supported by the National Key R&D Program of China, grant number 2020YFE0200900.

Institutional Review Board Statement: Not applicable.

Informed Consent Statement: Not applicable.

Data Availability Statement: All data that support the findings of this study are included within the article.

Conflicts of Interest: The authors declare no conflict of interest.

Nomenclature

SEM	Scanning Electron Microscopy
TEM	Transmission Electron Microscopy
FIB	Focused Ion Beam
XRD	X-ray Diffraction
BSE	Backscattered Electron

References

1. Yin, Y.; Li, J.; Ma, H. Friction and wear properties of sintered Fe_3Al intermetallic compound based friction materials. *Tribology* **2004**, *24*, 457–461.
2. McKamey, C.G.; DeVan, J.H.; Tortorelli, P.F.; Sikka, V.K. A review of recent developments in Fe_3Al-based alloys. *J. Mater. Res.* **1991**, *6*, 1779–1805. [CrossRef]
3. Stoloff, N.S. Iron aluminides: Present status and future prospects. *Mater. Sci. Eng. A* **1998**, *258*, 1–4. [CrossRef]
4. Stoloff, N.S.; Liu, C.T.; Deevi, S.C. Emerging applications of intermetallics. *Intermetallics* **2000**, *8*, 1313–1320. [CrossRef]
5. Chen, Y.; Wang, H.M. Microstructure and wear resistance of laser clad TiC reinforced FeAl intermetallic matrix composite coatings. *Surf. Coat. Technol.* **2003**, *168*, 30–36. [CrossRef]
6. Zhang, J.; Zhang, X.; He, Z.; Hao, X.; Wang, Y.; Li, J. Application of Nano-sized Fe_3Al/Al_2O_3 gradient coating on Steel Plate by laser cladding method. *Paint. Ind.* **2019**, *42*, 61–64.
7. Zhao, X. Study on Microstructure and Wear Resistance of Laser Cladding Ni_3Al/Cr_3C_2 Composite. Master's Thesis, Hebei University of Technology, Tianjin, China, 2019.
8. Chen, C.; Zhao, X.; Fang, Q.; Zhao, L.; Han, W.; Peng, Y.; Yin, F.; Tian, Z. Wear Resistance of Laser Cladding Ni_3Al/Cr_3C_2 Composites. *Rare Met. Mater. Eng.* **2020**, *49*, 1388–1394.
9. Li, M.; Han, B.; Song, L.; He, Q. Enhanced surface layers by laser cladding and ion sulfurization processing towards improved wear-resistance and self-lubrication performances. *Appl. Surf. Sci.* **2020**, *503*, 144226. [CrossRef]
10. Hemmati, I.; Ocelík, V.; De Hosson, J.T.M. Effects of the alloy composition on phase constitution and properties of laser deposited Ni-Cr-B-Si coatings. *Phys. Procedia* **2013**, *41*, 302–311. [CrossRef]
11. Xu, B.; Guo, F. A micromechanics method for transverse creep behavior induced by interface diffusion in unidirectional fiber-reinforced metal matrix composites. *Int. J. Solids Struct.* **2019**, *159*, 126–134. [CrossRef]
12. Colaço, R.; Vilar, R. Abrasive wear of metallic matrix reinforced materials. *Wear* **2003**, *255*, 643–650. [CrossRef]
13. Mendiratta, M.G.; Ehlers, S.K.; Chatterjee, D.K.; Lipsitt, H.A. Tensile Flow and Fracture Behavior of DO_3 and Fe—25 at.% Al and Fe—31 at.% Al Alloys. *Metall. Trans. A* **1987**, *18*, 283–291. [CrossRef]
14. Li, Z.; Peng, B. Formation mechanism and control measures of laser cladding cracks. *J. Mater. Prot.* **2016**, *49*, 61–66.
15. Zhao, Z. Wear of Engine Cylinder Liner and Its Preventive Measures. *J. North China Inst. Aerosp. Technol.* **2005**, *15*, 14–15.
16. Gao, C. The Relationship between Hardness and Wear Resistance of Metals. *Mech. Eng.* **1981**, *4*, 6–8.
17. Liu, Z.; Yin, F.; Chen, C.; Zhao, L.; Fu, L.; Feng, Y.; Cao, Y.; Peng, Y.; Tian, Z.; Li, C. Microstructure and Wear Resistance of a Cr7C3 Reinforced Ni_3Al Composite Coating Prepared by Laser Cladding. *Coatings* **2022**, *12*, 105. [CrossRef]
18. Yuan, Y.; Li, Z. Effects of rod carbide size, content, loading and sliding distance on the friction and wear behaviors of $(Cr, Fe)_7C_3$ reinforced α-Fe based composite coating produced via PTA welding process. *Surf. Coat. Technol.* **2014**, *248*, 9–22. [CrossRef]
19. Hu, J.; Li, D.; Llewellyn, R. Computational investigation of microstructural effects on abrasive wear of composite materials. *Wear* **2005**, *259*, 6–17. [CrossRef]

coatings

Article

Effect of Heat Treatment on Microstructure and Tribological Properties of Laser Cladding CeO_2/Ni60 Composite Coating on 35CrMoV Steel

Zhiming Gao [1], Shuqing Zhang [1], Zhongtang Gao [1,*], Haibo Ren [2] and Chuanwei Zhang [1]

[1] School of Mechanical Engineering, Xi'an University of Science and Technology, Xi'an 710054, China
[2] School of Mechanical Engineering, Yanshan University, Qinhuangdao 066004, China
* Correspondence: zhongtanggao@xust.edu.cn; Tel.: +86-29-85583157

Abstract: A Ni60 cladding layer with addition of 6.0% CeO_2 was prepared on 35CrMoV steel by laser cladding technology. The prepared sample was placed at 500 °C, 600 °C and 700 °C for 60 min to explore the effects of heat treatment on the tribological properties of the composite coating. The microstructure, phase composition, microhardness and tribological properties of the composite coating were characterized by optical microscopy and scanning electron microscopy (SEM), X-ray diffraction (XRD), micro-Vickers hardness tester and MicroXAM-800 optical surface photometer, respectively. According to the above experimental results analysis, the main components of 6.0% CeO_2/Ni60 cladding layer are γ-(Fe,Ni),Cr_7C_3,$Cr_{23}C_6$,CrB, CrFeB and Cr_2Ni_3. By calculating the FWHM value and the left shift of the XRD diffraction peak, it is found that the coating grains are remarkably refined and the microstructure uniformity is significantly improved under the condition of heat treatment at 500 °C. The experimental results show that the Ni60 composite coating with 6.0% CeO_2 has the best friction and wear performance at 500 °C. The wearing quality of the composite coating at 500 °C was reduced by 43%.

Keywords: laser cladding; heat treatment; CeO_2; wear resistance

1. Introduction

To work under high stress conditions, 35CrMoV alloy steel is often manufactured as a key component [1,2]. The working life of the parts cannot be guaranteed because the parts are in service under harsh conditions for a long time [3]. The resultant damage of important parts will not only delay the working time, but also make it difficult to guarantee the repair quality even if the damaged parts are repaired by traditional methods [4].

Since laser cladding technology has significant advantages for metal surface modification, the cladding coating on the surface of parts can significantly improve the hardness, wear resistance and corrosion resistance compared with the matrix [5–8]. The use of laser cladding technology can effectively improve the service life of parts and reduce the production cost [9,10]. Nickel-based alloys have high hardness, good wear resistance and high temperature oxidation resistance, and are widely used in laser cladding, butt welding and spraying surface strengthening technology [11,12]. The nickel-based coatings prepared by laser cladding technology on steel surfaces have excellent mechanical properties, such as high hardness, which is much higher than the matrix wear resistance and excellent metallurgical bonding performance [13]. Although nickel-based materials have excellent mechanical properties, it is difficult to avoid the occurrence of pores, cracks and other defects in nickel-based composite coatings of 35CrMoV matrix due to the distinct difference in physical properties and chemical composition between the steel substrate and nickel-based self-fusing alloy [14]. In recent years, it has been proved that rare earth oxides can effectively lessen the number of cracks and the formation of pores. Wang et al. prepared nickel base coatings with addition of La_2O_3, CeO_2 and Y_2O_3 by laser cladding on 6063 aluminum alloy. From the perspectives of micro morphology, dendrite density, wear morphology and wear amount, the mechanical properties of the cladding layer with addition of rare earth oxides result in being better than those without rare earth addition [15]. Oh et al. studied the separate action of four rare earth oxides on the refinement of microstructure for nickel-based alloys. With the result that the addition of La_2O_3 has the greatest refinement effect through sufficient oxide dispersion and delayed recrystallization, and the distribution homogeneity of microstructure is the highest [16]. Xu et al. prepared laser cladding 316L stainless steel coatings with addition of CeO_2. The results show that CeO_2 facilitates refined grains in the cladding layer. As a consequence, both the hardness and the mechanical properties of the coating were greatly improved [17]. Our previous study also confirmed that adding rare earth elements contributes to the improvement of mechanical properties of the coatings.

It is well known that heat treatment, as a traditional process, it is one of the key means to improve the structure, mechanical properties and internal stress state of materials [18,19]. In recent years, researchers have studied the mechanical properties of laser cladding coatings after heat treatment. Pan et al. investigated the effect of annealing on the mechanical properties of nickel-based coatings cladding on 430 stainless steel at heat treatment temperatures ranging from 0 to 800 °C for two hours. The research showed that with the increase in heat treatment temperature, both the elastic modulus and the yield strength of the coating show downward trends [20]. Wang et al. studied the effects of rare earth element, Ce, and different heat treatment temperatures on the oxidation resistance of steel. The results show that with the increase in temperature, Ce can improve the oxidation resistance of steel and effectively inhibit the shedding of oxides [21]. Guo et al. studied the mechanical properties of iron base laser cladding coatings under the influence of heat treatment. With the result that the optimal wear resistance of the coating can be obtained after 850 °C heat treatment [22]. The above research shows that heat treatment has a positive effect on the mechanical properties of the coating.

The authors' previous study confirmed the beneficial effect of CeO_2 on the microstructure mechanical properties of laser cladding Ni60 coating on 35CrMoV, but the performance variation of the cladding layer after heat treatment was not clear. Therefore, the main purpose of this work is to further explore the effect of the heat treatment process on the mechanical properties of laser cladding Ni60 coating on 35CrMoV steel with 6.0%CeO_2 addition on the basis of the original research.

2. Materials and Experimental Procedures

2.1. Materials

In this study, 35CrMoV steel was selected as the laser cladding coating substrate, and the size of the substrate was 100 mm × 60 mm × 20 mm. The cladding material is composed of sub-micron rare earth oxide CeO_2 (purity ≥ 99.99%) (20–70 μm) and Ni60 (60–150 μm). Its chemical element composition is shown in Table 1. Before laser cladding, CeO_2 powder and Ni60 powder were mixed by ball milling method. Mixing experiment parameters are as follows: ball powder mass ratio of 10:1, ball mill speed of 600 rpm and mixing time of 60 min. The mixed Ni60/6.0%CeO_2 (wt.%) powder was dried in a box resistance furnace at 100 °C for 120 min.

Table 1. Chemical composition of 35CrMoV steel and Ni60 powder (wt.%).

Samples	C	Si	Mn	Cr	Mo	V	P	S	Cu	Fe	B	Ni
35CrMoV	0.30–0.38	0.17–0.37	0.40–0.70	0.80–1.10	0.15–0.25	0.10–0.20	≤0.035	≤0.035	≤0.030	—	—	≤0.030
Ni60 powder	0.8	4.0	—	15.5	—	—	—	—	—	15.0	3.5	excess

2.2. Coating Preparation Process

The 35CrMoV substrate was cleaned with acetone before laser cladding to purge oil stains. The purpose of polishing the surface of the substrate is to remove the oxide layer and is cleaned with acetone and then dried to prevent being re-oxidized. The mixed power was spread evenly on the substrate. The thickness of powder was controlled by a special laying card groove, and the powder thickness was 1.5 mm.

BS-OF-3000-15-4L general laser heating equipment (Laser-300, Xi'an Beisheng Laser Technology Co., Xi'an, China) was used for laser cladding of multi-channel lapping. The optimal processing parameters are power of 3000 W, spot diameter of 3 mm, scanning speed of 120 mm/min, focal length of 16.0 mm, coating rate of 30% and high purity argon gas flow rate of 15 L/min.

2.3. Heat Treatment Method

The 35CrMoV substrate with laser cladding coating was cut into samples of 15 mm × 10 mm × 10 mm by wire-cutting method. The samples were then divided into three groups, which were placed in a resistance furnace at 500 °C, 600 °C and 700 °C for 1 h and then cooled to room temperature in the furnace. The main parameters of the high temperature sintering furnace are: maximum power 1.2 kW and maximum heating temperature 1100 °C. The temperature in the furnace can be effectively controlled by intelligent PID mode. The heating rate and cooling rate of the samples were 5 °C/min and 5 °C/min, respectively.

2.4. Detection Method

The macroscopic morphology of the cladding layer was characterized using a Nikon MA200 metallographic microscope. The microstructure of the selected cladding area was analyzed by scanning electron microscopy(JEOL/JSM-5610LV, Ishizuka Electronics Corporation, Mitaka, Japan). Energy dispersive spectrometer (EDS, Ishizuka Electronics Corporation, Mitaka, Japan) and X-ray diffractometer (XRD, BRUKER-AXS-D8, BRUKER AXS GMBH, Berlin, Germany) were used to determine the elements' distribution and phase constitution in the cladding layer. The microhardness of the coating was determined by a micro-Vickers hardness tester (HMAS-D1000SZ, Hardness Precision Instruments Co., Shanghai, China) with a loading force of 9.8 N for 15 s. An MWT-R4000 electrochemical corrosion friction and wear tester (MFT-R4000, Zhongke Kaihua Technology Co., Lanzhou, China) was used to conduct the dry sliding friction test to test the friction and wear properties of the cladding layer. In this friction and wear experiment, A Si_3N_4 ball with a diameter of 6 mm was used as the friction pair, the loading load was 5 N, the wear time was 20 min, the friction travel was 10 mm and the fixed frequency was 2 Hz. A Microxam-800 optical profilometer was used to measure the 3D morphology of the samples after friction and wear, and the friction coefficient of the cladding layer at different temperatures was characterized and analyzed.

3. Results and Discussion

3.1. Macroscopic Morphology of Composite Coating

Figure 1 shows the macroscopic morphologies of the surface of Ni60 composite coating with the addition of 6.0%CeO_2 at different heat treatment temperatures. Figure 1a shows the macroscopic morphology of the composite coating at room temperature. Thus, it may be deduced that the cladding layer is mostly composed of coarse columnar dendrites and the microstructure uniformity is poor. This is because the precipitation of Fe_5C_2 enhanced phase in the 6.0%CeO_2/Ni60 cladding layer is inhibited by excessive rare earth element, Ce, which leads to the reduction of the refining effect of Fe_5C_2 on the coarse dendritic eutectic in the coating. Under these conditions, the intrinsic microstructure of the cladding zone is mostly composed of Ni and Cr enriched phases, and the dendritic eutectic with increasing size of (Ni, Fe) + Fe_5C_2 + $Cr_{23}C_6$ eventually leads to the formation of microstructure defects [23]. Under the condition of heat treatment at 500 °C, the dendritic eutectic structure was uniform, and there is obvious tendency for the dendrite structure to be significantly refined compared with the non-heat-treated structure, as shown in Figure 1b. Compared with the cladding zone at 500 °C, the microcosmic organization and homogeneity of the composite coating at 600 °C have different changes, as shown in Figure 1c. First, coarse columnar dendrites coexist with tiny dendrites, and the dendrites have a certain tendency to grow compared with that of 500 °C. Secondly, it can be observed that the uniformity of the microstructure has decreased to a certain extent. When the heat treatment temperature reaches 700 °C, the microstructure is mainly composed of coarse eutectic, which presents a dispersed sparse distribution. According to the analysis of the experimental results, the microstructure of the cladding layer has been refined significantly under the 500 °C heat-treatment condition.

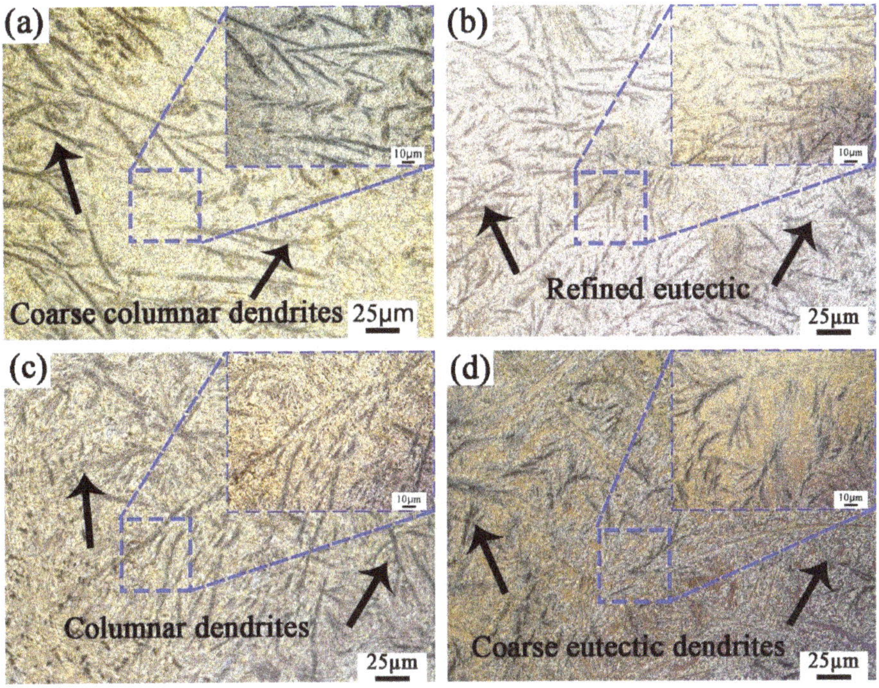

Figure 1. Macromorphology of Ni60 cladding layer with 6.0% CeO_2 at different annealing temperatures: (**a**) 25 °C; (**b**) 500 °C; (**c**) 600 °C; (**d**) 700 °C.

3.2. Phase Composition

Figure 2 indicates XRD patterns of composite coating with 6.0% CeO_2 additive amount at different annealing temperatures. From Figure 2 it can be seen that the main components of the coating are γ-(Fe,Ni),Cr_7C_3,$Cr_{23}C_6$,CrB, CrFeB and Cr_2Ni_3. Hard phases Cr_7C_3, $Cr_{23}C_6$ and CrB fully raise the hardness but also increase the brittleness of the cladding layer. The appearance of Cr_2Ni_3 phase is a result of the fusion of matrix materials. As we all know, the narrower the half-peak width of the XRD pattern, the larger the grain size. The left shift of the diffraction peak represents the increase in the half-peak width of the XRD pattern. From Figure 2 it can be seen that under the condition of heat treatment at 500 °C, the diffraction peak shifts significantly to the left, which proves that the width of the half-peak increases. It can be inferred that the grain size decreases under such experimental conditions. Under the conditions of 600 °C and 700 °C, the XRD patterns have different degrees of right movement, which shows that the grain size increases under these heat treatment conditions. The FWHM value measured by integral method can be used to compare the half-peak width at different heat treatment temperatures, which can indirectly assess the grain size variation at different heat treatment temperatures. The half-peak widths at 25 °C, 500 °C, 600 °C and 700 °C are 3.49, 3.91, 3.32 and 2.71, respectively. It can be concluded that the half-peak width is the maximum under the heat treatment condition of 500 °C, and it follows from what has been said that the particle size of the cladding coating is relatively small under this condition, which also confirms the conclusion drawn in Figure 1.

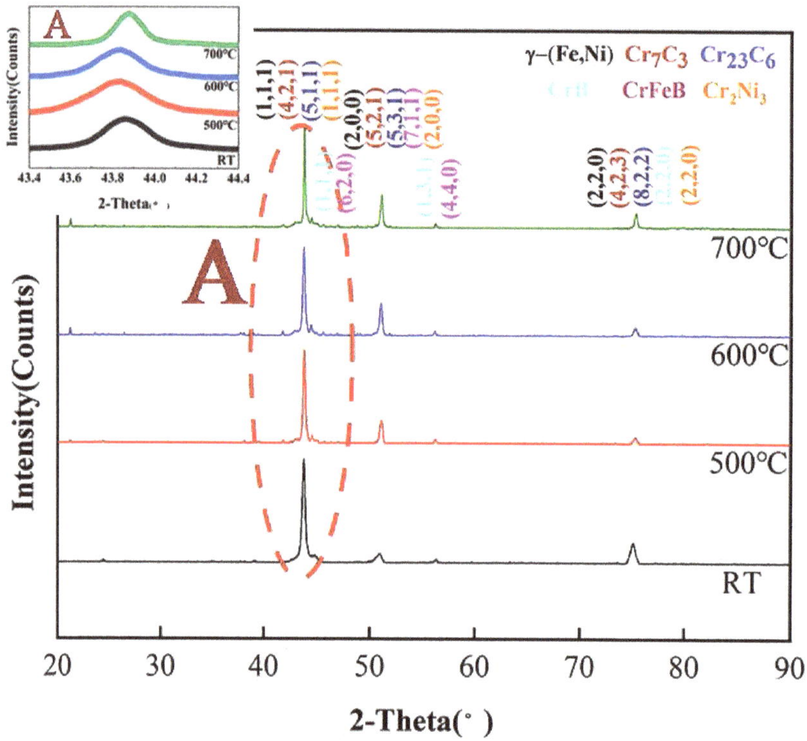

Figure 2. Micromechanism of the complex coating with different annealing temperatures.

3.3. Microstructures of Cladding Layer

Figure 3 shows the microstructure evolution after different heat treatment temperatures. The micromechanism of the complex coating under 25 °C, that is, without heat treatment, presents obvious coarse and sparse microstructure. The cladding layer is mainly composed of coarse columnar crystals, as shown in Figure 3a. Figure 3b shows the microstructure of the complex coating after 500 °C annealing. The micromechanism consists mainly of components of closely and evenly distributed dendrite crystals and the original coarse dendrites are fully refined. Figure 3c shows that under the experimental condition of heat treatment temperature of 600 °C, the microstructure refinement effect of the cladding layer can be seen to be significantly reduced. Under these conditions, relatively coarse dendrites reappear in the cladding coating, decreased the uniformity of the microstructure. When the heat treatment temperature is 700 °C, it can be found that the cladding layer has obviously coarse dendrites randomly distributed, and obvious segregation phenomenon can be seen in the tissue, as shown in Figure 3d. Combined with the above analysis and the microstructure of the cladding layer shown in Figure 3, it can be seen that the microstructure of the Ni60 cladding layer with 6.0%CeO_2 after the heat treatment experiment is improved to different degrees compared with that under 25 °C. When the heat treatment temperature is 500 °C, the particles' refining effect in the micro-organization is the best, and the homogeneity of the microstructure is obviously improved.

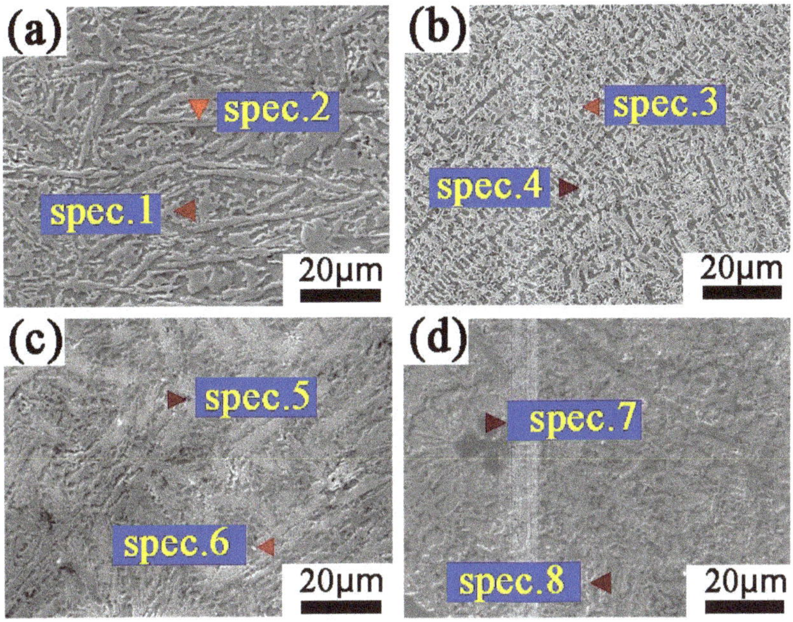

Figure 3. Microstructure of the complex coating with different annealing temperatures: (**a**) 25 °C; (**b**) 500 °C; (**c**) 600 °C; (**d**) 700 °C.

Combined with the XRD(Figure 2) and EDS(Table 2) results, it may be deduced that the coarse dendrites of the cladding layer are mainly composed of B, Cr, Fe, Ni and C elements, which are further confirmed as γ-(Fe,Ni) + Cr_7C_3 + $Cr_{23}C_6$ + CrFeB + Cr_2Ni_3 solution [3]. The fine compact grain region is consisted of Ni, Fe and Cr elements, which are γ-(Fe,Ni) + Cr_2Ni_3. The reinforced phase generated by B element can improve the hardness of the cladding layer. In addition, boron also has the effect of self-deoxidizing and reducing inclusions and pores in the cladding layer. Therefore, it can be inferred that the gradual transformation of fine dendrites into coarse columnar crystal in the clad coatings is related to the decrease in B content.

Table 2. EDS results of the coating(wt. %).

Spectrum	B	C	Si	Cr	Fe	Ni	Ce
Spec. 1	15.13	2.08	0.19	73.69	7.49	1.31	0.11
Spec. 2	1.10	2.28	3.11	15.81	26.72	50.81	0.19
Spec. 3	8.06	3.78	2.15	16.22	29.94	39.84	0.02
Spec. 4	0.83	1.25	2.31	15.51	35.06	45.34	-
Spec. 5	6.65	2.93	2.25	15.36	29.46	43.62	-
Spec. 6	-	4.50	2.14	13.71	26.23	57.47	0.14
Spec. 7	1.07	6.76	0.90	32.97	31.21	26.71	0.38
Spec. 8	-	1.62	3.49	8.56	30.41	56.06	0.40

3.4. Element Distribution Analysis

Figure 4 demonstrates the elements distribution of the Ni60 cladding layer containing 6.0%CeO_2 at several annealing temperatures. The degree of segregation of each element is different. In the four cases, the segregation severity of Ni, Fe, Si and Cr elements are more intense than that of B, C and Ce elements. In the cladding layer without heat treatment, it can be observed that the segregation of Si and Cr elements is obvious, as shown in Figure 4a_0–g_0. This is because the excessive addition of Ce in 6.0%CeO_2 resulted in the conversion of excess Ce from dissolution and segregation along the crystal boundaries during solidification [24]. The latent heat of phase transition emitted by the second phase compound at the crystal boundaries reduces the degree of supercooling and finally results in the columnar crystals. When the heat treatment temperature is 500 °C, it can be seen that the segregation of each element is significantly improved as can be seen in Figure 4a_1–g_1. The reason for the improvement of element segregation is that the reasonable temperature increases the diffusion of the elements in the sample, and the inhibition of CeO_2 on the segregation of elements in the clad coatings is further enhanced. When the heat treatment temperatures are 600 °C and 700 °C, it can be seen from Figure 4a_2–g_2 and a_3–g_3 that the segregation of Ni, Cr and Fe elements causes the uneven distribution of elements in the cladding layer, resulting in the decrease in the uniformity of the cladding layer structure. The above experimental results show that appropriate heat treatment temperature (500 °C) can effectively inhibit the segregation of elements, so as to effectively improve the uniformity of the distribution profile of elements in the clad coatings.

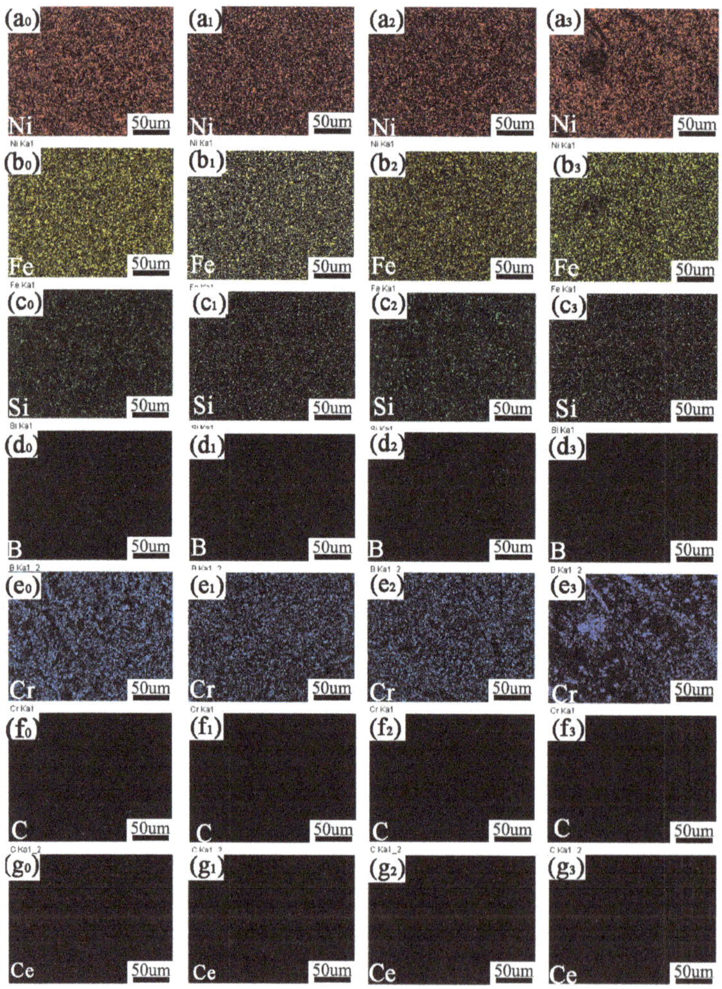

Figure 4. The element distribution of Ni60-6.0%CeO_2 complex coating at different heat treatment temperatures: (**a_0–g_0**) 25 °C; (**a_1–g_1**) 500 °C; (**a_2–g_2**) 600 °C; (**a_3–g_3**) 700 °C.

3.5. Microhardness of Cladding Layer

Figure 5 displays the microhardness allocation of 6.0%CeO_2 clad coatings at disparate annealing temperatures. The change of microhardness value directly affects the quality of cladding layer [25]. Generally, the microhardness distribution curve of the cladding layer can be divided into three sections, as shown in Figure 5. Along the cross section of the sample from top to bottom, it can be divided into the surface cladding region, the intermediate bonding region between the cladding layer and the matrix and the final 35CrMoV matrix region. In this experiment, the hardness of the coating was characterized by HMAS-D1000SZ micro-Vickers hardness tester. The loading force was 9.8 N and the duration was 15 s. For the accuracy of the test, three points were taken at equal distances to test the hardness. The distance was 0.5 mm apart, and the average value was taken as the final result. Average microhardness of samples after 25 °C, 500 °C, 600 °C and 700 °C heat treatments are 539.45 HV_1, 732.37 HV_1, 677.41 HV_1 and 658.19 HV_1, respectively. Microhardness of the clad coatings is the lowest under the heat treatment condition of 25 °C, and the hardness fluctuation is larger at the bonding zone. This is caused by the poor bonding performance. Additionally, the serious element segregation in non-heat-treated samples, which leads to the uneven structure, also contributes to the fluctuation hardness behavior. When the heat treatment temperature is 500 °C, the microhardness of the coverage shows an overall increase and the hardness curve is smoother than at other temperatures. At a reasonable heat treatment temperature, the grains of the coverage including 6.0%CeO_2 are obviously refined, and the microstructure uniformity reaches the optimal. As a consequence, the hardness curve of the cladding layer is relatively smooth. The reinforced phase formed by Cr and B in the cladding layer increase the hardness of the cladding layer obviously. Under the experimental conditions of the heat-treatment temperatures of 600 °C and 700 °C, it can be observed that the hardness curve of the cladding layer fluctuates obviously. Combined with microstructure and phase analysis, the effect of grain refinement in the cladding layer is reduced. It can be inferred from the EDS diagram that the reduction of B compounds is one of the reasons for the fluctuation of hardness values. Combined with the above conclusions, the heat treatment can uplift the rigidity of Ni60 coverage containing 6.0%CeO_2 to a certain extent. Under the condition of 500 °C, the rigidity of coverage obviously augmented, and the fluctuation range of microhardness is the smallest.

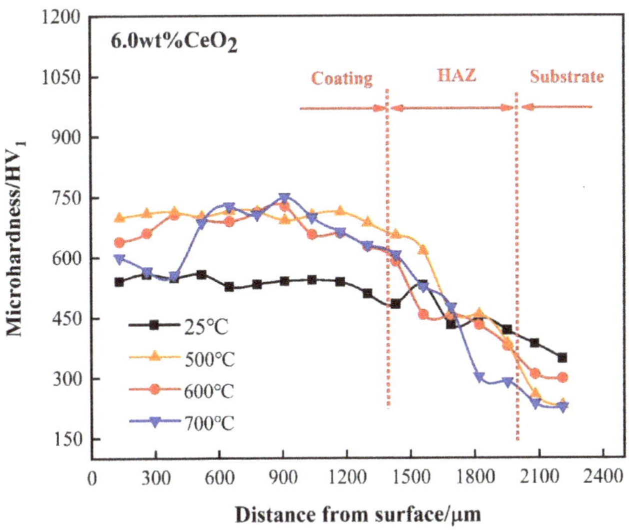

Figure 5. Microhardness distribution at different heat treatment temperatures.

3.6. Wear Morphology

In order to investigate the three-dimensional morphology change of the 6.0%CeO_2/Ni60 cladding layer, MicroXAM-800 optical profilometer was applied to represent the surface profile of the cladding wear marks. The measuring range is 71 × 55 μm, the measuring accuracy is 20 Pa and the resolution is 0.01. Deep grooves appear on the worn surface of the Ni60 cladding layer without heat treatment test, resulting in serious tearing and debris flaking, as shown in Figure 6a. The main wear mechanism that causes this phenomenon is abrasive wear. Figure 6b shows the wear morphology of the cladding layer under the condition of 500 °C heat treatment. The worn surface presents a clear groove track shape, and there is no obvious tearing and spalling phenomenon. The worn surface of the coating is smooth, and no bad slotting is generated after wear, indicating that the formation of the coating is correspondingly homogeneous. Under these experimental conditions the friction mechanism is mostly plastic deformation. The pattern of the coating after wear test when the heat treatment temperature reaches 600 °C is shown in Figure 6c. It can be seen that there are obvious differences in the flatness of the surface topography after wear. Tears and flakes occurred in the worn area. The friction mechanism is mainly adherence wear and grain-abrasion. After heat treatment at 700 °C, the wear morphology of the cladding layer deteriorates further after the wear test. The worn surface is heavily grooved and chipped. The wear mechanism is mainly abrasive wear and adhesive wear. Combined with these analyses, the Ni60 coatings of 6.0%CeO_2 can obtain excellent wear morphology and it has better abrasion resistance at the heat treatment temperature of 500 °C. This combined with the wear of the cladding layer shows that the Ni60 cladding layer of 6.0%CeO_2 is obviously refined after the heat treatment condition of 500 °C, and the microstructure uniformity is obviously improved, which is also a main cause for the preferable wearability of the coatings.

3.7. Friction Coefficient and Wear

The friction coefficient can also reflect the wear performance of the cladding layer. In order to better understand the friction coefficient change of the Ni60 cladding layer with 6.0%CeO_2 under different heat treatment experimental conditions, wear tests under unified test parameters (the load was 5 N and the test time was 20 min) were carried out on the samples after different heat treatment temperatures. The friction coefficient evolution can be seen in Figure 7. The friction factor of the cladding layer at 500 °C is significantly reduced compared with the other three heat treatment temperatures. The friction factors of coatings at 25 °C and 700 °C are almost the same. The friction factor of coatings at 600 °C shows instability, which is consistent with the characteristics of the three-dimensional morphology.

The change of wear performance presented by friction coefficient is also directly reflected in the wear quantity and wear depth. Figure 8 shows the wear quantity (a) and wear depth (b) of the Ni60 cladding layer with 6.0%CeO_2. Figure 8a clearly shows that the wear quantity of the cladding layer at 25 °C and 700 °C is almost the same, which is 28,646 μm^3 and 27,583 μm^3, respectively. The wear of the coating at 500 °C is only 16,583 μm^3, which is 43% less than that of the coating with no-annealing. The wear loss of coatings at 600 °C is 20,354 μm^3, which is slightly higher than that at 500 °C. The variation of wear depth is basically consistent with that of wear quantity, as shown in Figure 8b. In conclusion, the Ni60 cladding layer with 6.0%CeO_2 after 500 °C heat treatment has better tribological and wear performance than cladding layers at other heat treatment temperatures; the wear quantity of the coating without heat treatment is reduced by about 43%.

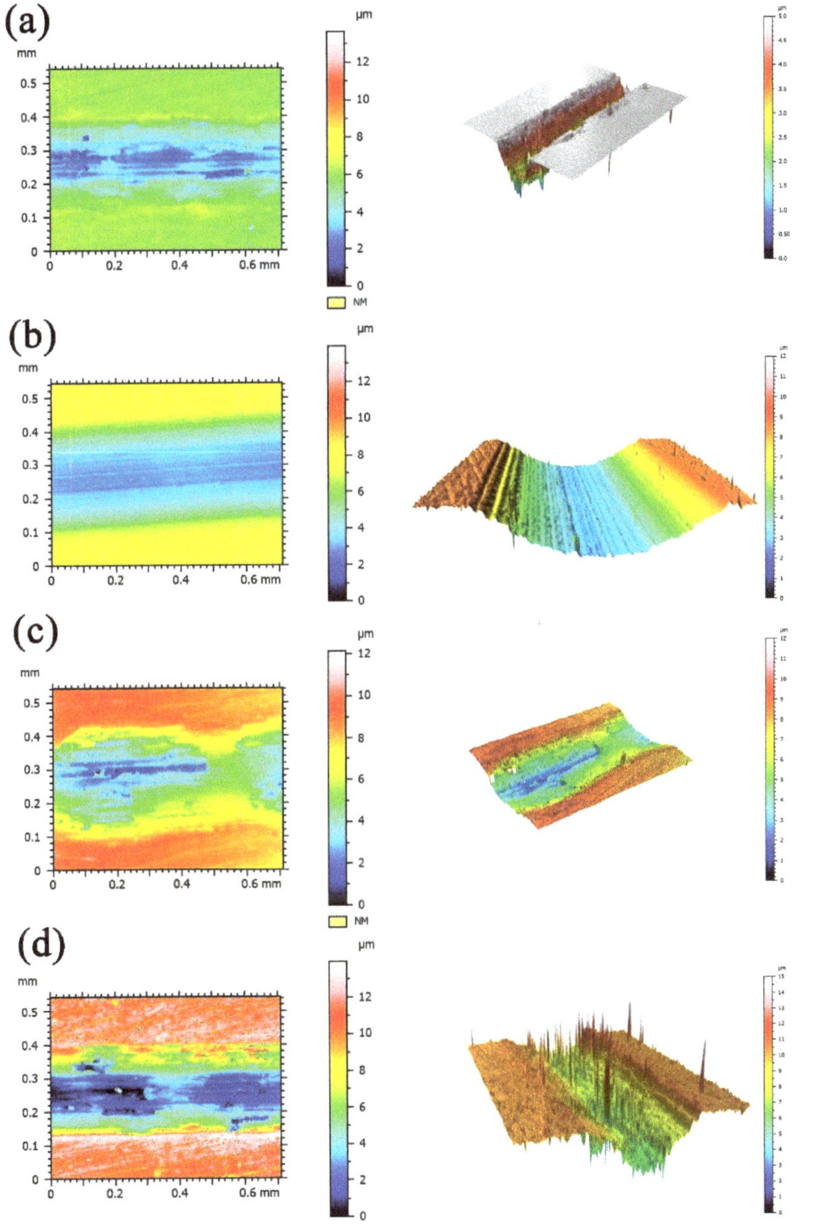

Figure 6. Wear morphology at different heat treatment temperatures: (**a**) 25 °C; (**b**) 500 °C; (**c**) 600 °C; (**d**) 700 °C.

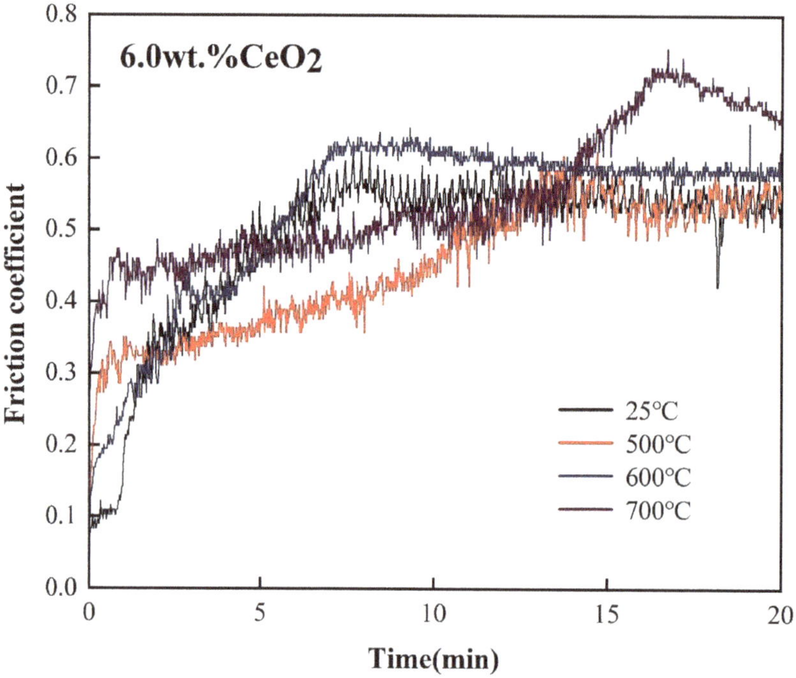

Figure 7. Friction coefficient of Ni60 coating with 6.0%CeO_2 at several annealing temperatures.

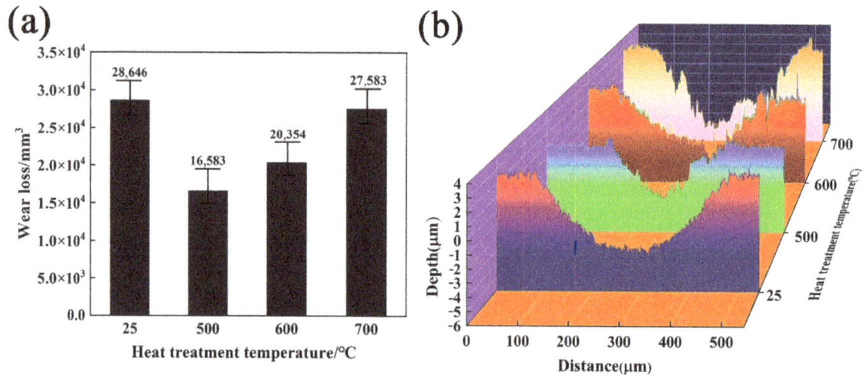

Figure 8. Attrition rate of Ni60 cladding layer under several annealing temperatures with 6.0% CeO_2 addition: (**a**) Wear quantity; (**b**) depth of wear.

4. Conclusions

(1) The tribological and wear properties of Ni60/6.0% CeO_2 cladding layers with annealing temperatures of 25 °C, 500 °C, 600 °C and 700 °C were investigated. The changes of the macroscopic morphology, microstructure, phase composition, microhardness, friction coefficient and wear quantity of the coatings at several annealing temperatures were surveyed and are briefly summarized as follows:

(2) The Ni60 cladding layer of 6.0%CeO_2 is comprised of γ-(Fe,Ni),Cr_7C_3,$Cr_{23}C_6$,CrB, CrFeB and Cr_2Ni_3. Hard phases such as Cr_7C_3, $Cr_{23}C_6$ and CrB significantly raise the hardness and brittleness of the coatings. By calculating the FWHM value and the left shift of XRD diffraction peak at 500 °C, it is found that the grain of the composite coating is

obviously refined, the element segregation phenomenon is effectively inhibited and the homogeneity of the microstructure is greatly improved.

(3) The microhardness distribution of the CeO_2/Ni60 cladding layer is the most homogenous at 500 °C. The average microhardness of the cladding layer is about 732.37 HV_1.

(4) The wear quantities of Ni60 coatings at 25 °C, 500 °C, 600 °C and 700 °C are 28,646 μm^3, 16,583 μm^3, 20,354μm^3 and 27,583 μm^3, respectively. The wear quantity of the coating at 500 °C is only 16,583 μm^3, which is 43% less than that of the coating without annealing. The change the situation of wear depth and friction coefficient is basically consistent with that of wear quantity. Therefore, the 6.0%CeO_2/Ni60 cladding layer obtained the optimal tribological and wear properties at the annealing temperature of 500 °C.

Author Contributions: Writing—original draft, Z.G. (Zhiming Gao), S.Z., H.R. and C.Z.; Writing—review & editing, Z.G. (Zhongtang Gao) All authors have read and agreed to the published version of the manuscript.

Funding: This work was supported by the National Natural Science Foundation of China (52104384, 51804251), China Postdoctoral Science Foundation (2020M683675XB), NSF of Shaanxi Provincial Department of Education (21JK0761), Science Fund of Shandong Laboratory of Yantai Advanced Materials and Green Manufacturing (AMGM2021F04), and the Planning project of Xi'an Science and Technology (21XJZZ0041).

Institutional Review Board Statement: Not applicable.

Informed Consent Statement: Not applicable.

Data Availability Statement: All data that support the findings of this study are included within the article.

Conflicts of Interest: The authors declare no conflict of interest.

References

1. Gao, Z.; Ren, H.; Geng, H.; Yu, Y.; Gao, Z.; Zhang, C. Effect of CeO_2 on Microstructure and Wear Property of Laser Cladding Ni-Based Coatings Fabricated on 35CrMoV Steel. *J. Mater. Eng. Perform.* **2022**, *31*, 9534–9543. [CrossRef]
2. Jiang, X.; Zhou, Y.; Shi, C.; Mao, D. Effects of Ultrasonic-Aided Quenching on the Corrosion Resistance of GB 35CrMoV Steel in Seawater Environment. *Metals* **2018**, *8*, 104. [CrossRef]
3. Liu, J.; Li, X.; Bai, J.; Zhang, T.; Xu, Y.; Yu, Y. Effects of Thermal Treatment on Microstructure and Wear Properties of Ni60/CeO_2 Composite Coating 35CrMoV Steel by Laser Cladding. *Coatings* **2022**, *12*, 1575. [CrossRef]
4. Gao, Y.; Tong, Y.; Guohui, L.; Lu, P.; Zhang, D. Microstructure and Mechanical Properties of Ni-Based Alloy Composite Coating on Cr12MoV by Laser Cladding. *Coatings* **2022**, *12*, 1632. [CrossRef]
5. Cao, S.; Liang, J.; Wang, L.; Zhou, J. Effects of NiCr intermediate layer on microstructure and tribological property of laser cladding Cr_3C_2 reinforced Ni60A-Ag composite coating on copper alloy. *Opt. Laser Technol.* **2021**, *142*, 106963. [CrossRef]
6. Cai, Y.; Luo, Z.; Chen, Y.; Ao, S. Influence of CeO_2 on tribological behaviour of TiC/Fe-based composite coating. *Surf. Eng.* **2017**, *33*, 936–943. [CrossRef]
7. Liang, J.; Yin, X.; Lin, Z.; Chen, S.; Liu, C.; Yan, S.; Dong, S. Effects of LaB6 on microstructure evolution and properties of in-situ synthetic TiC+TiBx reinforced titanium matrix composite coatings prepared by laser cladding. *Surf. Coat. Technol.* **2020**, *403*, 126409. [CrossRef]
8. Quazi, M.M.; Fazal, M.A.; Haseeb, A.S.M.A.; Yusof, F.; Masjuki, H.H.; Arslan, A. Effect of rare earth elements and their oxides on tribo-mechanical performance of laser claddings: A review. *J. Rare Earths* **2016**, *34*, 549–564. [CrossRef]
9. Liu, J.; Yu, H.; Chen, C.; Weng, F.; Dai, J. Research and development status of laser cladding on magnesium alloys: A review. *Opt. Lasers Eng.* **2017**, *93*, 195–210. [CrossRef]
10. Chen, W.; Yang, X.; Li, X.; Chai, C.; Liu, W. Study on Microstructure and Properties of Nickel-Based Self-Lubricating Coating by Laser Cladding. *Coatings* **2022**, *12*, 753. [CrossRef]
11. Shu, D.; Cui, X.; Li, Z.; Sun, J.; Wang, J.; Chen, X.; Dai, S.; Si, W. Effect of the Rare Earth Oxide CeO_2 on the Microstructure and Properties of the Nano-WC-Reinforced Ni-Based Composite Coating. *Metals* **2020**, *10*, 383. [CrossRef]
12. Wang, C.; Gao, Y.; Wang, R.; Wei, D.; Cai, M.; Fu, Y. Microstructure of laser-clad Ni60 cladding layers added with different amounts of rare-earth oxides on 6063 Al alloys. *J. Alloys Compd.* **2018**, *740*, 1099–1107. [CrossRef]
13. Yinghua, L.; Xuelong, P.; Jiacai, K.; Yingjun, D. Improving the microstructure and mechanical properties of laser cladded Ni-based alloy coatings by changing their composition: A review. *Rev. Adv. Mater. Sci.* **2020**, *59*, 340–351. [CrossRef]

14. Liang, C.J.; Wang, C.L.; Zhang, K.X.; Liang, M.L.; Xie, Y.G.; Liu, W.J.; Yang, J.J.; Zhou, S.F. Nucleation and strengthening mechanism of laser cladding aluminum alloy by Ni-Cr-B-Si alloy powder based on rare earth control. *J. Mater. Process. Technol.* **2021**, *294*, 117145. [CrossRef]
15. Wang, C.; Gao, Y.; Zeng, Z.; Fu, Y. Effect of rare-earth on friction and wear properties of laser cladding Ni-based coatings on 6063Al. *J. Alloys Compd.* **2017**, *727*, 278–285. [CrossRef]
16. Oh, Y.; Han, C.H.; Wang, M.; Chun, Y.-B.; Han, H.N. Effect of rare earth oxide addition on microstructure and mechanical properties of Ni-based alloy. *J. Alloys Compd.* **2021**, *853*, 156980. [CrossRef]
17. Xu, Z.; Wang, Z.; Chen, J.; Qiao, Y.; Zhang, J.; Huang, Y. Effect of Rare Earth Oxides on Microstructure and Corrosion Behavior of Laser-Cladding Coating on 316L Stainless Steel. *Coatings* **2019**, *9*, 636. [CrossRef]
18. Lu, X.-L.; Liu, X.-B.; Yu, P.-C.; Zhai, Y.-J.; Qiao, S.-J.; Wang, M.-D.; Wang, Y.-G.; Chen, Y. Effects of heat treatment on microstructure and mechanical properties of Ni60/h-BN self-lubricating anti-wear composite coatings on 304 stainless steel by laser cladding. *Appl. Surf. Sci.* **2015**, *355*, 350–358. [CrossRef]
19. Liu, L.; Xu, H.; Xiao, J.; Wei, X.; Zhang, G.; Zhang, C. Effect of heat treatment on structure and property evolutions of atmospheric plasma sprayed NiCrBSi coatings. *Surf. Coat. Technol.* **2017**, *325*, 548–554. [CrossRef]
20. Pan, Y.; Cui, J.; Lei, W.; Zhou, J.; Ma, Z. Influence of Heat Treatment on the Mechanical Properties of Ni Films on 430 Stainless Steel Substrate. *High Temp. Mater. Process.* **2017**, *36*, 855–861. [CrossRef]
21. Wang, S.; Zheng, Z.; Zheng, K.; Long, J.; Wang, J.; Ren, Y.; Li, Y. High temperature oxidation behavior of heat resistant steel with rare earth element Ce. *Mater. Res. Express* **2020**, *7*, 016571. [CrossRef]
22. Guo, W.; Zhang, L.; Xu, C.; Chai, R.; Gao, Z.; Kogo, B.; Chizari, M.; Zhang, C.; Wang, B. Study on the wear resistance of laser cladding iron-base alloy by heat treatment. *Mater. Res. Express* **2018**, *6*, 2. [CrossRef]
23. Gao, Z.; Ren, H.; Yuan, Y.; Gao, Z.; Liu, E.; Zhang, C. Effect of CeO(2) on the microstructure and microhardensss of laser-cladded Ni60 on 35CrMoV alloys. *Micron* **2021**, *150*, 103146. [CrossRef]
24. Yang, L.; Li, Z.; Zhang, Y.; Wei, S.; Wang, Y.; Kang, Y. In-situ TiC-Al3Ti reinforced Al-Mg composites with Y_2O_3 addition formed by laser cladding on AZ91D. *Surf. Coat. Technol.* **2020**, *383*, 125249. [CrossRef]
25. Farahmand, P.; Liu, S.; Zhang, Z.; Kovacevic, R. Laser cladding assisted by induction heating of Ni–WC composite enhanced by nano-WC and La_2O_3. *Ceram. Int.* **2014**, *40*, 15421–15438. [CrossRef]

coatings

MDPI

Article

High Temperature Oxidation and Oxyacetylene Ablation Properties of ZrB_2-ZrC-SiC Ultra-High Temperature Composite Ceramic Coatings Deposited on C/C Composites by Laser Cladding

Kaijin Huang [1,2,*], Yahao Xia [1] and Aihua Wang [1]

1 State Key Laboratory of Materials Processing and Die & Mould Technology, Huazhong University of Science and Technology, Wuhan 430074, China
2 State Key Laboratory of Advanced Technology for Materials Synthesis and Processing, Wuhan University of Technology, Wuhan 430070, China
* Correspondence: huangkaijin@hust.edu.cn; Tel.: +86-027-87543676

Abstract: In order to improve the high temperature oxidation and ablation resistance of C/C composites, ZrB_2-ZrC-SiC ultra-high temperature composite ceramic coatings were prepared on C/C composites by laser cladding using Zr, B_4C, and Si as raw materials. The microstructure of the coating was characterized by scanning electron microscopy (SEM) and X-ray diffraction (XRD). Air isothermal oxidation (1600 °C, 80 min) and oxyacetylene flame ablation (2400 kW/m^2, 300 s) were used to evaluate the high-temperature oxidation and ablation properties of the coating, respectively. The results show that the microstructure of laser cladding coating is a totem of black and white. The white part is mainly the first solidified high melting point ZrB_2 phase, and the black part is the latter solidified eutectic structure, which is mainly composed of $ZrB_2(ZrB_{12})$-ZrC or $ZrB_2(ZrB_{12})$-SiC two phases. After oxidation at 1600 °C and 80 min, the coating is mainly composed of ZrO_2 and $ZrSiO_4$ phases, and $ZrSiO_4$ is basically distributed among ZrO_2 particles. The high temperature oxidation and ablation properties of the coating are better than the C/C composite matrix, and the mass ablation rate of the coating is about 1/4 of the latter.

Keywords: laser cladding; ultra-high temperature composite ceramic coatings; high-temperature oxidation; oxyacetylene flame ablation; C/C composites

1. Introduction

Carbon/carbon (C/C) composites, namely carbon fiber or graphite fiber reinforced carbon matrix composites, are a new type of materials with unique properties in the field of composite materials. They can show excellent high-temperature mechanical properties in the environment with no oxygen or very low oxygen content, maintain their advantages in the environment with 3000 °C or rapid temperature change, and when the temperature exceeds 1500 °C, their mechanical strength will increase with the increase of temperature. The preparation of C/C composites is formed by the densification of carbon matrix in carbon fiber. Therefore, people design carbon fiber preforms with different dimensions and sizes to meet the different requirements of material properties in the production process [1]. In addition, C/C composites have the advantages of low density (1.3–2.5 g/cm^3), high thermal conductivity (20–150 W/m.K), low thermal expansion coefficient (2×10^{-6}/K), and excellent high temperature mechanical properties. It has a wide range of applications in aerospace, nuclear energy, and automotive. However, its weakness of easy oxidation above 500 °C limits its long-term use in aerobic atmosphere and high temperature environment [2]. In order to solve the problem of surface oxidation resistance of C/C composites, there are two main types of external oxidation resistance coating technologies, which can isolate the contact between oxygen and matrix, and internal matrix modification technology, which can improve the oxidation resistance of carbon fiber and matrix by additive [3–6]. In other words, by preparing anti-oxidation coatings on C/C composites or introducing refractory borides, carbides and nitride phases such as ZrB_2, HfB_2, ZrC, TaC, HfC, and HfN into C/C composites through matrix modification technology, the ultra-high temperature ablative performance of C/C composites can be improved.

Among them, the main preparation technologies for anti-oxidation coatings include pack cementation, chemical vapor deposition (CVD), plasma spraying, slurry painting, sol–gel, reactive infiltration, chemical vapor reaction (CVR), in situ formation, laser cladding, etc. [5–12]. The main matrix modification technologies include chemical vapor infiltration/ deposition, precursor infiltration and pyrolysis, reactive melt infiltration method, slurry infiltration/impregnation, hot pressing method, etc. [3–5].

However, various coating preparation technologies have their own advantages and disadvantages [6]. For example, the coating prepared by CVD and plasma spraying has a weak binding force with the substrate and poor thermal shock resistance. The slurry painting and sol–gel method are easy to operate, but the compactness and thermal shock resistance of the coating are poor. Although the coating prepared by pack cementation and CVR has good bonding properties and thermal shock resistance, it is difficult to avoid the defects caused by preparation. At present, plasma spraying and laser cladding are the two main technologies that can achieve millimeter thickness coating. The former has the advantages of low requirements for coating materials, high deposition rate, simple operation, low preparation cost, controllable thickness, etc., but has the disadvantages of layered coating structure, high porosity in coating, weak interface bonding force between coating and substrate, etc. The latter has the advantages of fine microstructure of coating, metallurgical bonding between coating and substrate, controllable thickness, and easy automation. However, hypersonic weapons require a thickness of millimeter-level thermal protection coating for full-temperature long-term oxidation and ablative resistance [13]. Therefore, in this paper, the laser cladding technology was used to prepare the millimeter-thick ZrB_2-ZrC-SiC ultra-high temperature ceramic composite coating with oxidation and ablative resistance.

In fact, there are many literature reports on the high-temperature oxidation resistance modification of C/C composites [14] or the preparation of high-temperature oxidation resistance coating on its surface [15]. However, the preparation of ZrB_2-ZrC-SiC ultra-high temperature composite ceramic coatings with resistance to 1600 °C on C/C composites based on laser cladding technology has not been reported in the literature [7–12]. Wang et al. [7] prepared the inner Si-SiC-$MoSi_2$ and outer SiO_2-Nd_2O_3 double layer coating on C/C composites using slurry-based laser cladding technology. The ablative properties of the coating under 300/500 W fiber laser with wavelength of 1060–1090 nm were studied. The results show that the ablative resistance of the coating is improved. Wang et al. [8] prepared the inner Si-SiC and outer SiO_2-Nd_2O_3 double layer coating on C/C composites using slurry-based laser cladding technology. The ablative properties of 1–7 s with 300/500 W fiber laser with wavelength of 1060–1090 nm were studied. The results show that the ablative resistance of the coating is improved. Teng et al. [9] prepared Si-SiC coating on C/C composites using slurry-based laser cladding technique and studied the ablation properties of the coating under 300 W fiber laser with wavelength of 1060–1090 nm. The results show that the ablative resistance of the coating is improved. Wang et al. [10] prepared the inner Si-SiC and outer SiO_2-B_2O_3-Al_2O_3-Sm_2O_3 double layer coating on C/C composites using slurry-based laser cladding technique. The properties of the double-layer coating oxidized at 1500 °C for 10 h were studied. The results show that the coating containing Sm_2O_3 has better oxidation resistance than the coating without Sm_2O_3. Wang et al. [11] prepared Si-SiC coating on C/C composites using slurry-based laser cladding technique and studied the oxidation properties of the coating at 1500 °C for 10 h. The results show that the oxidation resistance of the coating is improved. Eric et al. [12] used laser cladding technique to prepare NiCr alloy (78%Ni/19Cr), alumina ceramic (95%Al_2O_3 + 3%TiO_2), and magnesium oxide/zirconia ceramic (76%ZrO2 + 22%MgO) coatings on C/C composites. However, only the alumina ceramic coating was successful, but its oxidation resistance was not tested.

In this paper, ZrB_2-ZrC-SiC ultra-high temperature composite ceramic coatings were prepared on C/C composites using Zr powder, B_4C powder, and Si powder as raw materials by laser cladding, and the microstructure characteristics, high temperature oxidation behavior at 1600 °C, and ablative behavior at 2400 kW/m^2 heat flux of the coatings were studied.

2. Materials and Methods

The C/C composites were provided by Yixing Hengli Aviation Technology Co., Ltd., Jiangsu Province, China. The sample size used in the experiment was length × width × thickness = 100 mm × 50 mm × 10 mm. Figure 1a is the XRD pattern of C/C composites. It can be seen from the figure that the main phase of C/C composites is graphite (JSPDS 96-120-0018). Figure 1b shows the SEM morphology of C/C composites. The figure shows the carbon fiber woven together, and the size of the single carbon fiber is about 7 μm.

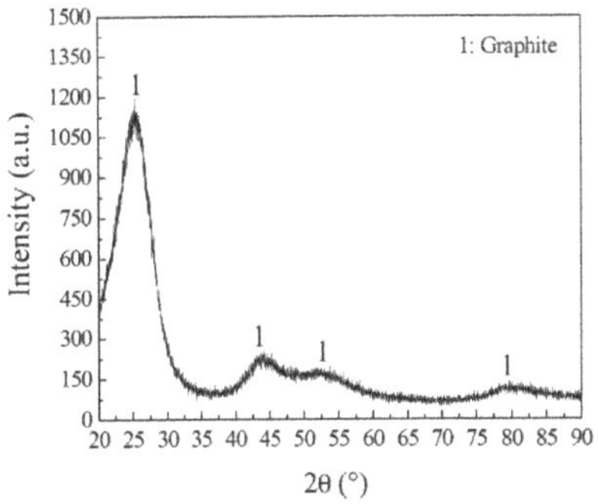

(a)

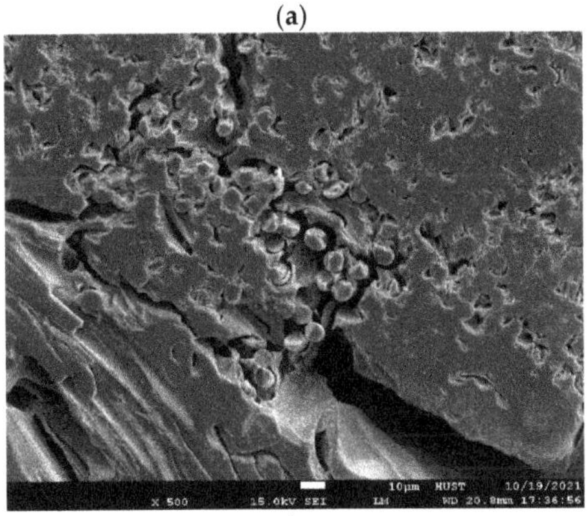

(b)

Figure 1. XRD pattern and SEM morphology of C/C composites. (**a**) XRD pattern; (**b**) SEM morphology.

Zr powder, Si powder, and B_4C powder are selected as raw materials for laser cladding. Among them, the purity of Zr powder is 99.5% and the particle size is 45 μm; the purity of Si powder is 99.8% and the particle size is 45 μm; the purity of B_4C powder is 98.0% and the particle size is 4.5 μm. They are irregular shapes. The first two were provided by Beijing High-Tech New Material Technology Co., Ltd., Beijing, China. The latter one was provided by Longlai Import & Export Trading Co., Ltd., Zhenzhou, China.

The ZrB_2-ZrC-SiC ultra-high temperature ceramic coating was prepared by laser cladding with 0.6 mm thick mixed powder layer on C/C composites. The binder used for pre-mixed powder layer was 4% PVA solution and the ratio of 4% PVA solution to powder mixture was approximately 1:10. The specific steps are as follows. First, according to the equation $2Zr + B_4C + Si = 2ZrB_2 + SiC$, the required mass of Zr powder, Si powder, and B_4C powder was calculated. Secondly, the corresponding Zr powder, Si powder, and B_4C powder were weighed by SOP Quintix124 (Sartorius, Gottingen, Germany) balance (measuring accuracy was 0.1 mg). Then it was ground in the ball mill for 8 h. After being mixed evenly, it was prepared with 4% PVA solution in advance to make it sticky. Finally, the mixed powder layer of about 0.6 mm thick was prefabricated on C/C composites, which was to be used after natural air drying.

The laser cladding experiment was carried out on the complete set of laser cladding equipment of RSF-C6000 manufactured by Wuhan Rayco Fiber Laser Technology Co., LTD (Wuhan, China). In the process of laser cladding, in order to prevent the oxygen in the air from entering the laser molten pool, argon gas with 99.99% high purity was used as the protective gas. The preliminarily optimized laser cladding parameters were as follows: laser power P = 2.5 KW, laser scanning speed V = 15 mm/s, laser spot size D = 5 mm, laser spot overlapping rate f = 30%, and high purity argon flow rate is 15 L/min.

The laser cladding samples and C/C composites matrix were cut into 10 mm × 10 mm × 10 mm, 20 mm × 20 mm × 1 mm, and 20 mm × 20 mm × 10 mm by wire cutting method, respectively. The sample of 10 mm × 10 mm × 10 mm was used for XRD analysis and SEM observation, while the sample of 20 mm × 20 mm × 1 mm was used for high-temperature oxidation experiment, and the sample of 20 mm × 20 mm × 10 mm was used for ablation experiment. All samples after wire-cut were cleaned by washing powder, ultrasonic cleaning, and drying oven before being placed in a drying dish for use.

The high temperature oxidation performance of laser cladding coating and C/C composites substrate was evaluated according to Chinese aviation industry standard HB5258-2000 "Test method for determination of oxidation resistance in steel and high temperature alloys". The 1600 °C isothermal oxidation experiment was carried out in KSL-1700X-A4 (Hefei Kejing Materials Technology Co., LTD, Hefei, China) type box furnace. The experimental process of isothermal oxidation was as follows. Firstly, the SOP Quintix124 electronic balance (accuracy is 0.1 mg) was used to weigh the total mass of the sample and crucible, and then the total mass of the sample and crucible after each oxidation period was weighed. In this paper, four oxidation periods of 20 min, 40 min, 60 min, and 80 min were set. Finally, the oxidation gain rate per unit area of the coated samples after oxidation in different time periods was calculated, and the relationship between the oxidation gain rate per unit area and the corresponding oxidation time was drawn, namely, the oxidation kinetics curve.

The oxyacetylene ablative performance of laser cladding coating and C/C composites substrate was evaluated according to Military standards of the People's Republic of China GJB 323B-2018 "Test method for ablation of ablators ". The specific test steps of oxyacetylene ablative performance were as follows. The oxyacetylene ablative sample was first weighed using SOP Quintix124 balance (accuracy is 0.1 mg) and the weight m_0 was recorded. It was weighed in the crucible and the weight m'_0 was recorded. Finally, the oxyacetylene ablative performance was tested. Table 1 lists the oxyacetylene ablative parameters. Three samples were made for each data point at the same time and their average values were taken.

Table 1. Oxyacetylene ablative parameters.

Parameters	Values
Ablation time/s	300
Oxygen pressure/MPa	0.8
Acetylene pressure/MPa	0.08
Nozzle to sample distance/mm	10
Nozzle diameter/mm	2
Angle between nozzle and sample/°	90
Heat flux/KW·m^{-2}	2400

The mass ablation rate of the sample was calculated according to the following formula:

$$R_m = \frac{m_0' - m_1'}{t}$$

R_m is the sample mass ablation rate; m_0' is the mass of the crucible containing the sample before ablation; m_1' is the mass of the sample containing the crucible after ablation; t is the ablation time.

The phase identification of samples was done in the X 'Pert PRO X-ray diffractometer produced by PANalytical B.V., Almelo, the Netherlands. The test parameters were Cu target with a wavelength λ of 0.15406 nm, tube pressure of 40 kV, tube flow of 30 mA, and 2θ scanning range of 20–90°. The sample morphology was analyzed by using Quanta 650 FEG scanning electron microscope produced by FEI Company, Hillsboro, USA. The composition and proportion of the corrosion solution used were HNO_3:HCl = 1:3, and the corrosion time was 30 s.

3. Results and Discussion

3.1. Phase and Microstructure of the Coating

Figure 2 shows the XRD pattern of laser cladding coating. As can be seen from the figure, the laser cladding coating is mainly composed of four phases. They are hexagonal ZrB_2 (JSPDS 96-151-0857), cubic SiC (JSPDS 96-101-1032), cubic ZrC (JSPDS 96-900-8778), and monoclinic ZrO_2 (JSPDS 96-230-0297). Compared with the design of the coating (equation 2Zr + B_4C + Si = $2ZrB_2$ + SiC), two phases ZrC and m-ZrO_2 appear in addition to the coating design ZrB_2 and SiC. Among them, the formation of m-ZrO_2 phase may be related to the incomplete protection of laser molten pool by high purity argon during laser cladding, because there is a very active elemental metal Zr in the original mixed powder, which is easy to react with O_2 in the air to produce ZrO_2. The formation reasons of ZrB_2, SiC, and ZrC phases are as follows:

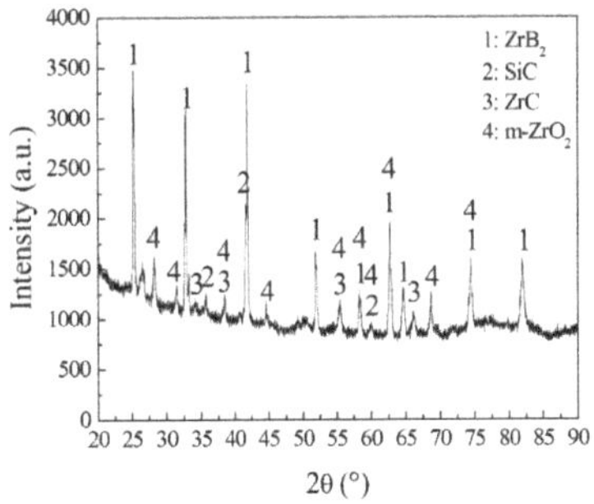

Figure 2. XRD pattern of laser cladding coating.

Three raw materials, Zr, Si, and B_4C, were used in laser cladding, and the possible reactions were as follows:

$$B_4C + Si = SiC + B \quad (1)$$

$$B_4C + Zr = ZrC + 4B \quad (2)$$

$$B_4C + 2Zr = 2ZrB_2 + C \quad (3)$$

$$2B + Zr = ZrB_2 \quad (4)$$

$$ZrC + Si = SiC + Zr \quad (5)$$

$$C + Zr = ZrC \quad (6)$$

$$C + Si = SiC \quad (7)$$

$$B_4C + 3Zr = 2ZrB_2 + ZrC \quad (8)$$

$$B_4C + 2ZrC + 3Si = 2ZrB_2 + 3SiC \quad (9)$$

$$B_4C + 2Zr + Si = 2ZrB_2 + SiC \quad (10)$$

The Gibbs free energy ΔG at room temperature ~3000 K of the above equations was found by HSC6.0 software (V6.0, 2006, Outokumpu Technology, Outokumpu, Finland), and the Gibbs free energy ΔG at different temperatures T was calculated and plotted, as shown in Figure 3.

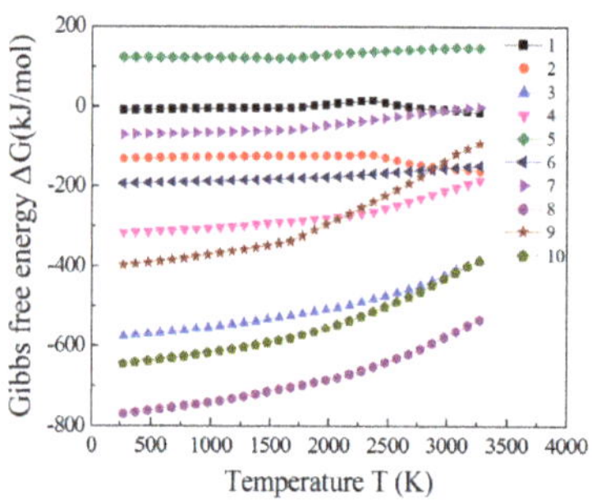

Figure 3. Gibbs free energy ΔG values at different temperatures for different reactions.

As can be seen from Figure 3, ΔG value of reaction (8) is the lowest at the same temperature, so from a thermodynamic point of view, reaction (8) will take precedence, resulting in the formation of ultra-high temperature ceramics ZrB_2 and ZrC phases. The Si in the original powder mixture will react with the generated part of ZrC and the remaining B_4C according to reaction (9) to form ZrB_2 and SiC phases. Therefore, the laser cladding coating will eventually get ZrB_2, ZrC, and SiC phases. This step-reaction mechanism has been reported in literature [16].

In addition, it can be seen from Figure 2 that the laser cladding coating contains four ultra-high temperature ceramic phases, ZrB_2, ZrC, SiC, and m-ZrO_2, but the relative intensity of diffraction peaks of each phase is significantly different, which indicates that the contents of the four phases are different in the laser cladding coating. In order to estimate their contents, the XRD-K value method reported in [17] was used in this paper. We calculated the mass fraction of different i and j phases in the coating according to the following formula:

$$\frac{w_i}{w_j} = \frac{w_i'/(1-w_c)}{w_j'/(1-w_c)} = \frac{I_i/K_c^i}{I_j/K_c^j}$$

where w_i and w_j represent the mass fraction of two phases i and j in the coating, respectively; w_i' and w_j' represent the mass fraction of phase i and phase j after adding the standard substance c, respectively; w_c represents the mass fraction of the added standard substance (such as α-Al_2O_3); I_i and I_j represent the intensity of the diffraction line of 2θ value corresponding to the strongest diffraction peak of the two phases i and j; K_c^i and K_c^j represent the reference intensity values (RIR values) of the i and j phases, respectively.

The strongest diffraction peaks of ZrB_2, SiC, ZrC, and m-ZrO_2 are (011), (111), (111), and $(11\bar{1})$, respectively. The corresponding 2θ values of the above four phases are 41.732°, 35.655°, 33.106°, and 28.179°, respectively. The corresponding RIR values of the above four phases are 9.25, 3.77, 9.80, and 4.98, respectively. According to Figure 2 and the above formula, the mass fractions of ZrB_2, SiC, ZrC, and m-ZrO_2 phases are 31.16%, 28.84%, 11.73%, and 28.27%, respectively. Obviously, the different proportions of ZrB_2-SiC-ZrC ultra-high temperature composite coating will affect its high temperature oxidation and ablation properties.

Figure 4 is the SEM image of the cross section of the laser cladding coating. As can be seen from the figure, the maximum thickness of the laser cladding coating is about 1.97 mm (Figure 4a). The laser cladding layer on the upper part of the coating is dense, but there are porosity defects of different scales in the lower part of the coating near the substrate (Figure 4a). There is local cracking at the interface of the coating and C/C composite matrix (Figure 4a). The internal microstructure of the coating presents a totem shape of black and white (Figure 4b–e). Among them, the pores in the coating with different sizes may be related to the H_2O and CO_2 formed by the decomposition of binder PVA during the laser cladding preset powder layer. Because the solidification rate of the laser cladding melt pool is very fast, a large number of gases cannot escape in time, resulting in the remaining pores in the coating, which has also been reported in the literature [18]. The local cracking at the interface of coating and C/C composite matrix should be caused by the difference of thermal expansion coefficient between coating containing ZrB_2, ZrC, SiC, ZrO_2, and C/C composite matrix. The thermal expansion coefficients of ZrB_2, ZrC, SiC, ZrO_2, and C at room temperature are 6.88×10^{-6} k^{-1}, 6.7×10^{-6} k^{-1}, 4.5×10^{-6} k^{-1}, 10.5×10^{-6} k^{-1}, and (0.5–1.5) $\times 10^{-6}$ k^{-1} [19].

Table 2 shows the EDS results of the five positions in Figure 4c. According to the XRD calibration results in Figure 2 and Table 2, the off-white positions 1 and 5 in the coating should be mainly large chunks of ZrB_2. Totemic black position 2 contains SiC. ZrB_{12} and ZrC were found in both the lying black patch site 3 and the erect off-white patch site 4.

Table 3 shows the EDS results of the six positions in Figure 4d. According to the XRD calibration results in Figure 2 and Table 3, the gray position 2 in the coating should be mainly large block ZrB_2. ZrB_{12} and ZrC were found at locations 1, 4, and 6. The dimpled black position 3 contains ZrB_{12} and SiC. The totemic black position 5 contains ZrB_{12} and ZrC.

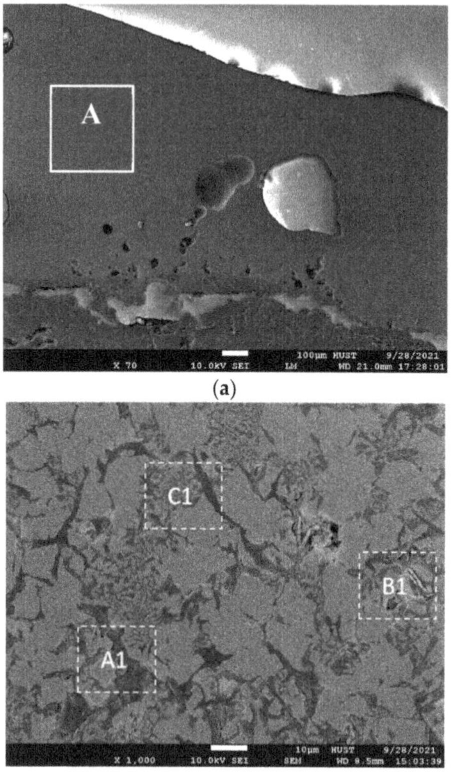

(a)

(b)

Figure 4. *Cont.*

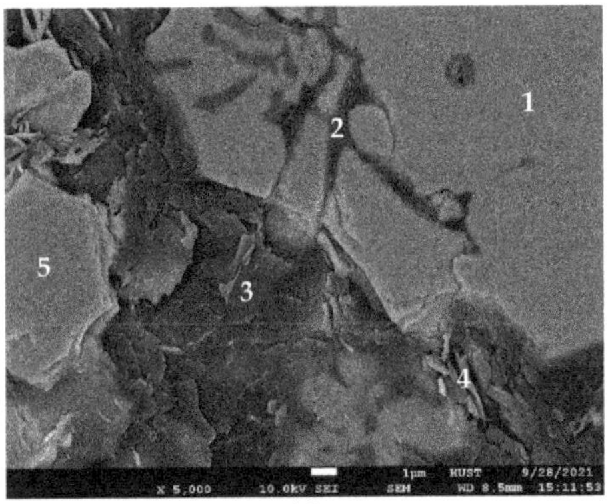

(**c**)

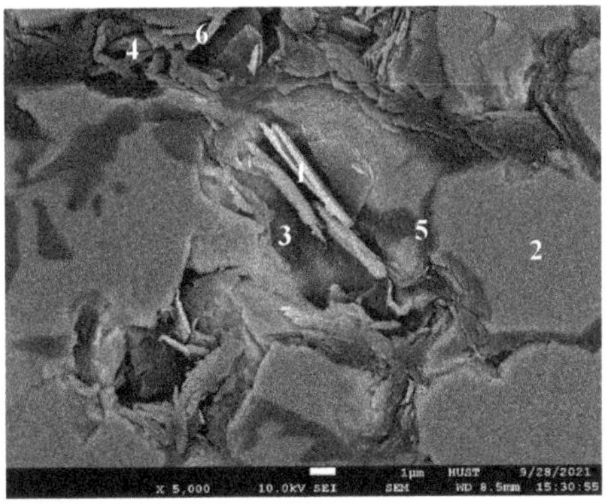

(**d**)

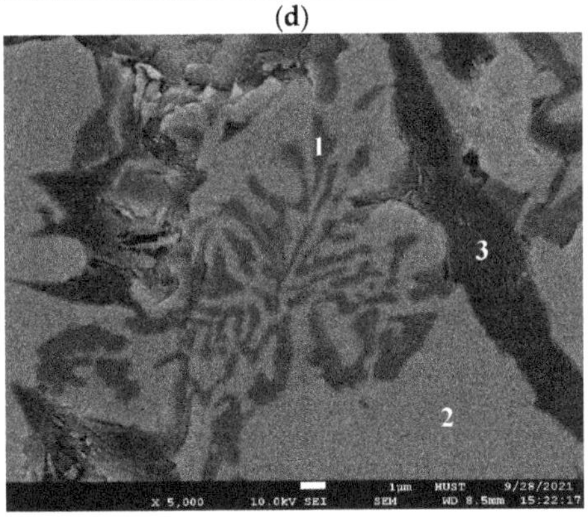

(**e**)

Figure 4. SEM image of cross section of laser cladding coating. (**a**) Whole SEM image of the coating; (**b**) Locally amplified SEM image at area A in (**a**); (**c**) Locally amplified SEM image at area A1 in (**b**); (**d**) Locally amplified SEM image at area B1 in (**b**); (**e**) Locally amplified SEM image at area C1 in (**b**). A, A1, B1, C1 represent different regions in the coating, while 1,2,3,4,5,6 represent different position points.

Table 4 shows the EDS results of the three positions in Figure 4e. According to the XRD calibration results in Figure 2 and Table 4, the totemic black position 1 in the coating contains ZrB_{12} and SiC. The grey and white position 2 should be mainly large block ZrB_2. The long black position 3 contains ZrB_{12} and ZrC.

Table 2. EDS results of five positions in Figure 4c.

Element (at.%)	Point 1	Point 2	Point 3	Point 4	Point 5
B	50.67	46.18	53.07	48.16	49.07
C	0.57	10.57	19.14	8.24	0.54
O	19.73	12.40	13.10	31.42	21.51
Si	-	3.64	-	-	-
Zr	29.03	27.22	14.69	12.18	28.89
Possible phases	ZrB_2-ZrO_2	ZrB_2-SiC-ZrO_2	ZrB_{12}-ZrC-ZrO_2	ZrB_{12}-ZrC-ZrO_2	ZrB_2-ZrO_2

Table 3. EDS results of six positions in Figure 4d.

Element (at.%)	Point 1	Point 2	Point 3	Point 4	Point 5	Point 6
B	42.73	47.04	54.81	52.24	49.16	54.14
C	11.43	0.33	14.91	20.28	9.93	22.90
O	33.07	25.42	18.31	19.42	26.08	16.91
Si	0.87	-	4.49	-	0.23	0.17
Zr	11.91	27.21	7.48	8.06	14.60	5.88
Possible phases	ZrB_{12}-ZrC-ZrO_2	ZrB_2-ZrO_2	ZrB_{12}-SiC-ZrO_2	ZrB_{12}-ZrC-ZrO_2	ZrB_{12}-ZrC-ZrO_2	ZrB_{12}-ZrC-ZrO_2

Table 4. EDS results of three positions in Figure 4e.

Element (at.%)	Point 1	Point 2	Point 3
B	43.64	48.10	54.76
C	1.32	0.53	19.29
O	11.67	21.94	18.53
Si	36.63	-	-
Zr	6.74	29.42	7.42
Possible phases	ZrB_{12}-SiC-ZrO_2	ZrB_2-ZrO_2	ZrB_{12}-ZrC-ZrO_2

It should be pointed out that in addition to the four phases of ZrB_2, ZrC, SiC, and ZrO_2 calibrated by XRD, ZrB_{12} phase also appears in the coating, which may be related to the fast solidification of laser cladding pool. According to the B-Zr binary phase diagram (Figure 4 in Ref. [20]), we know that under the slow solidification balance condition and low Zr content area, the peritectic reaction of L + ZrB_2→ZrB_{12} can happen at 2092 °C. The eutectic reaction L→ZrB_{12} + β-B can happen at 2026 °C. The eutectoid transformation of ZrB_{12}→ZrB_2 + β-B can happen at 1695 °C. Therefore, under the condition of equilibrium, cooling to room temperature should obtain ZrB_2 and β-B. However, it is well-known that the solidification rate of laser cladding pool is very fast, which is much higher than the equilibrium solidification rate. Therefore, the above eutectoid transformation may not be complete at room temperature, and ZrB_{12} phase can be obtained at room temperature. In addition, when Zr content is low and B content is high, this is consistent with the low Zr and high B in Tables 2–4.

To sum up, the microstructure of the coating is basically a totem of black and white, in which the white part is mainly the high melting point ZrB_2 phase solidified first, and the black part is mainly the eutectic reaction region solidified after, which is mainly composed of ZrB_2 (ZrB_{12})-ZrC or ZrB_2 (ZrB_{12})-SiC phases.

3.2. High Temperature Oxidation Properties of the Coating

Figure 5 shows the weight gain curves of C/C composite matrix and laser cladding coating after oxidation at 1600 °C and for 20 min, 40 min, 60 min, and 80 min. As can be seen from the figure: (1) The laser cladding coating is in the state of weight gain before 40 min, and the coating is in the state of weight loss after 40 min, which indicates that before 40 min, the weight gain rate of the oxidation products ZrO_2 and SiO_2 of the ultra-high temperature ceramic phase ZrB_2, ZrC, and SiC in the coating is greater than the weight loss rate of the gas phase B_2O_3, SiO, and CO in the oxidation products (because at

1600 °C, the oxidation of SiC changes from inert oxidation to active oxidation, that is, from $2SiC(s) + 3O_2(g) = 2SiO_2(s) + 2CO(g)$ to $SiC(s) + O_2(g) = SiO(g) + CO(g)$ [21]; the melting point and boiling point of B_2O_3 are 450 °C and 1500 °C, respectively; the melting and boiling points of ZrO_2 are 2680 °C and 4300 °C, respectively; the melting point and boiling point of SiO_2 are 1713 °C and 2230 °C, respectively). After the oxidation time exceeds 40 min, the content of ultra-high temperature ceramic phases ZrB_2, ZrC, and SiC in laser cladding coatings decreases, the weight gain rate of oxidation products ZrO_2 and SiO_2 is lower than the weight loss rate of gas phase B_2O_3, SiO, and CO in the oxidation products, and the result is a weight loss phenomenon. (2) At 40 min, the oxidation weight gain rate of laser cladding coating is 0.51 g/cm^2, while the oxidation weight loss rate of C/C composite is 2.0 g/cm^2, the former is about $\frac{1}{4}$ of the latter. (3) The C/C composite matrix was in a state of weight reduction before 40 min, and after 40 min, all of them were oxidized into CO_2 gas. (4) At 1600 °C, the laser cladding coating can protect the C/C composite matrix for 80 min.

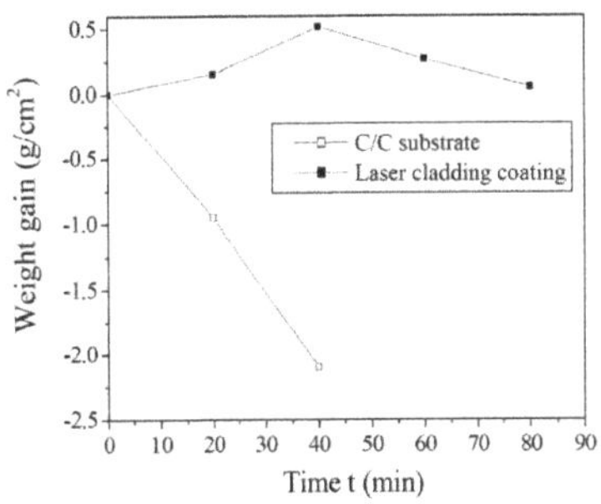

Figure 5. Weight gain curves of C/C matrix and laser cladding coating after oxidation at 1600 °C and at different time.

Figure 6 shows the XRD pattern of the laser cladding coating after oxidation at 1600 °C and 80 min. As can be seen from the figure, the oxidized products of laser cladding coating are composed of m-ZrO_2 (JSPDS 00-007-0343) and $ZrSiO_4$(JSPDS 01-071-0991) phases. This indicates that the ZrB_2, ZrC, and SiC phases in the laser cladding coating are completely oxidized after oxidation at 1600 °C and 80 min. That is, $2ZrB_2(s) + 5O_2(g) = 2ZrO_2(s) + 2B_2O_3(g)$, $2ZrC(s) + 3O_2(g) = 2ZrO_2(s) + 2CO(g)$, $2SiC(s) + 3O_2(g) = 2SiO_2(s) + 2CO(g)$ or $SiC(s) + O_2(g) = SiO(g) + CO(g)$, $2SiO(g) + O_2(g) = 2SiO_2(s)$ [20]. According to the ZrO_2-SiO_2 phase diagram (Figure 2 in Ref. [22]), the $ZrSiO_4$ phase can be generated by the peritectoid reaction ($ZrO_2 + SiO_2 = ZrSiO_4$) at 1676 °C.

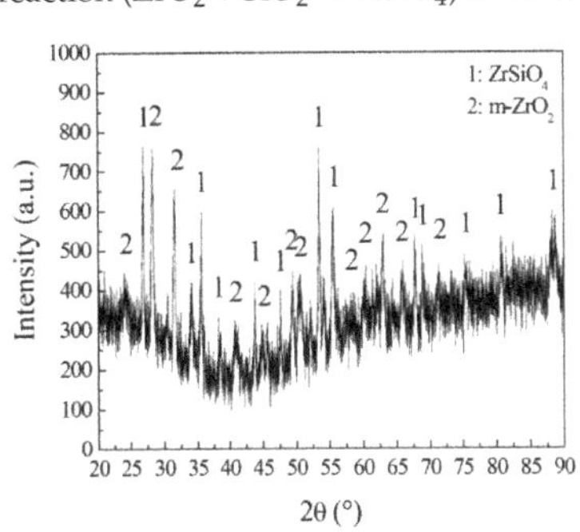

Figure 6. XRD pattern of laser cladding coating after oxidation at 1600 °C and 80 min.

Figure 7 shows the SEM image of the laser cladding cross section coating after oxidation at 1600 °C and 80 min. As shown in Figure 7 and Table 5, the coating appears hollow after oxidation at 1600 °C and for 80 min, and the maximum thickness of the coating after oxidation is about 1.162 mm. After the coating is oxidized, $ZrSiO_4$ phase mainly appears between ZrO_x oxides, among which ZrO_x oxides include both fully oxidized and coarse ZrO_2 phase and may also include incomplete oxidized ZrO and Zr_2O phases, which have been reported in the literature [23–25].

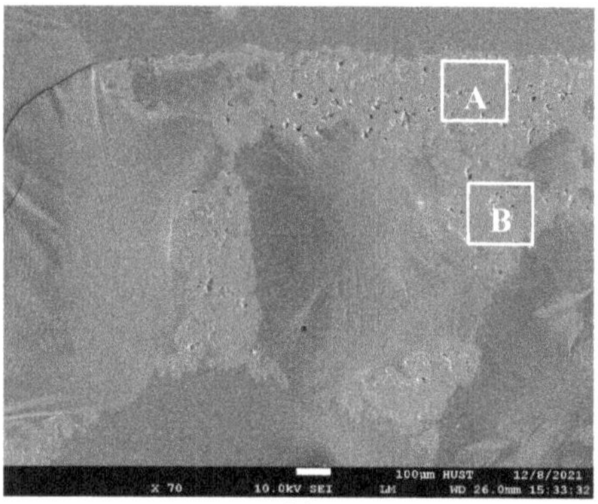

(**a**)

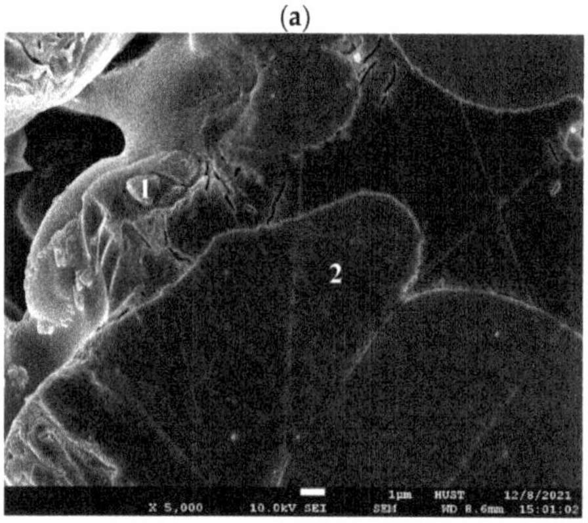

(**b**)

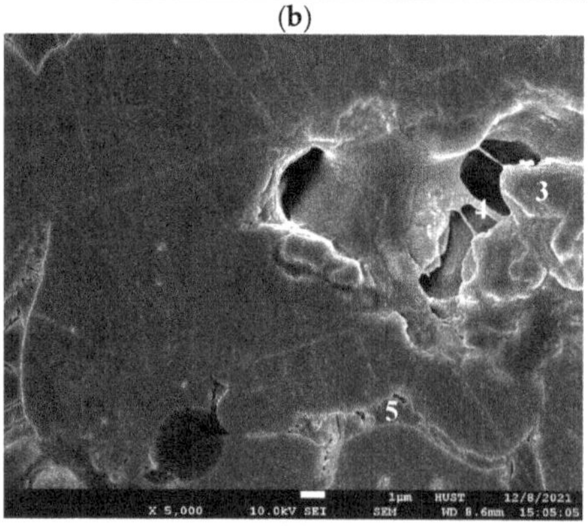

(**c**)

Figure 7. SEM image of the cross section of the coating after oxidation. (**a**) Whole SEM image; (**b**) Locally amplified SEM image at area A in (**a**); (**c**) Locally amplified SEM image at area B in (**b**). A, B represent different regions in the coating, while 1,2,3,4,5 represent different position points.

Table 5. EDS results at different parts of the cross section of the coating after oxidation.

Element (at.%)	Point 1	Point 2	Point 3	Point 4	Point 5
O	58.62	50.27	56.96	29.93	66.25
Si	0.36	1.08	-	-	10.26
Zr	41.03	48.65	43.04	70.07	23.49
Possible phases	ZrO_2	ZrO	ZrO_2	Zr_2O	$ZrSiO_4$-ZrO_2

3.3. Ablative Properties of the Coating

Figure 8 is a comparison of mass ablation rates of laser cladding coating and C/C composite matrix after 300 s ablation with oxyacetylene flame with a heat flux of 2400 kW/m^2. As can be seen from the figure, the mass ablation rate of the coating (0.504 mg/s) is much lower than that of the C/C composite matrix (2.045 mg/s), about 1/4 of the latter. This indicates that the laser cladding coating has a strong ability to resist the flame ablation of oxyacetylene. This should be due to the existence of ultra-high temperature ceramic phases ZrB_2, ZrC, and SiC in the laser cladding coating (Figure 2).

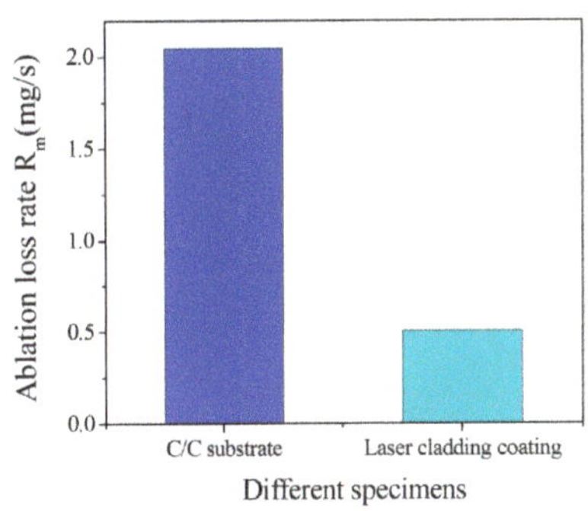

Figure 8. Comparison of oxyacetylene ablation rate between C/C composite matrix and laser cladding coating.

Figure 9 shows the XRD pattern of the C/C composite matrix after 300 s ablation with oxyacetylene flame with a heat flux of 2400 kW/m^2. It can be seen from the figure that the surface of C/C composite matrix is still graphite phase (JSPDS 96-120-0018) after being ablated by oxyacetylene flame. Figure 10 is the XRD pattern of the laser cladding coating surface after 300 s ablation with oxyacetylene flame with a heat flux of 2400 kW/m^2. It can be seen from the figure that the surface of laser cladding coating is mainly composed of hexagonal α-quartz (JSPDS 96-900-5032), monoclinic β-cristobalite (JSPDS 96-901-5393), and tetragonal $ZrSiO_4$ (JSPDS 96-900-0687) phases.

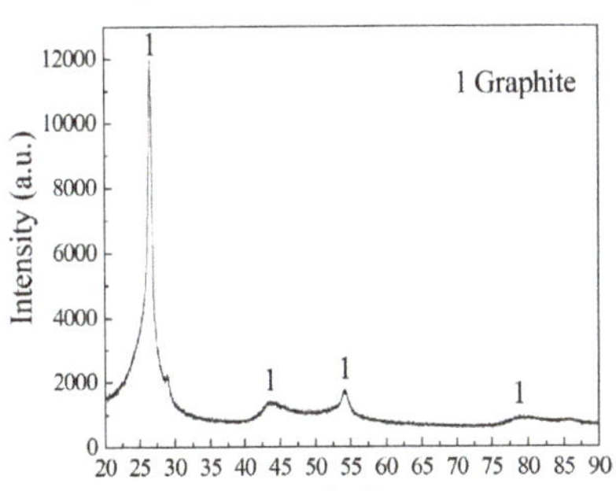

Figure 9. XRD pattern of oxyacetylene ablated surface of C/C composite matrix.

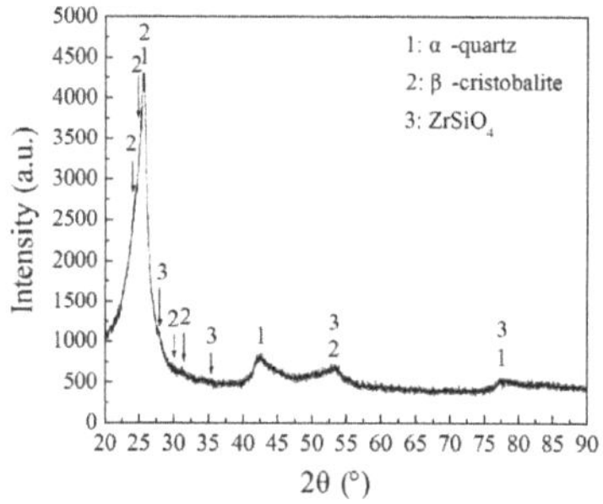

Figure 10. XRD pattern of oxyacetylene ablated surface of laser cladding coating.

In this paper, the heat flux of oxyacetylene flame ablation is 2400 kW/m^2, according to a report in the literature [26], the corresponding ablation temperature is 2000 ± 200 °C. The melting and boiling points of B_2O_3 are 450 °C and 1500 °C, respectively. The melting and boiling points of ZrO_2 are 2680 °C and 4300 °C, respectively. The melting and boiling points of SiO_2 are 1713 °C and 2230 °C, respectively. Therefore, in the process of ablation, the following five kinds of reaction may occur and generate α-SiO_2, β-SiO_2, and $ZrSiO_4$ phases. Among them, the phase of high temperature α-cristobalite (1470–1713 °C) can exist in the form phase of low temperature β-cristobalite (<180–270 °C). Its main reason is that the molten pool cooling rate of laser cladding is very fast.

$$2ZrB_2(s) + 5O_2(g) = 2ZrO_2(s) + 2B_2O_3(g)$$

$$2ZrC(s) + 3O_2(g) = 2ZrO_2(s) + 2CO(g)$$

$$2SiC(s) + 3O_2(g) = 2SiO_2(l) + 2CO(g)$$

$$SiC(s) + O_2(g) = SiO(g) + CO(g)$$

$$2SiO(g) + 2ZrO_2(s) + O_2(g) = 2ZrSiO_4(s)$$

Figure 11 shows the macroscopic morphology of C/C composite matrix and laser cladding coating after 300 s ablation by oxyacetylene flame with a heat flux of 2400 kW/m^2. As can be seen from the figure, obvious ablative rings and pits appear after the ablation of C/C composite matrix, indicating that the C/C composite matrix is subjected to the obvious ablation of oxyacetylene flame and the scouring force of oxyacetylene flame. Obvious white material appears on the surface of laser cladding coating after ablation. Combined with Figure 10, these white materials should be SiO_2.

Figure 12 shows the SEM image of the center of the C/C composite matrix after ablation, as shown in Figure 12 and EDS results (10.17 at.% O + 89.83 at.% C) for area S. The surface of C/C composite matrix after ablation is mainly composed of two elements, C and O, which may be C-CO_x after the ablation of the C/C composite matrix.

Figure 13 shows the SEM image of the center of the coating after ablation. As can be seen from Figure 10, Figure 13, and Table 6, after 300 s ablation by oxyacetylene flame with a heat flux of 2400 kW/m^2, the ZrB_2-ZrC-SiC ultra-high temperature composite ceramic coating on the C/C composite matrix basically ablated, forming $ZrSiO_4$ and SiO_2.

Table 6. EDS results of the center of the coating after ablation.

Element (at.%)	Point 1	Point 2	Point 3	Point 4
B	10.04	-	-	12.79
C	0.45	0.32	0.13	1.77
O	67.75	59.62	56.73	41.29
Si	2.33	4.11	13.90	5.91
Zr	19.44	35.94	29.24	38.25
Possible phases	$ZrSiO_4$-SiO_2-ZrB_2	$ZrSiO_4$-SiO_2	$ZrSiO_4$-SiO_2	$ZrSiO_4$-ZrB_2-SiO_2

(a)

(b)

Figure 11. Macroscopic morphology of C/C composite matrix and laser cladding coating after ablation. (**a**) C/C composite matrix; (**b**) laser cladding coating.

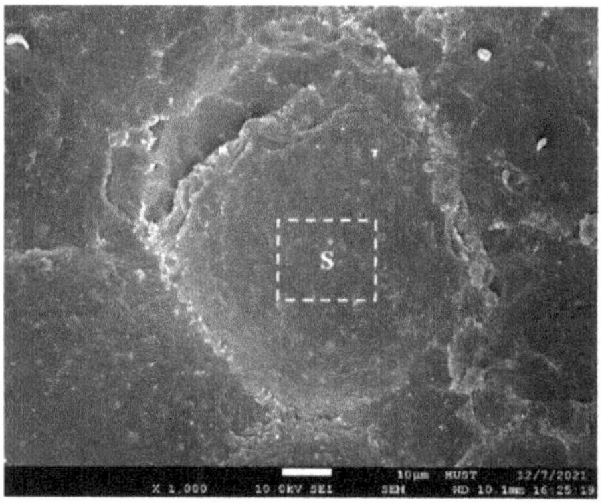

Figure 12. SEM image of the center of the C/C composite matrix after ablation. S represents region in the C/C composite matrix after ablation.

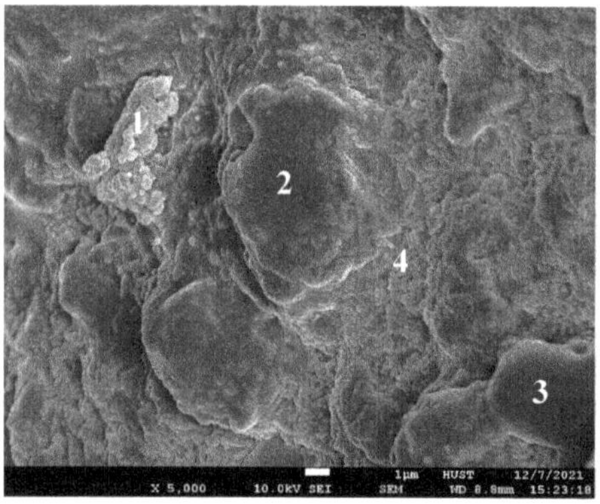

Figure 13. SEM image of the center of the coating after ablation. 1,2,3,4 represent different position points.

4. Conclusions

Using Zr powder, B_4C powder, and Si powder as raw materials, ZrB_2-ZrC-SiC ultra-high temperature composite ceramic coating was successfully prepared on C/C composites by laser cladding. The coating can protect the C/C composites matrix at 1600 °C for 80 min, and the oxidation weight gain rate at 1600 °C and 40 min is 0.51 g/cm^2. In addition, the ablation rate of the coating after 300 s ablation by oxyacetylene flame with a heat flux of 2400 kW/m^2 is about 1/4 of that of C/C composite matrix. These results indicate that ZrB_2-ZrC-SiC ultra-high temperature composite ceramic coating prepared by laser cladding has strong oxidation resistance at 1600 °C and oxyacetylene flame ablation resistance at 2400 kW/m^2. This provides an alternative technical approach for surface modification of C/C composites.

Author Contributions: Investigation—K.H., Y.X. and A.W.; writing—original draft preparation, Y.X.; writing—review and editing, K.H. All authors have read and agreed to the published version of the manuscript.

Funding: This work was supported by the Open Project Program (No. 2022-KF-19) of the State Key Laboratory of Advanced Technology for Materials Synthesis and Processing, Wuhan University of Technology.

Institutional Review Board Statement: Not applicable.

Informed Consent Statement: Not applicable.

Data Availability Statement: The data of this study are available from the corresponding author upon reasonable request.

Acknowledgments: The authors are also grateful to the Analytical and Testing Center of Huazhong University of Science and Technology.

Conflicts of Interest: The authors declare no conflict of interest.

References

1. Devi, G.R.; Rao, K.R. Carbon-carbon composites—An overview. *Def. Sci. J.* **1993**, *43*, 369–383. [CrossRef]
2. Windhorst, T.; Blount, G. Carbon-carbon composites: A summary of recent developments and applications. *Mater. Des.* **1997**, *18*, 11–15. [CrossRef]
3. Li, J.; Yang, X.; Liu, H.W.; Huang, Q.Z.; Su, Z.A. Research process of ultra-high temperature ceramics modified carbon/carbon composites for ablation resistance. *Chin. J. Nonferrous Met.* **2015**, *25*, 1731–1743.
4. Fu, Q.G.; Li, H.J.; Shen, X.T.; Li, K.Z. Domestic research process of matrix modification for carbon/carbon composites. *Mater. China* **2011**, *30*, 6–12.
5. Jin, X.C.; Fan, X.L.; Lu, C.S.; Wang, T.J. Advances in oxidation and ablation resistance of high and ultra-high temperature ceramics modified or coated carbon/carbon composites. *J. Eur. Ceram. Soc.* **2018**, *38*, 1–28. [CrossRef]
6. Yang, X.; Huang, Q.Z.; Su, Z.A.; Chang, X. Review of recent progress on oxidation protection for C/C composites at high temperature. *Aerosp. Mater. Tech.* **2014**, *44*, 1–15.
7. Wang, H.H.; Kong, J.A.; Xu, M.; Zhang, P.F.; Yang, L.; Shi, X.H.; Li, H.J. Laser ablation behavior of SiO_2-Nd_2O_3/Si-SiC-$MoSi_2$ coated C/C composites repaired by laser cladding. *Corros. Sci.* **2022**, *198*, 110132. [CrossRef]
8. Wang, H.H.; Kong, J.A.; Ge, J.; Teng, L.; Shi, X.H.; Li, H.J. Effects of reflectivity on laser-ablation resistance of the laser-cladding repaired Nd_2O_3 modified SiO_2 coatings on C/C composites. *J. Eur. Ceram. Soc.* **2021**, *41*, 6548–6588. [CrossRef]
9. Teng, L.; Shi, X.H.; Wang, H.H.; Li, W.; Yin, X.M.; Li, H.J. A new method to improve the laser-ablation resistance of Si-SiC coating on C/C composites: Laser cladding. *J. Eur. Ceram. Soc.* **2022**, *42*, 6425–6434. [CrossRef]
10. Wang, H.H.; Teng, L.; Kong, J.A.; Liu, X.S.; Shi, X.H.; Li, H.J. Enhancing anti-oxidation and thermal-radiation performance of the repaired borosilicate glass coating on C/C composites by Sm-doping. *J. Materiomics* **2022**, *8*, 417–426. [CrossRef]
11. Wang, H.H.; Li, H.J.; Shi, X.H.; Liu, X.S.; Kong, J.A.; Zhou, H. Repair of SiC coating on carbon/carbon composites by laser cladding technique. *Ceram. Int.* **2020**, *46*, 19537–19544. [CrossRef]
12. Eric, J.J.; Hull, R.J. Laser cladding of oxidation protection coatings on carbon-carbon composites. In Proceedings of the Laser Materials Processing Conference (ICALEO(R) 2000), Dearborn, MI, USA, 2–5 October 2000.
13. Wu, Q.; Ji, X.J.; Peng, H.R.; Shen, J.; Ren, X.J. Study on compound modified ZrB_2 based ultrahigh temperature oxidation resistant coatings. *China Ceram.* **2016**, *52*, 6–11.
14. Lin, H.J.; Li, H.J.; Shen, Q.L.; Shi, X.H.; Feng, T.; Guo, L.J. 3C-SiC nanowires in-situ modified carbon/carbon composites and their effect on mechanical and thermal properties. *Nanomaterials* **2018**, *8*, 894. [CrossRef] [PubMed]
15. Chu, Y.H.; Fu, Q.G.; Li, H.J.; Li, K.Z. Advancement of high temperature oxidation resistant ceramic coatings for Carbon/ Carbon composites. *Mater. Eng.* **2010**, *11*, 86–91.
16. Wu, W.W.; Zhang, G.J.; Kan, Y.M.; Wang, P.L. Reactive hot pressing of ZrB_2-SiC-ZrC ultra high-temperature ceramics at 1800 °C. *J. Am. Ceram. Soc.* **2006**, *89*, 2967–2969.
17. Zheng, Q.; Wang, X.H.; Xing, J.J.; Gu, H.; Zhang, G.J. Quantitative analysis for phase compositions of ZrB_2-SiC-ZrC ultra-high temperature ceramic composites. *J. Inorg. Mater.* **2013**, *28*, 358–362. [CrossRef]
18. Yang, Z.H. The Synthesis In-Situ of Metal-Ceramic Coatings Containing ZrB_2 by PTA. Master Dissertation, Tianjin University, Tianjin, China, 2009.
19. Liu, Y.M. Study of Preparation and Oxidation Resistance of ZrB_2 Based Ceramic Coatings for C/C Composites. Master Dissertation, Dalian University of Technology, Dalian, China, 2021.
20. Okamoto, H. Supplemental literature review of binary phase diagrams: Ag-Ca, Al-Yb, As-Fe, B-Zr, Co-U, Cu-Se, Cu-Th, La-Mo, Mg-Sn, Mo-Th, Sn-Ta, and Te-Ti. *J. Phase Equilib. Diff.* **2017**, *38*, 929–941. [CrossRef]
21. Ping, H.; Wang, G.L.; Zhi, W. Oxidation mechanism and resistance of ZrB_2-SiC composites. *Corros. Sci.* **2009**, *51*, 2724–2732.
22. Kaiser, A.; Lobert, M.; Telle, R. Thermal stability of zircon ($ZrSiO_4$). *J. Eur. Ceram. Soc.* **2008**, *28*, 2199–2211. [CrossRef]
23. Arroyave, R.; Kaufman, L.; Eagar, T.W. Thermodynamic modeling of the Zr-O system. *Calphad* **2002**, *26*, 95–118. [CrossRef]
24. Puchala, B.; Ven Van der, A. Thermodynamics of the Zr-O system from first-principles calculations. *Phys. Rev. B* **2013**, *88*, 094108. [CrossRef]
25. Chen, M.H.; Puchala, B.; Ven Van der, A. High-temperature stability of δ'-ZrO. *Calphad* **2015**, *51*, 292–298. [CrossRef]
26. Li, H.J.; Yao, X.Y.; Zhang, Y.L.; Li, K.Z.; Guo, L.J.; Liu, L. Effect of heat flux on ablation behaviour and mechanism of C/C–ZrB_2–SiC composite under oxyacetylene torch flame. *Corros. Sci.* **2013**, *74*, 265–270. [CrossRef]

coatings

Article

Effects of Process Parameters on Microstructure and High-Temperature Oxidation Resistance of Laser-Clad IN718 Coating on Cr5Mo Steel

Zelin Xu [1], Fengtao Wang [1,*], Shitong Peng [1], Weiwei Liu [2] and Jianan Guo [1]

1 Department of Mechanical Engineering, College of Engineering, Shantou University, Shantou 515063, China
2 School of Mechanical Engineering, Dalian University of Technology, Dalian 116024, China
* Correspondence: ftwang@stu.edu.cn; Tel.: +86-0754-86502941

Abstract: Cr5Mo steel with high thermal strength is frequently applied as the material for hydrocracking furnace tubes. Nonetheless, Cr5Mo tubes are prone to material failure in a high-temperature environment, threatening production safety. Considering that the IN718 nickel-base superalloy has favorable high-temperature oxidation resistance, the IN718 coating was fabricated on Cr5Mo substrate through laser cladding. The effect of process parameters on the high-temperature oxidation resistance of laser cladding IN718 coating was investigated. The results confirm that laser power and scanning speed affected the eutectic quantity precipitation of this layer, and the eutectic quantity precipitation was positively correlated with the mass gain of the coating. The high-temperature behavior of the coating could be divided into surface oxidation, intergranular corrosion, and material shedding. The scanning speed has a more significant impact on the high-temperature oxidation resistance. When the scanning speed is 15 mm/s, cracks originating in the heat-affected zone could exert a negative impact on the high-temperature oxidation resistance.

Keywords: laser cladding; Cr5Mo; IN718; HAZ crack; delamination crack; high-temperature oxidation resistance

1. Introduction

Cr5Mo tubes are a part of the furnace that directly heats the gas or liquid flowing in the tube by flame [1,2]. In general, the petrochemical furnace works continuously under high temperatures, high pressure, and corrosive mediums. The heated gas or liquid in the tube is commonly a flammable and explosive hydrocarbon [3]. Overheating and flame licking could damage the furnace tube, which adversely affects production safety and efficiency. Therefore, the high-temperature oxidation resistance of furnace tubes is an essential property for the safe as well as efficient service of the tubular furnace.

Laser cladding is one of the promising methods for manufacturing and repairing different industrial components such as pump shafts and turbine blades [4–7]. IN718 nickel-base superalloy has high strength, satisfying toughness, and corrosion resistance below 650 °C [8,9]. Thus, laser cladding IN718 coating is able to contribute to the improvement of the service life of Cr5Mo substrate in a high-temperature corrosion environment. Xie et al. [10] demonstrated that the increment of the cooling rate could inhibit undercooling, reduce the secondary dendrite arm spacing, and refine the Laves phase in the coating. Zhang et al. [11] prepared IN718 coating on the deformed IN718 alloy plate and investigated the effects of distinct laser cladding speeds on the microstructure and Nb segregation of the coating. These studies indicated that the process parameters vary in the microstructure of the coating, which may further affect the high-temperature oxidation resistance of the IN718 coating. Jia et al. [12] studied the high-temperature oxidation behavior and mechanism of IN718 components after selective laser melting (SLM) in the near-surface area. The results illustrated that increasing the laser energy density could facilitate the high-temperature oxidation resistance of the IN718 components. However, the laser cladding process differs from the SLM process in being exposed to the atmosphere, and the local temperature difference leads to shrinkage, residual stress, and deformation. Therefore, the effects of process parameters on coating quality and high-temperature oxidation resistance are worth investigating. Under specific process conditions, despite the fact that the surface quality of the laser cladding coating is intact, defects such as cracks and pores potentially occur inside [13]. Mazzarisi et al. mentioned [14] that a high number of cracks was found in areas with higher temperature and thermal gradient variations. Such defects have negative impact on the performance and behavior of AM-produced alloys [15]. In addition to the high-temperature oxidation behavior of the top area of the aforementioned laser cladding coating in the existing research [16], Thouless [17] reported that cracks and delamination through the thickness of the coating would accordingly damage the protective coating. It should be pointed out that there is an absence of research on the effects of process parameter adjustment and defects on the high-temperature oxidation resistance of laser cladding IN718 coating.

This work will study the microstructural evolution and defect generation mechanism of coating under different process parameters, analyze the correlation between process parameters and high-temperature oxidation resistance, and supplement the high-temperature behavior of laser cladding IN718 coating in a high-temperature oxidation environment. Investigating the high-temperature oxidation resistance of laser cladding IN718 coating is beneficial for improving the high-temperature oxidation resistance and service life of IN718 coating. Lastly, an engineering application example of remanufactured Cr5Mo hydrocracking furnace tubes is given.

2. Materials and Methods

2.1. Substrate and Laser Cladding Powder

This study adopted the Cr5Mo tube as substrate with an external diameter of 180 mm, a thickness of 15 mm, and a length of 250 mm and IN718 powder with a mesh size of 45–100 μm as the laser cladding material. The chemical compositions of these materials are listed in Table 1. The morphology of IN718 powder was nearly spherical, as shown in Figure 1a. The particle size was measured using a DynaPro NanoStar Dynamic Light Scattering (DLS) instrument (Wyatt Technology Corporation, Santa Barbara, CA, USA). The particle-size distribution of IN718 powder is shown in Figure 1b.

Table 1. Chemical composition of Cr5Mo substrate and IN718 powder.

Substrate	Cr	Mo	Mn	C	Si	Nb	Ni	Fe
Cr5Mo	5.02	0.54	0.35	0.10	0.54	0.02	-	Bal.
IN718	21.0	3.3	0.35	0.08	0.35	5.50	Bal.	14.50

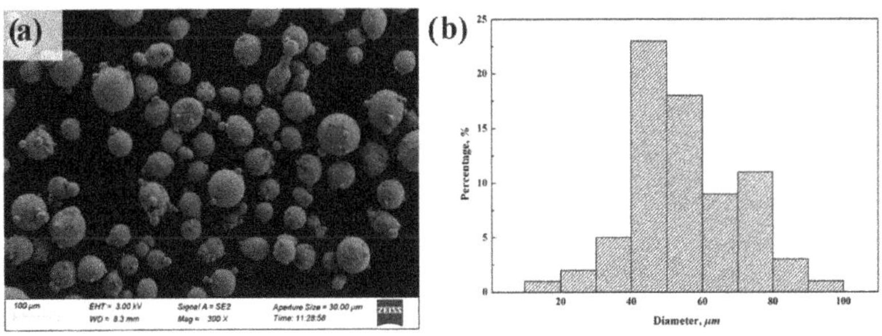

Figure 1. IN718 powder characterization: (**a**) SEM image of nearly spherical powder and (**b**) particle-size distribution.

2.2. Preparation of Laser Cladding Layer and Test Sample

A TRUMPF 3600 laser machine (spot size of 4 mm) equipped with coaxial powder feeding was used for laser cladding. The parameters of the coating preparation are listed in Table 2. Two critical operational parameters, namely, the laser power and scanning speed, were varied to study their effects on the physical properties of laser cladding coating. Other parameters include an overlap rate of 50% and a powder feeding rate of 15 g/min. As sketched in Figure 2, a total of five groups of the IN718 coatings with different parameters were prepared on Cr5Mo tubes, with average coating thickness of 3 mm and coating width of 15 mm. The distance between the coatings was 20 mm. Samples for testing were prepared from the deposited layer by wire electrical discharge machining (WEDM) (Makino, Tokyo, Japan), which were equipped with a size of 10 mm × 10 mm × 10 mm. A TASi TA603 thermometer (TASI, Suzhou, China) was used to measure the temperature of the substrate about 10 mm from the laser-clad track. The temperature of measured positions after laser cladding was mostly lower than 150 °C.

Table 2. Parameters for studying the effect of the laser cladding coating.

Experimental Group	Laser Power (*W*)	Scanning Speed (mm/s)
1600-5	1600	5
1600-10	1600	10
1600-15	1600	15
2000-10	2000	10
2400-10	2400	10

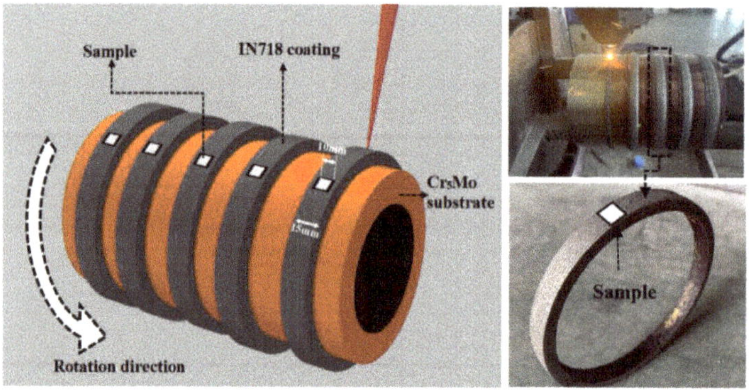

Figure 2. Preparation of laser cladding samples with different parameters on Cr5Mo.

2.3. Analysis and Testing

For microstructural observation, samples were initially polished and then etched with Marble's reagent (10 g $CuSO_4$ + 50 mL HCl + 50 mL H_2O). Microstructure and microhardness were analyzed using an MR5000 metallography microscope and a TMVP-1 microhardness tester. The load for the hardness test was set to 0.5 kg, and the dwell time was set to 10 s. Zeiss Sigma 300 scanning electron microscope (SEM) (Carl Zeiss AG, Oberkochen, Germany) was equipped with energy dispersive X-ray spectroscopy (EDS) for microstructural and chemical analysis. This paper adopted a Panalytical X'Pert'3 powder X-ray diffraction system (XRD) (PANalytical, Almelo, Netherlands) to observe the crystalline structure. Samples were heated in an SX2-5-12A muffle furnace. The increased oxidation weight (mass gain) tested by an EX1035 electronic analytical balance was employed to quantify the high-temperature oxidation resistance [18]. The oxidation resistance of IN718 coatings was quantified at high temperature. The samples were heated at 900 °C for 100 h. Moreover, Minitab software (Minitab 2021) was used for data fitting.

3. Results

3.1. Microstructure

Figure 3 indicates the XRD spectra of IN718 coating with different process parameters. The phases of the IN718-cladded layer were mainly γ-Ni (PDF-# 65-0380), γ′-Ni_3(Al, Ti) (PDF-# 18-0872), and γ″-Ni_3Nb (PDF-# 15-0101), which were consistent with the research of Sumit et al. [19] without the appearance of new phases when changing the process parameters. However, the diffraction peak intensity of the samples did change. When the power was 1600 W and the scanning speed was 5 mm/s (sample 1600-5), the peak intensities of γ-Ni and γ′-Ni_3(Al, Ti) were higher than that of the samples 1600-10 and 1600-15, indicating the more significant proportion of γ-Ni and γ′-Ni_3(Al, Ti). The decrease in scanning speed reduced the cooling rate, which was conducive to the growth of austenite dendrites. When the scanning speed was 10 mm / s and the laser power increased from 1600 W to 2400 W (samples 1600-10, 2000-10, and 2400-10), the intensity of the second peak significantly increased. As can be seen from the results, the higher laser power can promote the growth of dendrites.

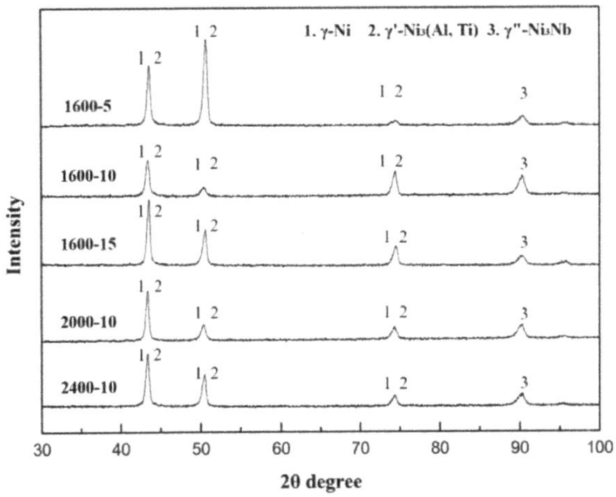

Figure 3. XRD spectra of IN718 coating under different process parameters.

Figure 4 displays the effects of power and scanning speed on the primary dendrite arm spacing (PDAS) of five samples. The dendrites among all the samples grew opposite to the heat transfer direction. The refinement degree of the microstructure was compared by measuring the distance between primary dendrites. With an increment in the laser power from 1600 W to 2400 W, the PDAS of IN718 laser cladding layer increased from 19.47 μm to 27.83 μm. Nevertheless, the PDAS decreased from 25.23 μm to 7.16 μm when the scanning speed increased from 5 mm/s to 15 mm/s. These results were in line with prior studies on the influence of laser power and scanning speed on the microstructure [20]. The size of the martensite and lower bainite increased during the process of low alloy steel fabrication through laser melting deposition. With the increase in laser power, the tempered martensite gradually coarsens. In the study of laser selective melting of 316 L stainless steel, the increase in scanning speed could induce grain refinement [21]. In terms of IN718 laser cladding coating, the variation of heat input and cooling rate led to the difference of PDAS.

Figure 5 presents the microstructural evolution of the top region of the laser cladding IN718 coating under different process parameters. It can be seen from Figure 5a–c that when the scanning speed increased from 5 mm/s to 15 mm/s, the top region's dendrites refined, and eutectic quantity increased. As shown in Figure 5b,d,e, when the laser power increased from 1600 W to 2400 W, the dendrite in the top region coarsened, and the eutectic quantity decreased. Figure 6 shows the microstructure evolution of the bottom region of laser cladding IN718 coating under different process parameters. Du et al. [22] mentioned that columnar dendrites were generated in the bottom region, and equiaxed dendrites were formed in the top region. It can explain the difference of microstructure morphology between Figures 5 and 6.

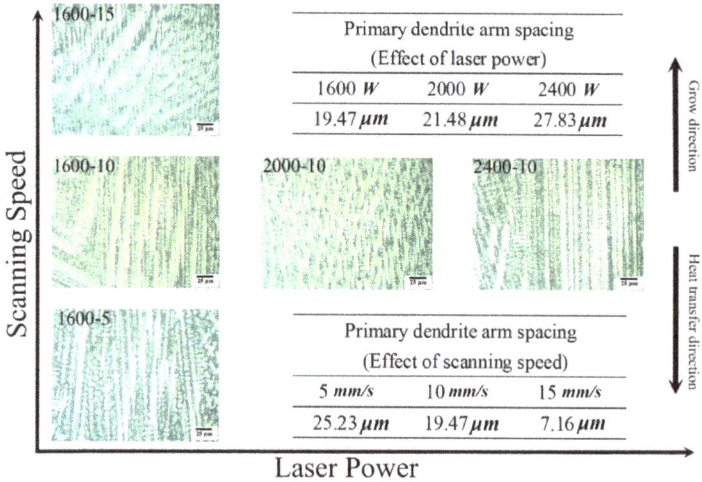

Figure 4. Effects of laser power and scanning speed on sample PDAS.

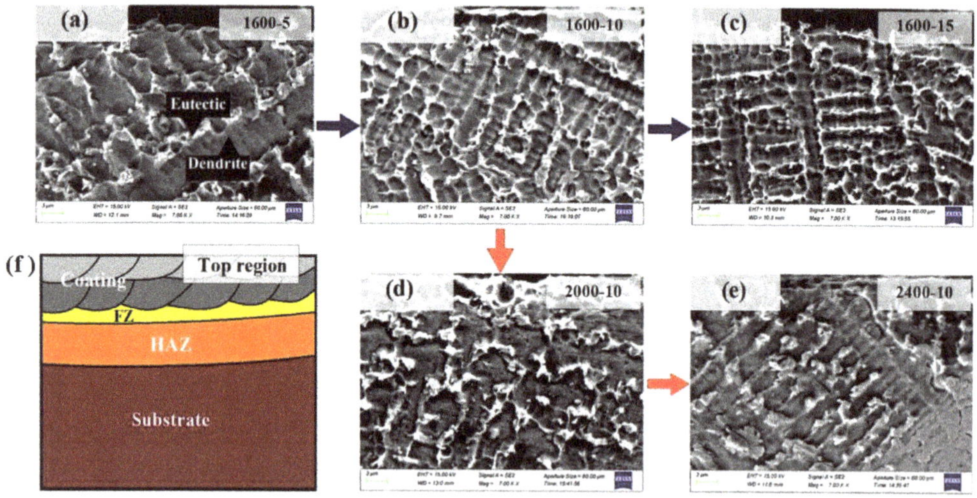

Figure 5. Microstructural evolution in the top region of laser cladding IN718 coating under different process parameters (**f**): Scanning speed increased from 5 mm/s to 15 mm/s (blue route (**a**–**c**), and laser power increased from 1600 W to 2400 W (red routes (**b**,**d**,**e**).

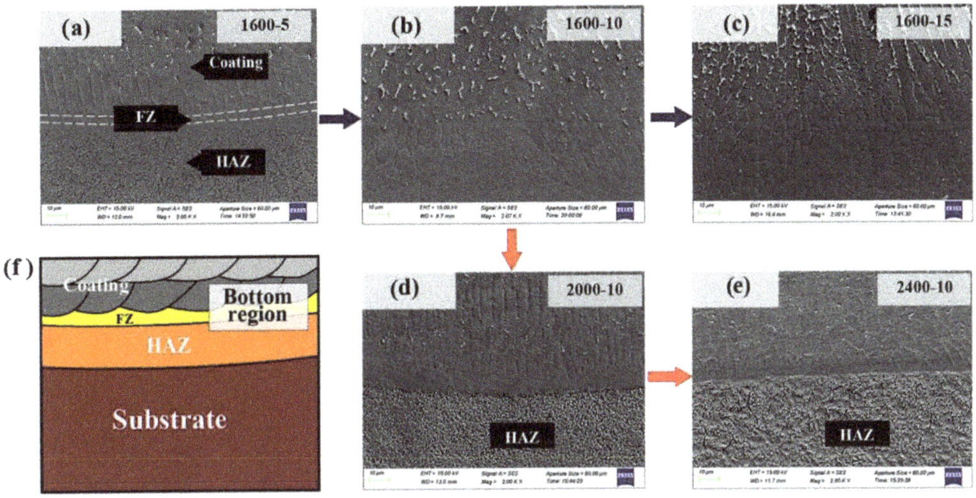

Figure 6. Microstructural evolution in the bottom region of laser cladding IN718 coating under different process parameters (**f**): Scanning speed increased from 5 mm/s to 15 mm/s (blue route (**a**–**c**), and laser power increased from 1600 W to 2400 W (red routes (**b**,**d**,**e**)).

In Figure 6a, the bottom region of the coating, the fusion zone (FZ), and the heat-affected zone (HAZ) were observed. As indicated in Figure 6a–c, there was an increased eutectic quantity in the top region when the scanning speed increased from 5 mm/s to 15 mm/s. In Figure 6b,d,e, the eutectic quantity decreased gradually when the laser power increased from 1600 W to 2400 W. Higher cooling rates could result in increased precipitation of eutectic elements. Lower power and faster scanning speed improve the cooling rate. Additionally, the higher cooling rates could eventually increase eutectic compounds while decreasing austenite dendrites. Such phenomena have also been observed in previous work [23].

3.2. Hardness Testing

Microhardness testing was measured from the coating surface to the substrate at room temperature, with seven testing locations in a straight line, whose results are listed in Table 3. Figure 7 indicates that the laser power and scanning speed affect the microhardness from the coatings to the substrate. In Figure 7a, the scanning speed (invariant) was 10 mm/s, while the laser power (variable) was increased from 1600 W to 2400 W. In Figure 7b, the laser power (invariant) was 1600 W, while the scanning speed (variable) was increased from 5 mm/s to 15 mm/s. The hardness of the coating and HAZ in Figure 7 are the average values at different positions listed in Table 3.

Table 3. Microhardness (room temperature) from coating to the substrate, $HV_{0.2}$.

-	1600-5	1600-10	1600-15	2000-10	2400-10
Coating	300.8 249.8	344.3 276.5	345.6 367.9	279.8 238.7	306.4 251.5
FZ	218.3	225.7	304.2	186.5	162.4
HAZ	284.9 307.3 269.5	324.1 340.8 281.3	400.9 370.8 314.2	335.4 441.4 291.8	325.2 333.8 283.8
Substrate			Average 252.9		

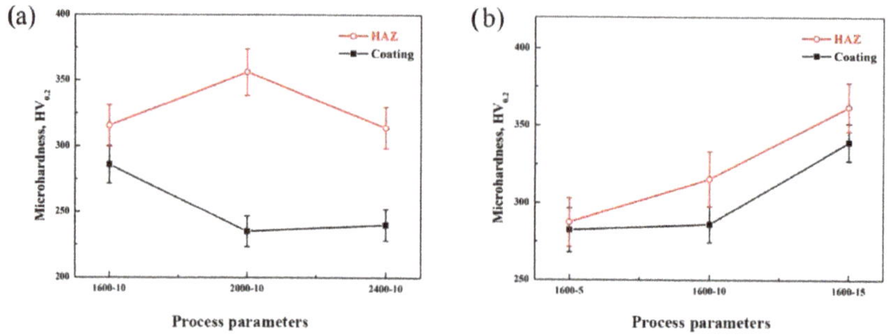

Figure 7. Factors affecting the microhardness of coating-substrate: (**a**) effects of laser power and (**b**) effects of scanning speed.

As shown in Figure 7a, with the increase in laser power, the average hardness of the coating decreases from 310.4 $HV_{0.2}$ to 278.9 $HV_{0.2}$, and the hardness of HAZ firstly shows an initial increment and then decreases. When the scanning speed increased (Figure 7b), the average hardness of the coating increased from 275.3 $HV_{0.2}$ to 356.7 $HV_{0.2}$, and the average hardness of HAZ increased from 287.2 $HV_{0.2}$ to 361.9 $HV_{0.2}$. Mankins et al. [24] stated that the primary strengthening mode of nickel-based superalloy stemmed from the precipitation strengthening of the $M_{23}C_6$ phase. Therefore, the change of laser process parameters affected the number of strengthening phases precipitated in the parent phase and subsequently changed the hardness of the coating.

Notably, the hardness of HAZ was generally higher than that of the coating and substrate. Adomako et al. [25] reported that the hardness of IN718 coating decreased continuously from the surface to the FZ due to the depletion of strengthening elements (Nb, Mo, and Ti) in the microstructure, which was also consistent with the present study. Under the condition of laser heating and rapid cooling, martensite was generated in the HAZ region [26]. The martensite had high hardness in the HAZ region.

It can be seen in Figure 7a that the hardness of HAZ first increased and afterwards decreased with the increase in laser power from 1600 W to 2400 W. As shown in Figure 6d,e, the HAZ microstructures of samples with different powers exhibit apparent differences. The SEM images of HAZ in samples with laser powers of 2000 W and 2400 W are observed in Figure 8. With the increase in laser power, the temperature of laser cladding track rose, and the cooling rate decreased, resulting in ferrite recrystallization and grain coarsening (Figure 8b). Microhardness reduction could also be proved when laser power increased from 2000 W to 2400 W.

The HAZ of these two samples was analyzed by EDS elemental mapping. In Figure 8c,d, the EDS elemental mapping results of HAZ in both samples with laser power of 2000 W and 2400 W showed a stable composition. Thus, the possibility of interference from other elements was excluded. It has been proved that changing the power produces two opposite factors in HAZ: With the increase in power, carbide precipitation can increase the hardness of HAZ; grains coarsen, and the hardness of HAZ decreases. Before discussing the high-temperature oxidation resistance of IN718 coating, it is necessary to study the microstructural evolution in HAZ, which determines the service life of IN718 coating.

3.3. Defects

Figure 9 provides various types of defects in the IN718 laser cladding coating. It is evident from Figure 9a that a crack starts from the bottom region of the coating and grows towards the surface of the coating. The crack grows along the boundary of dendrites. Combined with the analysis in Figure 6, more eutectic elements, including Laves phases and carbide, were generated in the bottom region of the coating under the condition of low laser power. Carbide and Laves phases promoted the segregation of impurity elements to grain boundaries, leading to low melting compounds [27], which induced intergranular microfissures due to grain boundary liquation. The increase in HAZ hardness was attributed to the precipitation hardening of the martensite, while it brought HAZ a higher brittleness and a higher crack sensitivity. A crack starting from HAZ can be seen in Figure 9b, which then extended into the coating. In Sample 1600-15, a thermal fatigue crack in HAZ initiated under thermal cyclic loading. The crack started from the HAZ also proved the necessity of studying the microstructural evolution of the coating–substrate bonding region. As shown in Figure 9c, a pore was observed in Sample 1600-15. Reactive and non-reactive gases exist in the process of laser cladding. In Sample 1600-15, the coating did not release the gas promptly due to the rapid solidification at high scanning speed. Therefore, the unreleased gas caused pore defects in the samples.

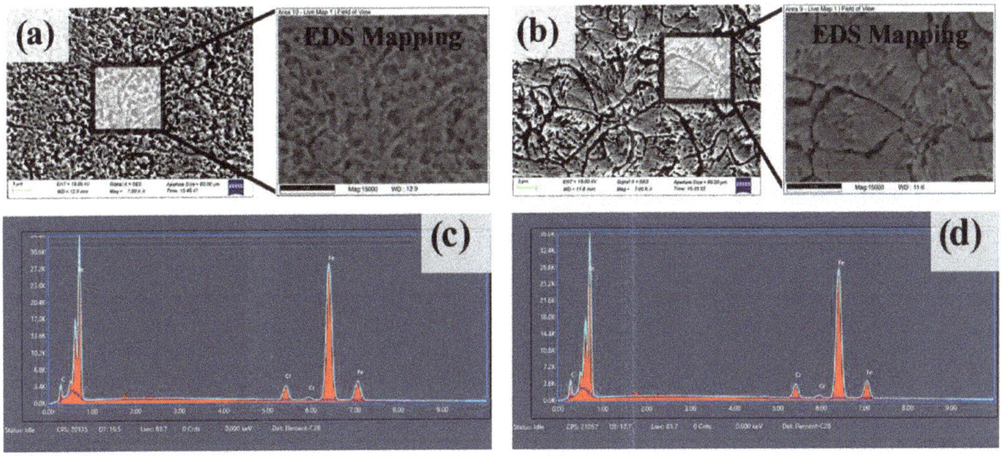

Figure 8. SEM images and EDS mapping results of HAZ: samples with laser power of 2000 W (**a**,**c**) and samples with laser power of 2400 W (**b**,**d**).

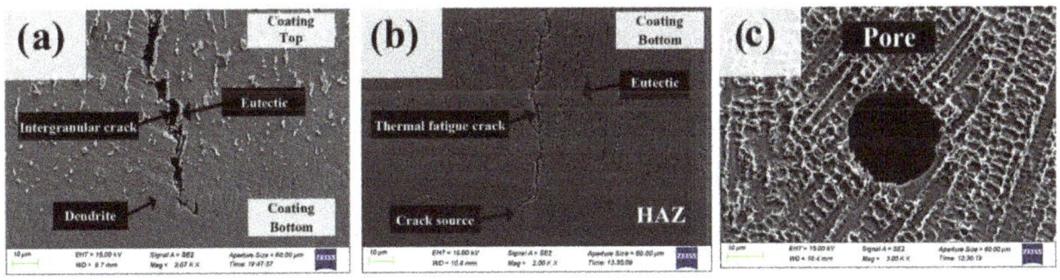

Figure 9. Different types of defects in IN718 coating: (**a**) intergranular crack in Sample 1600-10; (**b**) thermal fatigue crack in HAZ of Sample 1600-15; (**c**) pore in Sample 1600-15.

3.4. High-Temperature Oxidation Resistance

Nickel-based materials could be oxidized at a high temperature of 900 °C, and an oxide film was formed on the coating surface [28,29]. Figure 10 is the EDS mapping results of the Sample 1600-10 after the high-temperature oxidation treatment at 900 °C. The enrichment of element O in region (I) of the coating indicates that the final oxidation products are some oxides of element Ni, such as $NiFe_2O4$, NiO, and $NiCr_2O_4$. These phases have also been observed by Wu et al. [30]. After high-temperature oxidation treatment, the eutectic products in region (II) below the oxide film were significantly reduced. The content of Nb in region (III) increased with the increase in the eutectic content. The formation of oxide was closely related to the distribution of the Ni3Nb phase in the top region of the coating. Element Nb tended to be enriched below the oxide film of IN718 after high-temperature treatment, which was attributed to the eutectic generation near the surface and the movement of the element Nb.

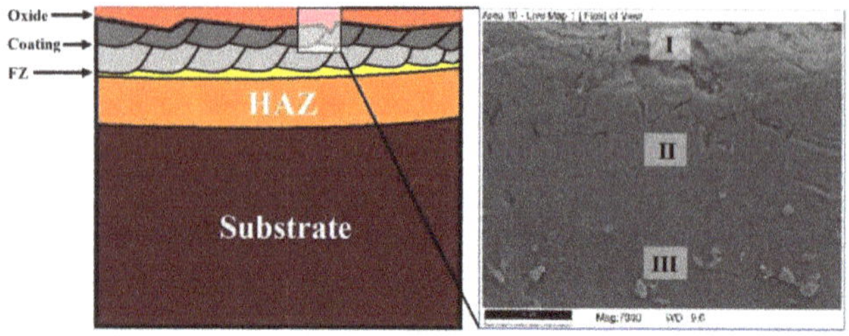

(a)

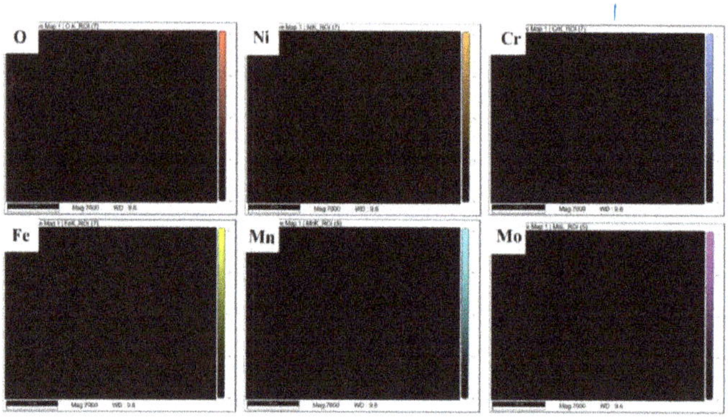

(b)

Figure 10. SEM (**a**) and EDS mapping result (**b**) of Sample 1600-10 after high-temperature oxidation treatment.

In Figure 11a, the enriched eutectic particles led to the cracking of the Sample 1600-10. As stated by Prakash et al. [31], the element Nb and the distribution of second-phase particles affect the crack growth mechanisms and rates in IN718 alloy. Figure 11b implies the lamellar distribution of eutectic particles in the coating. The cracks growing along the direction of the X–Y axis in this region could be called delamination cracks. Figure 11c is the diagram of the delamination in the IN718 coating. The bending was drawn to make delamination more visible. It can be seen from Table 4 that the content of the element Nb decreases significantly from the enrichment region to the surface. The element Nb plays a role in improving intergranular corrosion resistance. More serious intergranular corrosion occurred above the delamination when the element Nb decreased. In Sanviemvongsak's study [16], intergranular corrosion was observed in the top region of the IN718 coating after high-temperature oxidation. Additionally, it can be found in Figure 10 that the oxygen element diffused from region (I) to region (III), which indicates the intergranular corrosion occurred in the high-temperature oxidation environment. With the decrease in mass gain in Sample 1600-10 after 80 h of high-temperature oxidation treatment, it can be concluded that the material shedding occurred under the combined action of delamination cracks and intergranular corrosion (Figure 11d). In the process of high-temperature oxidation, Nb-rich and second-phase particles were formed, which increased the crack sensitivity of the coating. Hence, the cracks formed in HAZ were more simply extended to the surface of the coating, and consequently, the cracks passed through the thickness of the IN718 coating (Figure 9b).

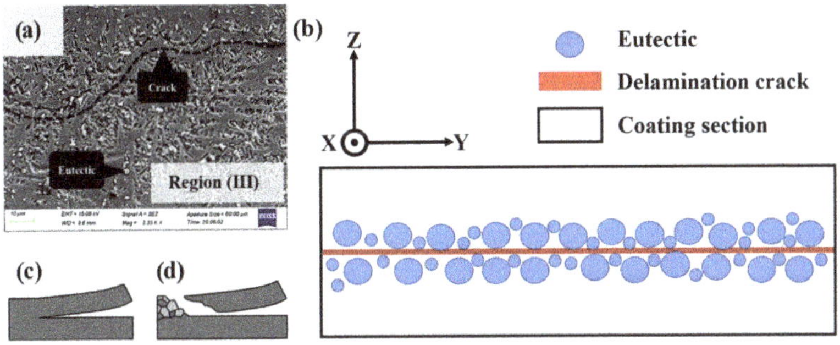

Figure 11. Delamination crack below oxide film: (**a**) dense eutectic particles and a crack in region III; (**b**) schematic of lamellar distribution of eutectic particles and a delamination crack; (**c**) delamination in the coating; (**d**) intergranular corrosion and material shedding.

Table 4. EDS analysis of Sample 1600-10 after high-temperature oxidation treatment (*wt*%).

Marked Locations	O	Mo	Cr	Mn	Fe	Ni	Nb
I	6.37	8.13	7.49	0.11	28.01	48.59	1.30
II	0.86	4.71	18.83	0.37	27.50	45.51	2.22
III	1.01	25.13	2.38	0.26	5.06	45.34	20.83

Figure 12 provides the mass gain variations of the samples with different process parameters under the high-temperature oxidation for 100 h. The mass gain of all samples increased with the increasing treatment time before 80 h. The increase in mass gain contributed to the oxide formation. On the contrary, after 80 h, the mass gain decreased in samples 1600-10 and 1600-15, which indicated the occurrence of material shedding in the IN718 coating, leading to the mass decrease at 100 h.

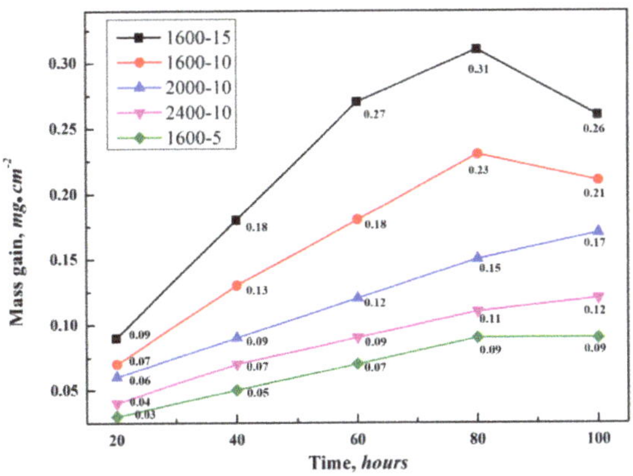

Figure 12. Mass gain variations of the samples under high-temperature oxidation for 100 h.

Figure 13 provides the final mass gain of the samples with different process parameters after high-temperature oxidation for 100 h. With the scanning speed increasing from 5 mm/s to 15 mm/s, the mass gain of the sample oxidized at high-temperature increased from 0.09 mg/cm^2 to 0.26 mg/cm^2, which indicated that the oxidation products of the sample increased, and the high-temperature oxidation resistance decreased. When the laser power increased from 1600 W to 2400 W, the mass gain of the sample decreased from 0.21 mg/cm^2 to 0.12 mg/cm^2, indicating the improvement of high-temperature oxidation resistance. The analysis of the oxide film and the microstructural evolution nearby presented a positive correlation between eutectic content and oxide. When the eutectic quantity in the top region of the IN718 coating increased, the oxide on the coating surface also increased after high-temperature oxidation treatment. Combined with the microstructure analysis in the top region (Figure 5,) when the laser power increased, the dendrites in the top region of the coating increased, while the eutectic quantity decreased. Higher laser power reduced the mass gain of the final oxide and enhanced the high-temperature oxidation resistance of IN718 coating. However, with the increase in scanning speed, more eutectic elements precipitated, and more oxides formed in the top region of the coating. In addition, the high-temperature oxidation resistance of the coating decreased.

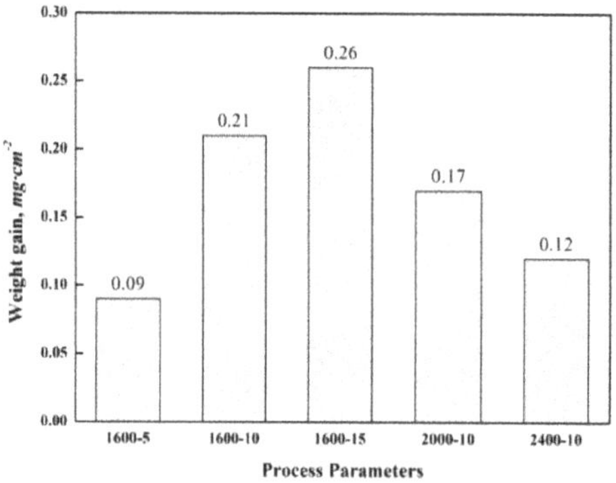

Figure 13. Final mass gain of the samples after high-temperature oxidation for 100 *h*.

Two linear regression equations of mass gain and process parameters are shown in Figure 14. When the laser power (*P*) increases by 50%, the mass gain (*M*) decreases by about 43%. When the scanning speed (*V*) is increased by 200%, the mass gain (*M*) is increased by about 188%. The data are also nonlinear-fitted by software (Minitab 2021), and the equation is as follows:

$$\text{Mass gain } (\text{mg}\cdot\text{cm}^{-2}) = 0.40\ P^{0.0087}\ V^{0.25} - 0.64.$$

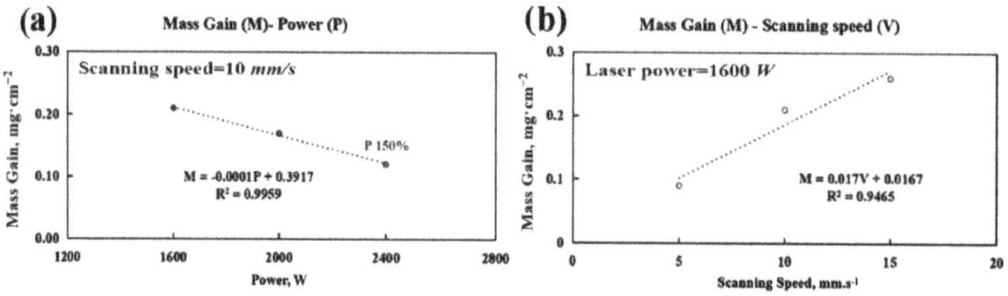

Figure 14. Linear regression equation of mass gain and process parameters; power (**a**), and scanning speed (**b**).

This equation is used for the preliminary evaluation of high-temperature oxidation resistance. The scanning speed has a more significant impact on the mass gain of the sample. It can be explained as the effect of crack. In the study of Xing et al. [32], oxides grew on the inner wall of the crack. As a result, the inside of the crack was filled with oxide.

Figure 15 shows the schematic of the high-temperature oxidation mechanism of the surface cracks in the IN718 coating. The inside of the crack was gradually filled with oxide during the high-temperature oxidation treatment (Figure 15a–c). It was considered that when the laser power reduced or the scanning speed increased, the dendrite was refined, and the eutectic compounds increased. Those changes negatively affected the high-temperature oxidation resistance of the IN718 coating. In the range of process parameters selected in this experiment, with the increase in scanning speed, the coating had higher cracking sensitivity under stress. The crack originated in HAZ in Figure 15d and extended to the surface. The crack extension could increase the surface contact area of the IN718 coating in the oxidation environment and further improve the mass gain of the IN718 coating.

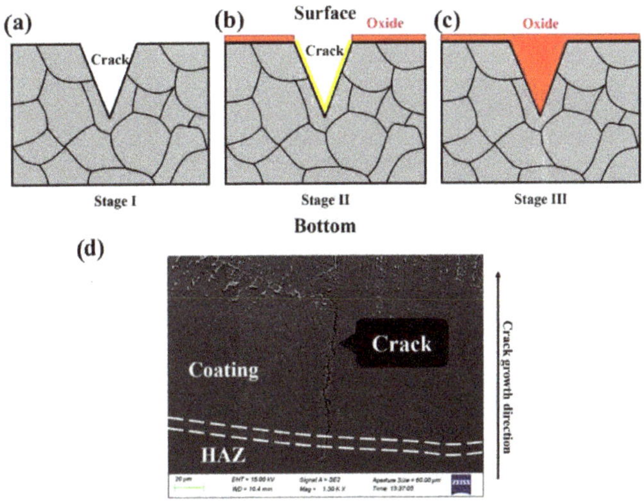

Figure 15. Schematic (**a–c**) and SEM (**d**) of the high-temperature oxidation mechanism of the surface cracks in the IN718 coating.

The high-temperature behavior of the IN718 coating could be divided into surface oxidation, intergranular corrosion, and material shedding. The purpose of changing process parameters was to adjust the precipitation of eutectic elements and reduce the mass gain following high-temperature oxidation. Cracks were the leading cause of coating failure. The microstructure evolution of coating and substrate should be considered comprehensively in the preparation of high-temperature coating to prevent cracks in the coating and HAZ. Figure 16 shows the relationship between process parameters, microstructure, and high-temperature oxidation resistance, where the positive sign (+) indicates an increase, while the negative (−) sign indicates a decrease.

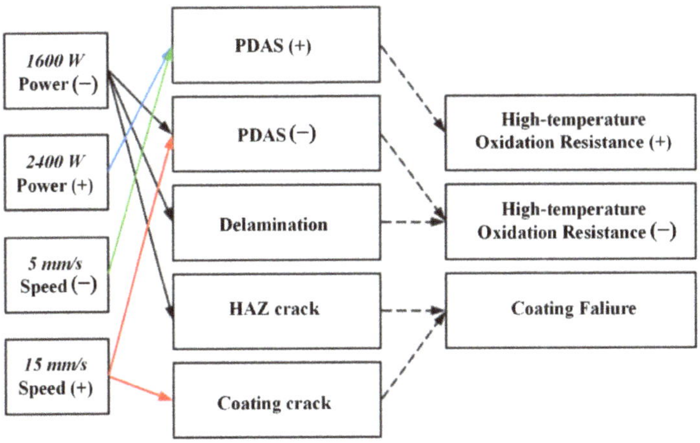

Figure 16. Effects of process parameter adjustment on high-temperature oxidation resistance.

3.5. Engineering Application

Figure 17 demonstrates a remanufacturing process of a Cr5Mo hydrocracking furnace tube for a petrochemical enterprise. The design life of the Cr5Mo furnace tubes was more than 15 years, and the actual service time was 15 years. In Figure 17a, severe oxidation has occurred on the surface of the Cr5Mo furnace tube. In Figure 17b, the surface of the furnace tube is processed to remove the oxide scale. After turning to remove the surface oxide layer and defects, the thickness of the original Cr5Mo tube complied with the standard ISO 13704. Penetrant testing was used to find surface defects, and the surface roughness value was Ra 6.3 (profile arithmetic average error, μm). Laser cladding IN718 coating on the surface of the Cr5Mo furnace tube is shown in Figure 17c. In Figure 17d, the surface of the furnace tube after laser cladding is processed to be smooth, resizing according to the requirements of engineering drawings. In the process of laser cladding, in regard to the experimental results, a selected laser power of 2400 W and scanning speed of 10 mm/s prepared the IN718 coating without significant defects after ultrasonic non-destructive testing and surface penetrant testing. The desired laser power of 2400 W can prevent the grain boundary liquefaction at low laser power. Scanning speed has a more significant impact on the high-temperature oxidation resistance of IN718 coating. Hence, the selected scanning speed of 10 mm/s can reduce the crack sensitivity of HAZ. Afterwards, this paper excluded the scanning speed of 5 mm/s for the purpose of ensuring the laser cladding efficiency. According to the experimental results, the Sample 2400-10 exhibited favorable high-temperature oxidation resistance. When the power was 2400 W and the scanning speed was 10 mm/s, there were no defects in FZ and HAZ. The laser cladding process could be divided into several stages to prevent the thermal deformation of the tube in excessive heating. The temperature of the Cr5Mo substrate during the laser cladding process ought to be strictly controlled below 600 °C.

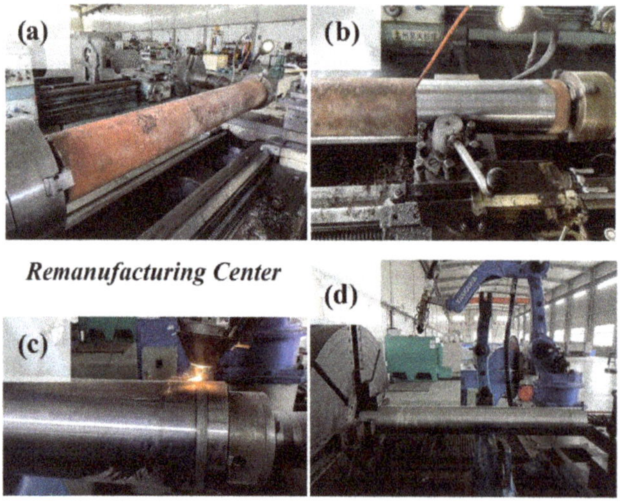

Figure 17. Remanufacturing of Cr5Mo hydrocracking furnace tube for a petrochemical enterprise: (**a**) serious oxidation of the Cr5Mo furnace tube; (**b**) surface processing of the furnace tube to removed oxide scale; (**c**) laser cladded IN718 on the surface of the Cr5Mo furnace tube; (**d**) surface processing and resizing after laser cladding.

The Larson–Miller equation LMP = T1 [C + log (t)] can be used to evaluate the remaining life of the tubes [33], where LMP is the Larson–Miller parameter, T1 is the temperature (in Kelvin), and t is the time to rupture. The value of the C-parameter locates between 15 and 30. It can be seen from the equation that the remaining life of the furnace tubes is significantly affected when overtemperature occurs. The high-temperature properties of materials play a vital role in the service life of furnace tubes. In the high-temperature creep test, when the Cr5Mo is performed at 550 °C and 190 MPa [34] and IN718 is performed at 700 °C and 560 MPa [35], both have the same stress rupture time of approximately 70 h. As IN718 has higher stability above 600 °C, it is helpful to improve the service life of remanufactured Cr5Mo furnace tubes in high-temperature oxidation environments. Figure 18 records the overhaul period of the Cr5Mo furnace tube in a certain petrochemical enterprise since 2019, which partly proves this conclusion. Deep cracks were found on the original Cr5Mo furnace tube without IN718 protective coating in the last testing. Thus, the remanufactured Cr5Mo furnace tube has longer service life than the original Cr5Mo furnace tube under the same operating conditions.

Type of furnace tube	Temperature	Examine and repair period			
		2019.10	2020.10	2021.10	2022.10
(A) Original Cr5Mo	500°C–700°C	Normal	Normal	Normal	Failure

(B) Remanufactured Cr5Mo		Normal	Normal	Normal	Normal

Figure 18. Overhaul period of the (**A**) original and (**B**) remanufactured Cr5Mo furnace tube in a certain petrochemical enterprise.

4. Conclusions

This study revealed the relationship between microstructure and high-temperature oxidation resistance in laser cladding IN718 coating, which proposed a loss mechanism of the coating under the combined action of delamination cracks and high-temperature corrosion. The experimental results were applied in a practical application. However, the application requires feedback after sufficient running time because various factors occur in the service environment. The main conclusions of the current study can be summarized as follows:

1. When the scanning speed increased from 5 mm/s to 15 mm/s, the top region's dendrites refined, and eutectic quantity increased. When the laser power increased from 1600 W to 2400 W, the dendrite in the top region coarsened, and the eutectic quantity decreased. Lower power and faster scanning speed improve the cooling rate, resulting in increased precipitation of eutectic elements. The change of laser process parameters affected the number of strengthening phases precipitated in the parent phase and subsequently changed the hardness of the coating.
2. When the scanning speed was increased from 5 mm/s to 15 mm/s, the average hardness of HAZ was increased by about 26 %. Under the condition of laser heating and rapid cooling, martensite with high hardness was generated in the HAZ region. Nevertheless, the brittleness of martensite also increased the crack sensitivity of HAZ, and the thermal fatigue cracks in HAZ initiated under thermal cyclic loading. The HAZ cracks along the thickness of the coating, resulting in coating failure.
3. When the scanning speed was increased from 5 mm/s to 15 mm/s, the mass gain was increased by about 188%. When the laser power increased from 1600 W to 2400 W, the mass gain decreased by about 43%. The formation of oxide was closely related to the distribution of the Ni3Nb phase in the top region of the coating. The less eutectic precipitation, the better the high-temperature oxidation resistance of IN718 coating. The scanning speed had a more significant impact on the high-temperature oxidation resistance.
4. The high-temperature behavior of the IN718 coating can be divided into surface oxidation, intergranular corrosion, and material shedding. The combined action of delamination cracks and intergranular corrosion led to material shedding. Crack

extension increased the surface contact area in the oxidation environment, which negatively affected the high-temperature oxidation resistance of IN718 coating.

Author Contributions: Conceptualization, Z.X. and F.W.; methodology, Z.X.; software, Z.X. and J.G.; validation, Z.X., W.L., and S.P.; formal analysis, Z.X.; investigation, Z.X.; resources, F.W. and W.L.; data curation, Z.X.; writing—original draft preparation, Z.X.; writing—review and editing, F.W. and S.P.; visualization, J.G.; supervision, F.W.; project administration, F.W.; funding acquisition, F.W. All authors have read and agreed to the published version of the manuscript.

Funding: This work was supported by the National Natural Science Foundation of China (Grant No. 51875075), the Natural Science Foundation of Guangdong Province (Grant No. 2021A1515011989), the Guangdong Provincial University Innovation Team Project (Grant No. 2020KCXTD012), the 2020 Li Ka Shing Foundation Cross-Disciplinary Research (Grant No. 2020LKSFG01D), the Key Projects of Universities in Guangdong Province (Grant No. 2019KZDXM046), and Shantou University Research Startup Funding Project (Grant No. NTF19030), the Equipment Pre-research Foundation (No. 80923010401).

Institutional Review Board Statement: Not applicable.

Informed Consent Statement: Not applicable.

Data Availability Statement: Not applicable.

Conflicts of Interest: The authors declare no conflict of interests.

References

1. Chen, S.; Ma, Y. The Research of the Chemistry Composition of High Temperature Oxidization under Loads for the Cr5Mo Alloy. *Mater. Sci. Forum* **2011**, *694*, 729–732. [CrossRef]
2. Chen, S. Study on the Micro-Structure of High Temperature Oxide Scales Applied Loads on Cr5Mo Alloy. *Adv. Mater. Res.* **2011**, *197–198*, 1086–1091. [CrossRef]
3. Muth, W.A. Effect of Pressure on the Flammable Limits of Some Hydrocarbon-Air Mixtures. Ph.D. Thesis, Iowa State University of Science and Technology, Ames, IA, USA, 1963. [CrossRef]
4. Yuan, W.; Li, R.; Chen, Z.; Gu, J.; Tian, Y. A comparative study on microstructure and properties of traditional laser cladding and high-speed laser cladding of Ni45 alloy coatings. *Surf. Coat. Technol.* **2018**, *405*, 126582. [CrossRef]
5. Song, M.; Wu, L.; Liu, J.; Hu, Y. Effects of laser cladding on crack resistance improvement for aluminum alloy used in aircraft skin. *Opt. Laser Technol.* **2021**, *133*, 106531. [CrossRef]
6. Tang, Z.; Liu, W.; Zhu, L.; Liu, Z. Investigation on coaxial visual characteristics of molten pool in laser-based directed energy deposition of AISI 316L steel. *J. Mater. Process. Technol.* **2021**, *290*, 116996. [CrossRef]
7. Mazzarisi, M.; Campanelli, S.L.; Angelastro, A.; Palano, F.; Dassisti, M. In situ monitoring of direct laser metal deposition of a nickel-based superalloy using infrared thermography. *Int. J. Adv. Manuf. Technol.* **2020**, *112*, 157–173. [CrossRef]
8. Lv, H.; Li, Z.; Li, X.; Yang, K.; Li, F.; Xie, H. Investigation on the Microporosity Formation of IN718 Alloy during Laser Cladding Based on Cellular Automaton. *Materials* **2021**, *14*, 837. [CrossRef]
9. Rezaei, A.; Kermanpur, A.; Rezaeian, A.; Badrossamay, M.; Foroozmehr, E.; Sadeghi, F.; Han, J.; Park, T.M. Contribution of hot isostatic pressing on densification, microstructure evolution, and mechanical anisotropy of additively manufactured IN718 Ni-based superalloy. *Mater. Sci. Eng. A* **2021**, *823*, 141721. [CrossRef]
10. Xie, H.; Yang, K.; Li, F.; Sun, G.; Yu, Z. Investigation on the Laves phase formation during laser cladding of IN718 alloy by GA-FE. *J. Manuf. Process.* **2020**, *52*, 132–144. [CrossRef]
11. Zhang, Y.; Yang, L.; Dai, J.; Liu, J.; Zhang, W.; Chen, H.; Wang, Z.; Shi, K. Microstructure and mechanical properties of pulsed laser cladded IN718 alloy coating. *Surf. Eng.* **2016**, *34*, 259–266. [CrossRef]
12. Jia, Q.; Gu, D. Selective laser melting additive manufactured Inconel 718 superalloy parts: High-temperature oxidation property and its mechanisms. *Opt. Laser Technol.* **2014**, *62*, 161–171. [CrossRef]
13. Jian, Y.; Shi, Y.; Liu, J. Effect of laser pretreatment and laser cladding on the interface bond strength and surface properties of vermicular graphite. *J. Mater.* **2022**, *57*, 6929–6942. [CrossRef]
14. Mazzarisi, M.; Angelastro, A.; Latte, M.; Colucci, T.; Palano, F.; Campanelli, S.L. Thermal monitoring of laser metal deposition strategies using infrared thermography. *J. Manuf. Process.* **2023**, *85*, 594–611. [CrossRef]
15. Korkmaz, M.E.; Gupta, M.K.; Waqar, S.; Kuntoglu, M.; Krolczyk, G.M.; Maruda, R.W.; Pimenov, D.Y. A short review on thermal treatments of Titanium & Nickel based alloys processed by selective laser melting. *J. Mater. Res. Technol.* **2022**, *16*, 1090–1101. [CrossRef]
16. Sanviemvongsak, T.; Monceau, D.; Macquaire, B. High temperature oxidation of IN 718 manufactured by laser beam melting and electron beam melting: Effect of surface topography. *Corros. Sci.* **2018**, *141*, 127–145. [CrossRef]
17. Thouless, M.D. Cracking and delamination of coatings. *J. Vac. Sci. Technol. A* **1991**, *9*, 2510–2515. [CrossRef]

18. Fang, Y.; Chen, N.; Du, G.; Zhang, M.; Zhao, X.; Cheng, H.; Wu, J. High-temperature oxidation resistance, mechanical and wear resistance properties of Ti(C,N)-based cermets with Al0.3CoCrFeNi high-entropy alloy as a metal binder. *J. Alloys Compd.* **2020**, *815*, 152486. [CrossRef]
19. Sumit, S.K.; Biswas, K.; Nath, A.K.; Manna, I.; Majumdar, J.D. Microstructural Change during Laser Welding of Inconel 718. *Optik* **2020**, *218*, 165029. [CrossRef]
20. Kang, X.; Dong, S.; Wang, H.; Yan, S.; Liu, X.; Xu, B. Effect of laser power on gradient microstructure of low-alloy steel built by laser melting deposition. *Mater. Lett.* **2020**, *262*, 127073. [CrossRef]
21. Liu, J.; Song, Y.; Chen, C.; Wang, X.; Li, H.; Zhou, C.; Wang, J.; Guo, K.; Sun, J. Effect of scanning speed on the microstructure and mechanical behavior of 316L stainless steel fabricated by selective laser melting. *Mater. Des.* **2020**, *186*, 108355. [CrossRef]
22. Du, L.; Gu, D.; Dai, D.; Shi, Q.; Ma, C.; Xia, M. Relation of thermal behavior and microstructure evolution during multi-track laser melting deposition of Ni-based material. *Opt. Laser Technol.* **2018**, *108*, 207–217. [CrossRef]
23. Savitha, U.; Gokhale, H.; Reddy, G.J.; Venkataramana, A.; Gokhale, A.A.; Sundararaman, M. Effect of Process Parameters on Porosity in Laser Deposited IN625 Alloy. *Trans. Indian Inst. Met.* **2012**, *65*, 765–770. [CrossRef]
24. Mankins, W.L.; Hosier, J.C.; Bassford, T.H. Microstructure and phase stability of INCONEL alloy 617. *Metall. Mater. Trans. B* **1974**, *5*, 2579–2590. [CrossRef]
25. Adomako, N.K.; Park, H.J.; Cha, S.C.; Lee, M.; Kim, J.H. Microstructure evolution and mechanical properties of the dissimilar joint between IN718 and STS304. Mater. *Sci. Eng. A* **2021**, *799*, 140262. [CrossRef]
26. Abdollah-Zadeh, A.; Salemi, A.; Assadi, H. Mechanical behavior of CrMo steel with tempered martensite and ferrite-bainite-martensite microstructure. *Metall. Mater. Trans. A* **2008**, *483–484*, 325–328. [CrossRef]
27. Lingenfelter, A. Welding of Inconel alloy 718: A historical overview. *TMS* **1989**, *718*, 673–683. Available online: https://www.tms.org/Superalloys/10.7449/1989/Superalloys_1989_673_683.pdf (accessed on 1 January 2021).
28. Cao, J.; Zhang, J.; Hua, Y.; Rong, Z.; Chen, R.; Ye, Y. High temperature oxidation behavior of Ni-based superalloy GH586 in air. *Rare Met.* **2017**, *36*, 878–885. [CrossRef]
29. Cao, J.; Zhang, J.; Hua, Y.; Chen, R.; Ye, Y. Improving the high temperature oxidation resistance of Ni-based superalloy GH202 induced by laser shock processing. *J. Mater. Process. Technol.* **2017**, *243*, 31–39. [CrossRef]
30. Wu, L.; Xiao, Y.; Xia, X.; Qian, J.; He, Y. High temperature oxidation resistance of porous Ni-Cr-Fe materials. *Int. J. Mater. Res.* **2020**, *111*, 126582. [CrossRef]
31. Leo Prakash, D.G.; Walsh, M.J.; Maclachlan, D.; Korsunsky, A.M. Crack growth micro-mechanisms in the IN718 alloy under the combined influence of fatigue, creep and oxidation. *Int. J. Fatigue* **2009**, *31*, 1966–1977. [CrossRef]
32. Xing, L.; Gu, C.; Lv, Z.; Bao, Y. High-temperature internal oxidation behavior of surface cracks in low alloy steel bloom. *Corros. Sci.* **2022**, *197*, 110076. [CrossRef]
33. Cao, L.; Du, C.; Shao, S.; Xie, G. Remaining life prediction of ethylene cracking tubes under carburizing and creep joint action. *ASME PVP* **2015**, *56994*, V06AT06A040. [CrossRef]
34. Zheng, Y.; Yang, S.; Ling, X. Creep life prediction of small punch creep testing specimens for service-exposed Cr5Mo using the theta-projection method. *Eng. Fail. Anal.* **2017**, *72*, 58–66. [CrossRef]
35. Xu, Z.; Cao, L.; Zhu, Q.; Guo, C.; Li, X.; Hu, X.; Yu, Z. Creep property of Inconel 718 superalloy produced by selective laser melting compared to forging. *Mater. Sci. Eng. A* **2020**, *794*, 139947. [CrossRef]

coatings

MDPI

Article

Gradient Coating of Laser Cladding TiB_2/Ti-Based Alloy on Titanium Alloy Surface

Yinghua Lin [1], Haibo Wang [1], Mingxing Zhang [1], Hui Lin [2,*], Dengqiang Yan [3], Qinghua Lin [2], Xin Kang [2] and Xinlin Wang [1,*]

1 Hunan Province Key Laboratory for Ultra-Fast Micro/Nano Technology and Advanced Laser Manufacture, School of Mechanical Engineering, University of South China, Hengyang 421001, China
2 School of Mechanical and Electrical Engineering, Putian University, Putian 351100, China
3 Hunan Special Equipment Inspection and Testing Institute, Hengyang Branch, Hengyang 421007, China
* Correspondence: hui.linca@gmail.com (H.L.); wxl_ly000@aliyun.com (X.W.)

Abstract: The technology of the TiB_2/TiB cladding layer addresses the issue of the insufficient wear resistance of cup-shaped parts composed of titanium alloy materials. In order to eliminate the cracking problem of laser cladding TiB_2/Ti-based alloy, 30%TiB_2/Ti-based alloy gradient coating was prepared on the surface of titanium alloy by laser cladding in this study. The results revealed that the microstructure of the matrix and the cladding layer is metallurgically bonded. The microstructures of the cladding layer appear as rod-like and coarse-grained features on the surface, and fine needle-like and small-grained morphologies inside. The fine needle-like TiB precipitated in situ from the melt has a flat interface with Ti and exhibits a low degree of interfacial mismatch, while the interface between small particle-like TiB and Ti is wavy and has a high degree of interfacial mismatch. The gradual increase in the amount of TiB is present from the surface to the bottom of the cladding layer, while the amount of unmelted TiB_2 particles decreases. The chemical structure of the cladding layer is mainly presented as TiB_2, TiB and α-Ti phases. The maximum hardness of the cladding layer is 725 $HV_{0.2}$, where it is more than twice the hardness of the substrate. The fretting wear resistance of the cladding layer is better than that of the titanium alloy substrate under low loads (50 N–100 N), while a high load (more than 150 N) triggers a reverse outcome.

Keywords: laser cladding; titanium-based alloys; TiB_2; microstructure; wear resistance

1. Introduction

Titanium alloy has the mechanical properties of low density, high specific strength, high specific modulus, corrosion resistance and non-magnetic. Recently, titanium alloys have been adopted as a substitute for structural steel in the construction of aircraft parts such as bulkheads, slide rails, landing gear beams, fans, compressor discs and blades [1,2]. Moreover, due to the high specific fracture toughness, good medium and low temperature toughness and high melting point, titanium alloys are also extensively used in the manufacturing of various accessories of aero-engine casings, such as cups, double hinges, anti-icing shells and afterburner impellers of fuel systems, etc. [3,4]. Titanium alloy has been widely used in ships, metallurgical equipment, chemical machinery and other fields. However, these alloys are sensitive to fretting wear during service [5,6] due to a large friction factor and serious adhesive wear in the material. A mechanical component composed of titanium alloy will be scrapped even though failure is present in a part of the component. It was reported [7–9] that the fatigue life of titanium alloys decreased by 60% due to the fretting effect, and the friction coefficient of a friction pair composed of titanium alloys without lubricants increased from 0.45 to 0.80. The main forms of fretting damage are fatigue delamination and adhesive wear. Therefore, how to improve the fretting fatigue and reduce the fretting wear of titanium alloys are the critical issues to be addressed.

The technology of the TiB_2/TiB cladding layer addresses the issue of the insufficient wear resistance of cup-shaped parts composed of titanium alloy materials. The application of TiB_2 cermet coating on titanium alloys by using laser cladding has been reported as a promising technology thanks to its distinct mechanical properties. Laser cladding employs a high-energy laser beam to rapidly melt, expand and solidify the surface material of a mechanical component, and further obtain a metallurgical bond with the substrate with a dense structure. In regard to the selection of cladding material, compared with commonly used carbide ceramics, nitride ceramics and their composite materials, TiB_2 has both strong Ti-B ionic bonds and B-B covalent bonds. Its physical and chemical advantages are present as follows [10–12]: ① High hardness. TiB_2 has great hardness both at room temperature and high temperature, with a Vickers hardness rating of up to 800 HV even at 600 °C. ③ Low friction coefficient. TiB_2 exhibits a lower friction coefficient than ceramics, such as WC and TiN. Vairamuthu et al. [13] concluded the higher hardness of TiB_2-M cermet composite coating compared with WC-Co, as well as the superior cutting ability for Fe-based materials. Weng et al. [14] also found that TiB_2-M metal–ceramic composite coating has higher wear resistance than Cr_3C_2-NiCr and other coatings. The application of TiB_2 cermet coating on titanium alloys by using laser cladding has been regarded as a promising technology thanks to its distinct mechanical properties.

TiB_2 and TiB have been used as reinforcing phases in titanium-based composites due to their reasonable elastic modulus, high hardness, favorable wear resistance, sound oxidation resistance, and admirable compatibility and stability with Ti. Furthermore, the residual stress in the composite material is reduced thanks to the similar density and thermal expansion coefficient of Ti substrate and cladding materials, TiB_2 and TiB [15,16]. However, the laser cladding TiB_2-reinforced titanium-based composite coating is sensitive to cracks [17]. The relatively weak surface hardness of the cladding layer was reported as a culprit to impair the anti-fretting wear performance [18,19]. This study aims to create a novel metallurgical coating exhibiting a decrease in hardness from the outside towards the inside layers to gain a favorable anti-fretting wear performance. The mixture of 500 nm of TiB_2 and Ti at a special composition ration will be developed as the laser cladding material, due to the high hardness characteristics generated from the aggregation of submicron TiB_2 particles during the solidification process of the laser molten pool.

2. Experimental Section

2.1. Preparation of Composite Coating

The TC4 titanium alloy was selected as the experimental substrate (Table 1), and the titanium alloy was cut into discs with a diameter of 30 mm and a length of 15 mm. The end faces of the alloy discs were first polished by water sandpaper into smooth surfaces, and further cleaned with anhydrous ethanol to ensure that the polished faces were free of impurities such as residual oxides and oil stains. TiB_2 powder with a mass fraction of 99.5% (grain size 0.4 μm~0.6 μm) and Ti powder with a mass fraction of 99.6% (grain size 4 μm~6 μm) were selected as cladding materials, as shown in Tables 2 and 3. In order to reduce the loss rate of the cladding material, a small amount of 2123 phenolic resin was added into the material as the adhesive. The TiB_2 powder and Ti powder were pre-placed on the end face of the sample disc at a ratio of 3:7 (volume fraction), and the pre-coating thickness was set as 0.30–0.35 mm. Then, the preset layer was placed in a drying box, and the temperature and drying time were set to 120 °C and 5 h, respectively. A YLS-6000 IPG fiber laser was employed as a laser device. The protective gas was Ar gas, and the flow rate was set to 15 L/h. The laser process parameters were as follows: 2.6 kW of the laser power P, 7 mm/s of the laser beam scanning rate V, and 5 mm of the spot size D (circular shape spot). Figure 1 illustrates the experimental system of laser cladding. The dilution modulation of the titanium alloy base material is crucial to the formation of the gradient coating. The appropriate dilution is able to produce an optimal increase in the content of the Ti matrix from the surface to the inside during the solidification process of the laser molten pool. Furthermore, the required gradient coating is created as a result of different in situ reactions of TiB_2 and Ti and varying dissolution conditions from the surface to the inside.

Table 1. Chemical composites of TC4 (wt.%).

Composition	Al	V	Fe	S	C	O	Ni	Ti
Content	6.01	3.84	0.3	0.15	0.1	0.1	0.15	89.35

Table 2. Chemical composites of TiB_2 (wt.%).

Composition	Ti	B	C	Ca	Co	Fe	O
Content	67.5	28.5	1.0	1.0	1.0	0.5	0.5

Table 3. Chemical composites of Ti (wt.%).

Composition	Ti	Fe	Si	Al	C	N	O	H
Content	98.95	0.5	0.03	0.01	0.05	0.05	0.4	0.01

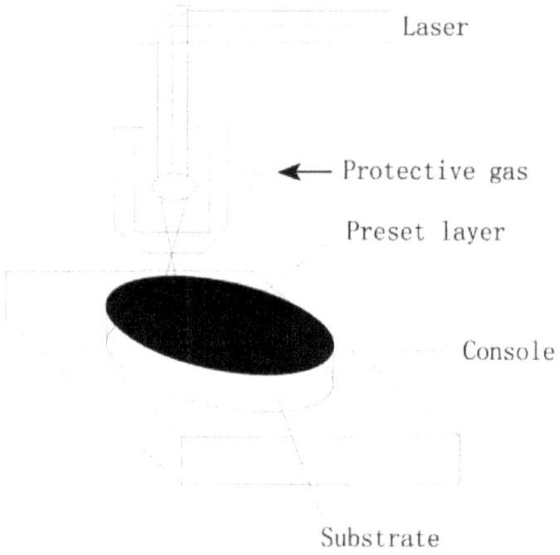

Figure 1. Schematic diagram of laser cladding test method.

2.2. *Composite Coating Characterization and Performance Testing*

A post-processing procedure of laser cladding included the following steps: cutting, grinding and polishing the sample component by using a wire electrical discharge machine; etching the component by an etchant comprising a solution of HNO_3, HF and H_2O in a volume ratio of 3:2:95. A scanning electron microscope (SEM, JSM-6490LA, Japan Electronics Co., Ltd., Tokyo, Japan) was used to study the microstructure of the cross section of the composite coating. D/max 2550 VL/PC, an X-ray diffractometer (XRD, D/max 2550 VL/PC), was utilized to identify the phase of the composite coating surface. Due to the low atomic number of B element, its content cannot be quantitatively detected by energy spectrum. In this study, a JXA-8100 electron probe (EPMA, JXA-800, Japan Electronics Co., Ltd., Tokyo, Japan) was used to analyze the phase types in the microstructure. The fine structure of the samples was investigated using a JEOL-2100 transmission electron microscope (TEM, JEOL-2100, Japan Electronics Co., Ltd., Tokyo, Japan) at operating voltage (200 kV). The TiB/Ti interface mismatch degree was calculated using the Bramfitt plane mismatch lattice model, i.e.,

$$\delta = 2 \times (d_{\text{whisker}} - d_{\text{matrix}})/(d_{\text{whisker}} + d_{\text{matrix}}) \times 100\% \quad (1)$$

where δ represents the degree of interfacial mismatch, d_{whisker} represents the TiBw lattice spacing near the interface and d_{matrix} represents the Ti lattice spacing near the interface.

The HDX-1000 digital microhardness tester was used for microhardness measurement under the experimental conditions, with 300 g of load and 10 s of the loading duration. The measurement was carried out from the outside to the inside along the cross section of the cladding layer at intervals of 100 μm. The FTM200 micro-vibration friction and wear tester was employed to conduct the measurements of fretting friction and wear of the composite coating, where line–surface contact was present in the friction process. The experimental parameters are shown in Table 4. The morphologies of the wear scars were observed with scanning electron microscopy (SEM, JSM-6490LA, Japan Electronics Co., Ltd., Tokyo, Japan). The wear losses of the cladding layers and the substrate were measured using a Mahr-M1 profilometer. The wear volume loss was calculated using the equation below:

$$V = S \times A \tag{2}$$

where V is wear volume (mm^3); S is wear scar cross-sectional area (mm^2); A is displacement value (mm).

Table 4. Experimental parameters of fretting wear test.

Test Temperature (°C)	Load (N)	Frequency (Hz)	Amplitude (μm)	Time (min)	Size (mm)
23~30	50, 100, 150	20	100	30	12 × 20 × 10

3. Results and Discussion

3.1. Microstructure Characteristics of Laser Cladding Layer on Titanium Alloy Surface

Figure 2a shows the cross-sectional morphology of the laser cladding 30%TiB_2/titanium-based alloy cladding layer on the titanium alloy surface. The thickness of the original preset layer is 0.3~0.35 mm, while the thickness of the cladding layer is 0.4~0.45 mm (Figure 2a), indicating that the substrate is partially melted under laser cladding, which is conducive to the formation of metallurgical bonding between cladding layer and the substrate. The combination layer of cladding layer and the substrate is divided into a melting zone (CZ), a bonding zone (BZ), a heat affected zone (HAZ) and a substrate (SUB) from the outside and inside. It can be seen from Figure 2a,b that there is a visible but not an excessively obvious demarcation line without any pores or cracks at the joint of the cladding layer and the substrate. This suggests the similarity of the thermal expansion coefficient values of TiB_2 and the titanium substrate, as well as no presence of cracks in the bonding area during the solidification of the molten pool. These microstructural morphologies strengthen the anti-spalling performance of the cladding layer during friction and wear.

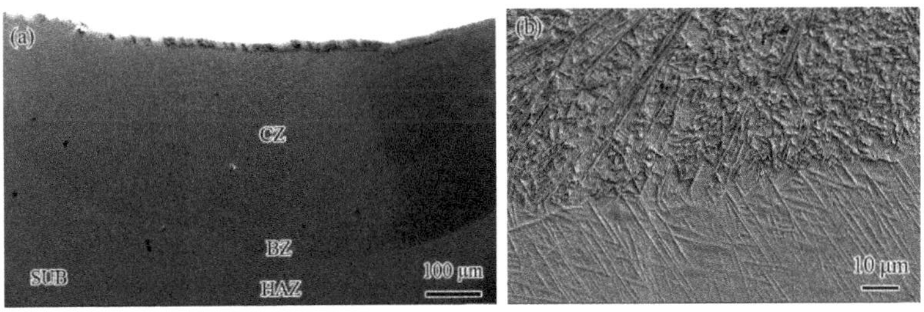

Figure 2. Macroscopic morphology (**a**) and microstructure (**b**) of the cross section of laser cladding layer (CZ—cladding zone, BZ—bonded zone, HAZ—heat affected zone, SUB—substrate).

Figure 3a shows the surface layer structure of the 30%TiB_2/Ti-based alloy cladding layer, which is mainly composed of thick rods (diameter 4 μm–6 μm), fine needles (diameter 0.5 μm–1 μm) and granular substances (diameter 2 μm–3 μm). Thick rods are formed by the agglomeration of particles. Figure 3b displays the middle layer structure of the cladding layer, which exhibits three types of morphological characteristics: coarse particle structure (3 μm–6 μm in diameter), fine needle-like structure (0.5 μm–1 μm in diameter) and small particle structure (1 μm–2 μm in diameter). The diameter of the thick rods is virtually unchanged, but the length visibly decreases. The particle structure in the middle layer has a decreasing size and an increase in amount. The microstructure of the bottom layer of the cladding layer shows three morphological features: dendrite-like crystal (3 μm–4 μm in diameter), fine needles (0.5 μm–1 μm in diameter) and small particles (1 μm–2 μm in diameter), as shown in Figure 3c. The diameter of the dendritic-like crystal diameter is smaller than that of the thick rod, but the length is significantly larger than that of the thick rod. The amount of small particles tends to increase compared with other layers. Figure 3d shows the structure of the bonding zone and heat-affected zone of the cladding layer. The microstructure of the bonding zone exhibits a fine needle-like and fine particle-like pattern, and thick rod-like and dendrites-like structures are not present in this layer (Figure 3d). The morphological characteristics of the heat-affected zone feature as a net fine needle-like pattern (Figure 3d).

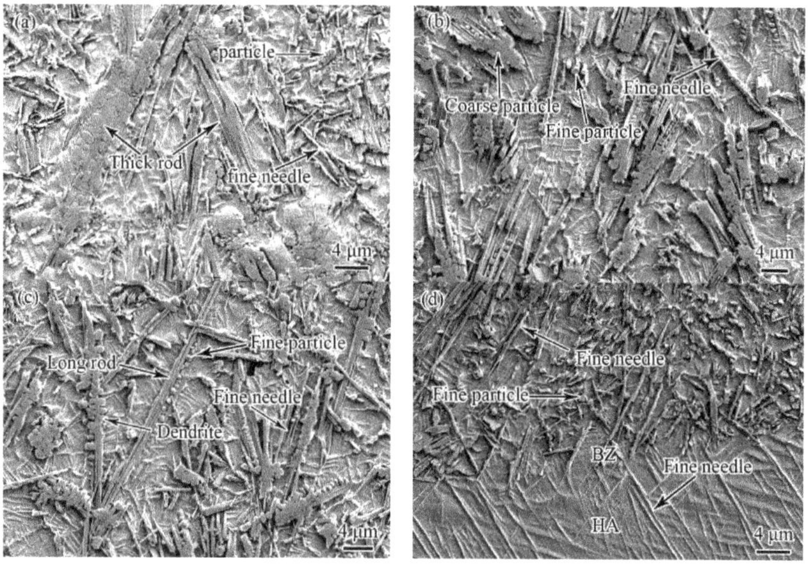

Figure 3. The cross-sectional microstructure of a laser cladding layer with 30%TiB_2/Ti-based alloy on titanium alloy surface: (**a**) surface layer; (**b**) middle layer; (**c**) bottom layer; (**d**) bonding area.

When a pre-set layer utilized TiB_2 and Ti particles with the size of 0.4 μm–0.6 μm, thick rod-like and coarse-grained structures would be formed during the laser preparation process. However, if a pre-set layer was created by using TiB_2 powder with a particle size of 5 μm, a thick rod-like and coarse-grained structure would not be present in the cladding layer [17,18], suggesting that the decrease in the particle size of TiB_2 to sub-micron level triggered the unstable surface action energy of TiB_2 particles. However, TiB_2 powder with a particle size of 0.4 μm–0.6 μm can exist in the form of single particles in a dry air medium at ambient temperature (Figure 4). This microstructural feature demonstrates that the agglomeration of TiB_2 particles at submicron-size level is induced in a multimedia physical environment.

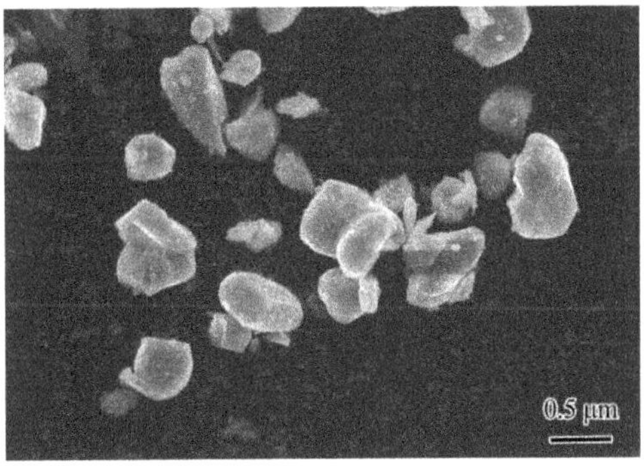

Figure 4. TiB_2 single particle with particle size of 0.4 μm–0.6 μm in dry air medium at ambient temperature.

There are three stages in the process of submicron TiB_2 forming coarse rod-like and coarse-grained structures:

Stage 1 Preset process. In the preparation process of the pre-set layer, ethanol is utilized as a medium to evenly mix a small amount of binder and TiB_2 powder. It has been demonstrated that the moisture absorption of particles generated liquid bridge force [20]; then, under the action of the stirring force exerted in the mixture it triggered the collision and coagulation of particles, further resulting in the agglomeration of some particles.

Stage 2 Drying process. Since there is a high fraction of structurally adsorbed water and physically adsorbed water in the presetting process, voids are likely to appear between the particles and develop a large capillary force [21] due to the evaporation of water during the drying process. Adhesive bonding generated from capillary forces between particles agglomerates some particles. The above two stages only appear in the water molecule medium and at a lower temperature (below 150 °C). The two stages simply develop soft agglomeration [22], since there is little presence of the effect of chemical bonds on the surface of TiB_2.

Stage 3 Laser cladding process. Due to the high melting point and melting enthalpy of TiB_2 particles, the particles are difficult to melt or dissolve in a short duration of laser processing, while the Ti melt can be generated from Ti matrix material in the short cladding process. In the stage of molten pool formation, the fluid-mechanical effects of temperature gradient, concentration gradient and gravity gradient trigger strong convection currents in the molten pool, thereby inducing more collision between TiB_2 particles. As a result, the formation of thick rods and coarse particles will be developed.

3.2. Phase Composition and Microstructure Analysis of 30%TiB_2/Ti-Based Alloy Laser Cladding Layer

The cladding layer is composed of α-Ti, TiB_2 and TiB, suggesting the formation of TiB created from the in situ reaction of Ti and TiB_2 under the laser cladding condition [17,18]. SEM and probe microanalysis (EPMA) identification was utilized to detect the microstructure of the cladding layer. The microstructure of TiB_2 and TiB exhibited rod-like and dendrite-like crystal structures as well as coarse particles, while the microstructure of TiB showed fine needle-like crystal and small particles (Figure 5 and Table 5). The surface and middle of the cladding layer presented thick rod-like and coarse-grained morphologies, and dendrite-like structures were formed at the bottom of the cladding layer, suggesting that the formation of dendritic-like structures in the cladding layer was greatly affected by the melt environment.

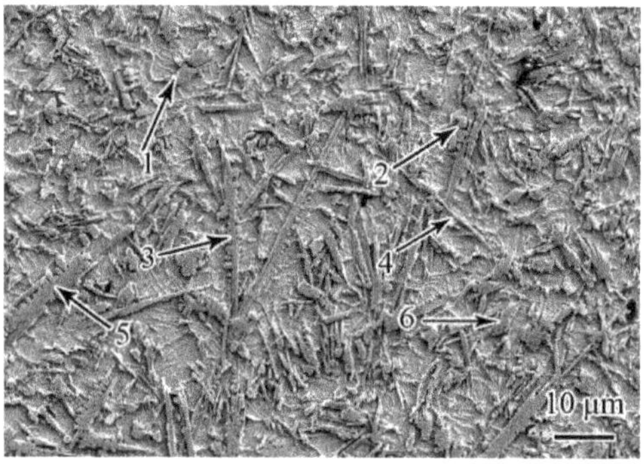

Figure 5. Microstructure of 30%/Ti-based alloy cladding layer.

Table 5. Measurements of EPMA point area for each location shown in Figure 5.

Position No.	Mass Fraction/%				
	Ti	B	Al	V	O
1	34.5	58.6	4.1	2.5	0.3
2	49.3	43.1	4.6	2.6	0.4
3	48.4	43.7	4.2	3.5	0.2
4	48.7	42.5	5.1	3.4	0.3
5	34.8	57.9	4.0	2.9	0.4
6	35.3	58.6	3.1	2.6	0.4

It was found that both the dendrite-like trunk and the dendrite arms in the cladding layer showed rod-like features, and the dendrite arms were perpendicular to the orientation of the trunk (Figure 6). The phase composition in the layer was identified by EPMA, and it was found that TiB_2 accounted for the primary portion in the central region of the trunk, while TiB particles were formed in the dendrite arms. Since TiB is formed by the in situ reaction between TiB_2 and Ti, the formation of different dendrite arms of TiB may possibly be created by obtaining B atoms from the TiB_2 particles existing in the dendrite-like trunk. However, the high melting point and melting enthalpy of TiB_2 particles inhibit the melting or dissolution of TiB_2. The solubility of TiB_2 particles can be improved by increasing the amount of Ti liquid or increasing the temperature of the molten pool. A large amount of Ti liquid existing in the bottom layer of the cladding layer facilitates the dissolution of TiB_2 particles.

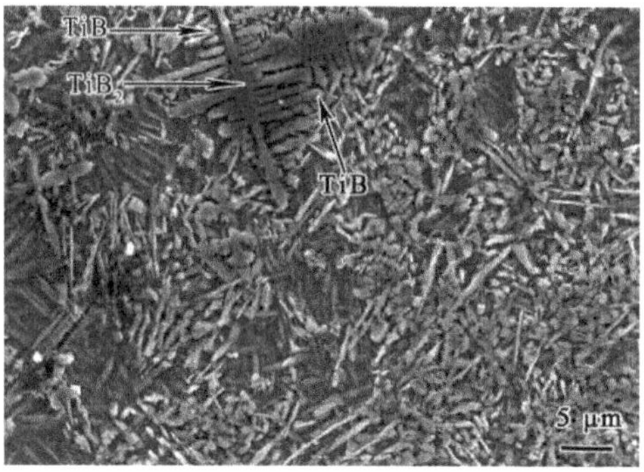

Figure 6. The dendritic-like morphology in the cladding layer.

It has been demonstrated that the highest dissolution rate is present on a TiB_2 crystal plane with the slowest crystal growth [23], and (0 0 0 1) and (1 0 1 0) TiB_2 crystal planes exhibit the slowest crystal growth [24]. These crystal planes exposed outside the surface layer will transform as dissolved crystal planes. As a result, these dissolved Ti and B atoms will re-precipitate and grow during the condensation process of the molten pool. Direct nucleation requires a large amount of supercooling under the limited concentration of dissolved Ti and B atoms. The bottom layer of the cladding layer facilitates nucleation thanks to its favorable physical environment.

Some published Ti-B phase diagrams revealed the low solubility of B element in titanium alloy [25]. Therefore, TiB tended to grow up by adsorption on the existing TiB_2 backbone. In addition, it was found that some dendrite arms did not cling to the dendrite-like trunk, since some defects were present in the TiB_2 trunk position. These defects accelerated the dissolution of the TiB_2 particle edge, and thus increased the dissolved B content compared to the condition without defects. As a result, the discharged B elements during solidification were concentrated at the interface front of the dendrite arm bulge to achieve effective compositional supercooling, which promoted nucleation in the liquid phase to induce TiB_2 crystal growth without attaching to the existing TiB_2 backbone.

3.3. TEM Detection of 30%TiB_2/Ti-Based Alloy Laser Cladding Layer

Figure 7 shows the interface and matrix defect results of two forms of TiB and α-Ti. The first form of TiB exhibits a long strip-like structure. The interface between TiB and α-Ti presents a flat shape without defects, while a few defects in the α-Ti matrix appear in a region near the interface (Figure 7a). The second form of TiB shows a granular structure, and a wave-like interface between TiB and α-Ti is present. In contrast to the first form, a relatively large amount of defects existing in the α-Ti matrix are present in the region near the interface (Figure 7b). TiB showed a needle-like structure that grows along the [0 1 0] optimal direction [26]. When B elements are freed from the surface of TiB_2 particles, they are affected by the turbulent flow of the molten pool or blocked by obstacles. The direction of movement of the B elements is shifted due to the presence of the turbulent flow or obstacles and further develops a wave-like interface. Moreover, the generation of a large number of defects in the α-Ti matrix near the interface is also related to the restricted free growth of TiB.

Figure 8 shows the results of the interfacial mismatch of two morphologies of TiB and α-Ti. The interfacial mismatch between fine needle-like TiB and α-Ti is 3.5%, while that of granular TiB and α-Ti is 11.2%. A large number of defects existing in the α-Ti matrix are present in the region near the granular TiB interface and concentrated in a single area. These morphological features result in higher stresses between granular TiB and Ti interfaces compared to that between the TiB and α-Ti interfaces. However, the pure fine needle-like structures are difficult to generate by the growth of TiB crystals during the nucleation and growth process. The following environmental conditions induce the modification of TiB morphology: the short duration of the existence of the laser melt pool, the intensive convection, and the high occurrence of collision and coalescence among particles [27].

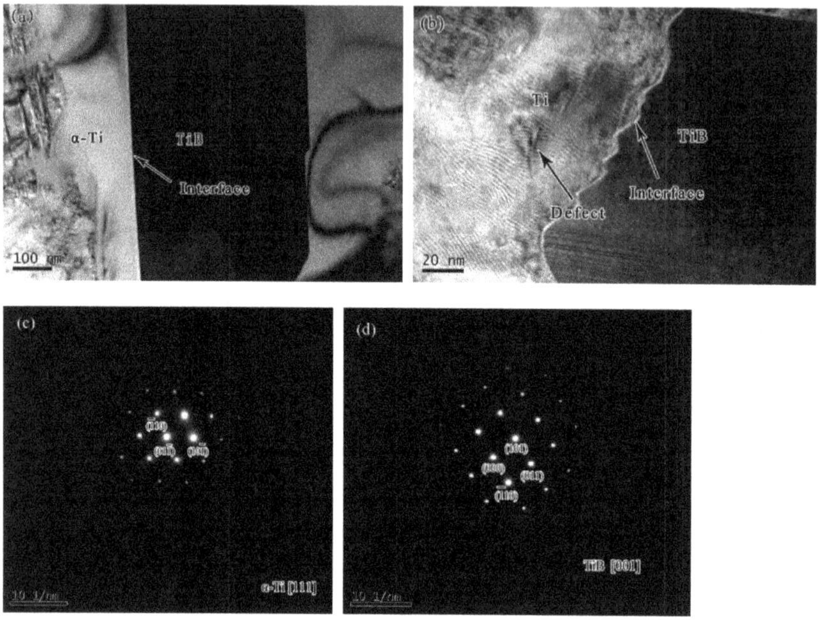

Figure 7. Interface and matrix defects of two different morphologies of TiB and α-Ti: (**a**) TiB and Ti matrix with long strip-like structures; (**b**) TiB and Ti matrix with granular structures of Figure b; (**c**) α-Ti diffraction spot; (**d**) TiB diffraction spot of Figure b.

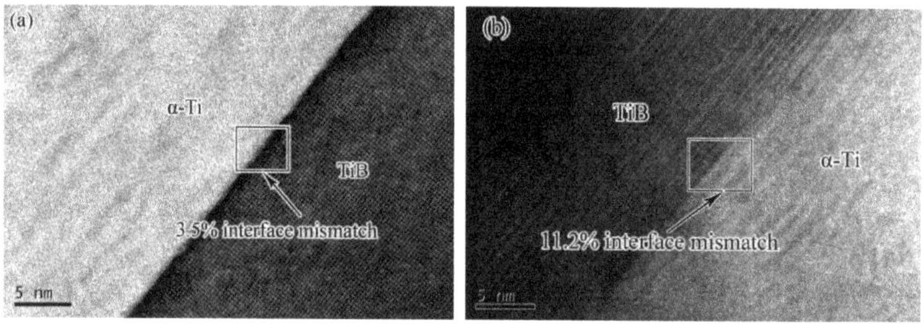

Figure 8. The interfacial mismatch degree of two morphologies of TiB and α-Ti: (**a**) the interfacial mismatch degree of fine needle-like TiB and α-Ti, 3.5%; (**b**) the interfacial mismatch degree for granular TiB and α-Ti, 11.2%.

3.4. Hardness of 30%TiB_2/Ti-Based Alloy Laser Cladding Layer

Figure 9 displays the microhardness distribution across the cross-section of the 30%TiB_2 /Ti-based alloy laser cladding layer. It can be seen that as the distance from the surface of the titanium alloy substrate increases, the hardness of the cladding layer gradually decreases. The microhardness variation presents three stages. The first stage is the microhardness of the cladding layer with a highest hardness of 725 $HV_{0.2}$ and an average hardness of 653 $HV_{0.2}$. The second stage is the microhardness of the heat-affected zone with a peak value of 521 $HV_{0.2}$ and an average hardness of 487 $HV_{0.2}$. The third stage is the hardness of the titanium alloy substrate with an average hardness of 332 $HV_{0.2}$. The cladding layer is triple the hardness of the titanium alloy substrate thanks to the enhancement of TiB_2 and TiB strengthening phase. There are two mechanisms to form TiB_2/TiB gradient coating. First, the substrate partially melts during the melting process of the TiB_2/Ti cladding material. As a result, the gradual increase in Ti content from the surface to the inside during the rapid solidification process of the molten pool is generated. The varying Ti content further affects the in situ reaction and dissolution/precipitation of TiB_2 and Ti. Second, the melting direction is from the surface down to the inside during the laser cladding process by using the preset method, while a reverse direction of solidification is present due to the cooling effect of the base material. Therefore, the temperature gradient and solidification rate vary from the surface down to the inside, resulting in the formation of a gradient coating during the cladding process.

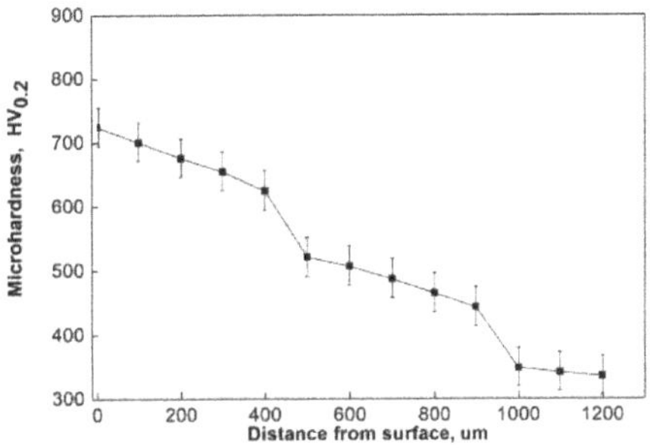

Figure 9. Microhardness distribution of cladding cross-section.

Under the chilling action of the substrate, the solidification rate of the laser molten pool gradually increases from the surface to the inside during the solidification process. Therefore, the laser cladding layer prepared by the preset method shows the gradual increase in residual stresses from the surface down to the inside. Consequently, the source of cracks is prone to formation inside the cladding layer, especially the non-gradient TiB_2/TiB cladding layer prepared by laser cladding. Technically, the inappropriate low content of Ti matrix inside the cladding layer is the culprit to significantly increase the risk of cracking of the cladding layer due to the unfavorable decrease in the internal plasticity, as shown in Figure 10.

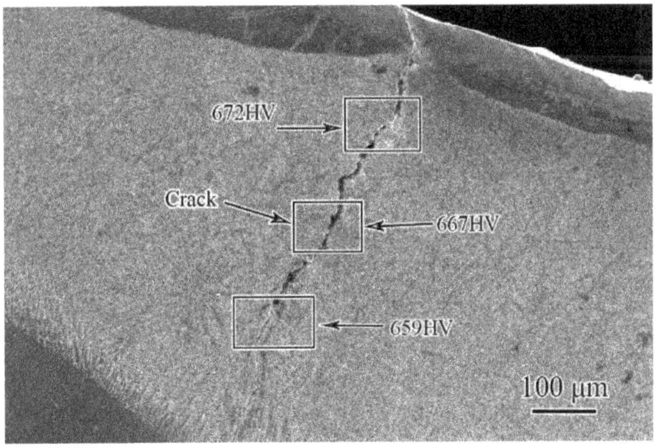

Figure 10. High hardness TiB_2/TiB cladding layer.

3.5. Wear Resistance of 30%TiB_2/Ti-Based Alloy Laser Cladding Layer

The fretting wear test was performed on the titanium alloy substrate and the cladding layer at room temperature. The detecting region for the experimental study was located at the contact area of the cladding layer with a hardness of 650 $HV_{0.2}$. A comparative analysis between titanium alloy substrate and the cladding layer was also conducted. The cladding layer is almost twice the microhardness of the titanium alloy substrate, but the friction coefficient of the cladding layer is slightly lower than that of the substrate (Table 6). The wear test results of titanium alloy substrate and cladding layer under different friction loads are shown in Figure 11. The wear amount of the cladding layer is lower than that of the titanium alloy substrate under low loads (50 N and 100 N), but a high load (150 N) reverses the wearing scenario.

Table 6. Fretting wear test results of contact zone.

Type	Average Microhardness ($HV_{0.2}$)	Average Friction Coefficient/100 N
Substrate	332	0.53
Coating	650	0.46

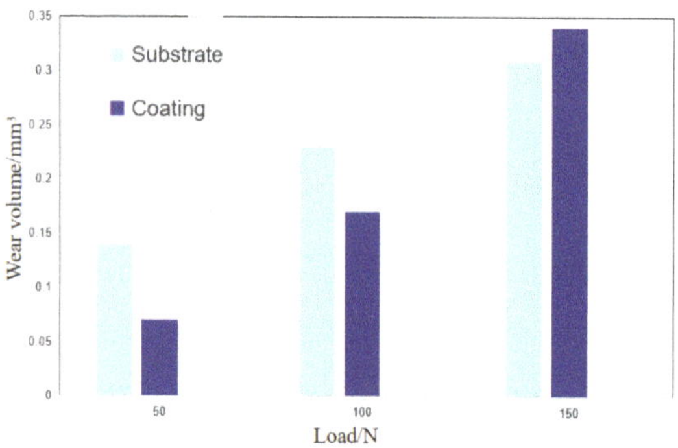

Figure 11. The wearing amounts of titanium alloy substrate and cladding layer under different loads.

To further investigate the fretting wear morphology of the cladding layer, SEM was utilized to examine the worn surface morphology of the substrate and the cladding layer under 50 N and 150 loads. Wearing damage zones appear on the surface of the titanium alloy substrate under all experimental loads (Figure 12a,b). The surface adhesive wear of the titanium alloy substrate is not obvious under a relatively low load (50 N), while a high load (150 N) induces visible adhesion pits on the surface of titanium alloy. Under a low load (50 N), the thickness of the wear damage zone of the cladding layer is significantly lower than that of the titanium alloy substrate. This morphological feature suggests that the strengthening phase in the cladding layer can improve the fretting wear resistance under a low load. However, under a high load (150 N), the depth of the wear damage zone of the cladding layer is significantly higher than that of the titanium alloy substrate.

The 30%TiB_2/Ti-based alloy laser cladding layer is composed of TiB_2 and TiB, which are both ceramic materials. The microstructures of these materials exhibit close-packed hexagonal and orthorhombic structures composed of B-B covalent bonds and Ti-B ionic bonds, respectively. It has been demonstrated that high bond energy and chemical stability are present in the two bonds [28]. Therefore, it is difficult for both TiB_2 particles and TiB short fibers to react chemically at relatively low temperatures, and the friction only triggers material wear or transfer. However, under high loads, the cladding layer is incapable of resisting the fretting wear. TiB_2 is a ceramic brittle phase, which develops weak bonding of the interface between TiB_2 particles and the matrix. Moreover, a higher particle interface stress is present as the increase in the particle size. A high load can trigger fatigue crack initiation and propagation. Consequently, the shedding of large particles will appear if cracks spread across the entire particle interface.

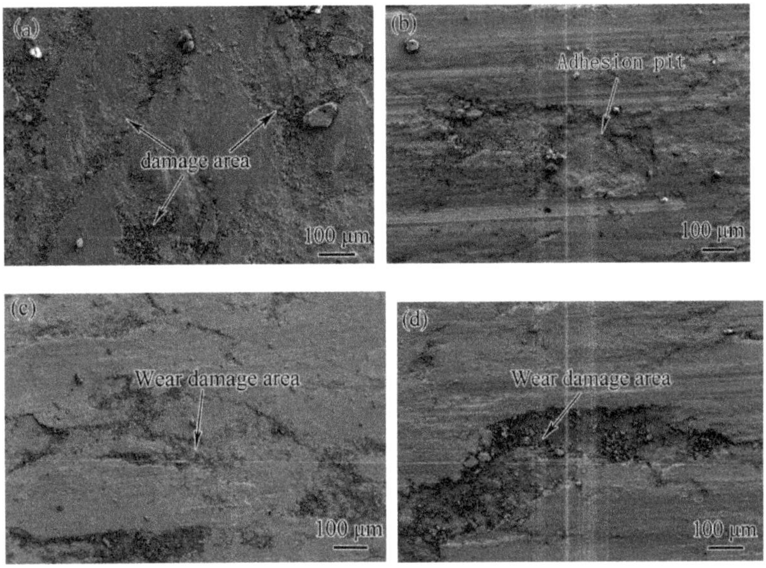

Figure 12. Wear morphologies of titanium alloy substrate and cladding layer under 50 N and 150 N loads: (**a**) titanium alloy substrate under 50 N load; (**b**) titanium alloy substrate under 150 N load; (**c**) cladding layer Under 50 N load; (**d**) cladding layer under 150 N load.

4. Conclusions

The TiB_2/titanium-based alloy laser cladding layer on the surface of titanium alloy is split into the melting zone, bonding zone and heat-affected zone. The microstructure of the substrate and cladding layer is present as a metallurgically bonded, well-formed surface structure, with no defects such as cracks and pores.

(1) The microstructures of the cladding layer show rod-like and coarse-grained features on the surface, and fine needle-like and small-grained morphologies inside. The TiB precipitated in situ from the melt exhibits fine needle-like and small-grained morphologies, and an increasing amount from the surface layer to the bottom of the cladding layer. The chemical structure of the cladding layer mainly presents TiB_2, TiB and α-Ti phases.

(2) The hardness of the cladding layer is gradually decreased with the increase in the distance from the surface of the cladding layer to the titanium alloy base. The average hardness of the cladding layer is 653 $HV_{0.2}$, and the maximum hardness is 725 $HV_{0.2}$. The laser cladding TiB_2/titanium-based alloy coating is around twice the hardness of the substrate.

(3) The fretting wear resistance of the cladding layer is better than that of the titanium alloy substrate under low loads (50 N–100 N), while a high load (more than 150 N) triggers a reverse outcome.

Author Contributions: Conceptualization, Y.L. and H.L.; methodology, X.W.; software, H.W.; validation, H.W., M.Z. and Q.L.; formal analysis, Y.L.; investigation, X.K.; resources, D.Y.; data curation, H.W.; writing—original draft preparation, Y.L. and H.L.; writing—review and editing, H.L.; visualization, X.K.; supervision, X.W.; project administration, Q.L.; funding acquisition, Y.L. All authors have read and agreed to the published version of the manuscript.

Funding: The authors are grateful for financial support from the Hunan Natural Science Foundation—Provincial and Municipal Joint Fund (2022JJ50019); Hengyang Guiding Plan Project (hyzdxjh202101); and University of South School Talent Introduction Project (210XQD017).

Institutional Review Board Statement: Not applicable.

Informed Consent Statement: Not applicable.

Data Availability Statement: We declare that the data in the paper is valid.

Conflicts of Interest: The authors declare no conflict of interest.

References

1. Li, G.Q.; Ma, F.C.; Liu, P.; Qi, S.C.; Li, W.; Zhang, K.; Chen, X.H. Review of micro-arc oxidation of titanium alloys: Mechanism, properties and applications. *J. Alloys Compd.* **2023**, *22*, 169773. [CrossRef]
2. Shu, L.S.; Li, J.H.; Wu, H.; Heng, Z. Optimization of Multi-Track Laser-Cladding Process of Titanium Alloy Based on RSM and NSGA-II Algorithm. *Coatings* **2022**, *12*, 1301. [CrossRef]
3. Akula, S.P.; Ojha, M.; Rao, K.L.; Gupta, A.K. A review on superplastic forming of Ti-6Al-4V and other titanium alloys. *Mater. Today Commun.* **2023**, *34*, 105343. [CrossRef]
4. Yue, Y.; Liu, S.G.; Qiu, W.W.; Wang, F.; Xue, Y.J.; Xia, C.Q.; Du, S.M. Comparative Study on Wear Behaviors of Monolayer and Heterogeneous Multilayer Ta Coatings in Atmospheric and SBF Environments. *Coatings* **2023**, *13*, 120. [CrossRef]
5. Zhou, K.; Xie, F.Q.; Wu, X.Q.; Wang, S.Q. Fretting wear behavior of nano ZrO_2 doped plasma electrolytic oxidation composite coatings on TC21 titanium alloy. *Surf. Coat. Technol.* **2021**, *421*, 127429. [CrossRef]
6. Wang, K.M.; Liu, W.; Hong, Y.X.; Shakhawat Sohan, H.M.; Tong, Y.G.; Hu, Y.L.; Zhang, M.J.; Zhang, J.; Xiang, D.D.; Fu, H.G.; et al. An Overview of Technological Parameter Optimization in the Case of Laser Cladding. *Coatings* **2023**, *13*, 496. [CrossRef]
7. Zhou, Y.; Shen, M.X.; Cai, Z.B.; Peng, J.F.; Zhu, M.H. Study on dual rotary fretting wear behavior of Ti6Al4V titanium alloy. *Wear* **2017**, *376–377*, 670–679. [CrossRef]
8. Xue, J.L.; Guo, W.; Yang, J.; Xia, M.S.; Xia, M.S.; Zhao, G.; Tan, C.W.; Wan, Z.D.; Chi, J.X.; Zhang, H.Q. In-situ observation of microcrack initiation and damage nucleation modes on the HAZ of laser-welded DP1180 joint. *J. Mater. Sci. Technol.* **2023**, *148*, 138–149. [CrossRef]
9. Wang, K.M.; Du, D.; Liu, G.; Pu, Z.; Chang, B.H.; Ju, J. A study on the additive manufacturing of a high chromium nickel-based superalloy by extreme high-speed laser metal deposition. *Opt. Laser Technol.* **2021**, *133*, 106504. [CrossRef]
10. Yang, S.; Han, Q.; Yin, Y.; Zhang, Z.; Wang, L.; Zhu, Z.; Liu, H.; Ma, T.; Gao, Z. Effects of TiB_2 content on the processability and mechanical performance of Hastelloy-X based composites fabricated by selective laser melting. *Opt. Laser Technol.* **2022**, *155*, 108441. [CrossRef]
11. Cui, Y.; King, D.J.M.; Horsfield, A.P.; Gourlay, C.M. Solidification orientation relationships between Al_3Ti and TiB_2. *Acta Mater.* **2020**, *186*, 149–161. [CrossRef]
12. Li, J.H.; Hage, F.S.; Ramasse, Q.M.; Schumacher, P. The nucleation sequence of α-Al on TiB_2 particles in Al-Cu alloys. *Acta Mater.* **2021**, *206*, 116652. [CrossRef]
13. Vairamuthu, J.; Senthil, A.K.; Sivakumar, G.D.; Naveen, S.N.; Kailasanathana, C. Mechanical properties of casting titanium alloy matrix composites reinforced by WC and TiB_2 ceramic particulates. *Mater. Today Proc.* **2022**, *59*, 1503–1507. [CrossRef]
14. Weng, F.; Yu, H.J.; Du, X.Y.; Tian, H.F.; Chen, C.Z. In situ formed TiB_2/TiC complex structure in laser-alloyed coatings with improved wear property. *Ceram. Int.* **2022**, *48*, 7056–7062. [CrossRef]
15. Wang, Z.; Lan, N.; Zhang, Y.; Deng, W.R. Microstructure and Properties of MAO-Cu/Cu-(HEA)N Composite Coatings on Titanium Alloy. *Coatings* **2022**, *12*, 1877. [CrossRef]
16. Jindal, J.D.V.; Sanders, A.P.; Chandran, K.S.R. CALPHAD-guided alloy design and processing for improved strength and toughness in Titanium Boride (TiB) ceramic alloy containing a ductile phase. *Acta Mater.* **2019**, *171*, 18–30.
17. Ouyang, D.L.; Hu, S.W.; Cheng, T.A.O.; Xia, C.U.I.; Zhu, Z.S.; Lu, S.Q. Experiment and modeling of TiB_2/TiB boride layer of Ti−6Al−2Zr−1Mo−1V alloy. *Trans. Nonferrous Met. Soc. China* **2021**, *31*, 3752–3761. [CrossRef]
18. Lin, Y.H.; Jiang, C.C.; Lin, Z.H.; Chen, Q.T.; Lei, Y.P.; Fu, H.G. Laser in-situ synthesis of high aspect ratio TiB fiber bundle reinforced titanium matrix composite coating. *Opt. Laser Technol.* **2019**, *115*, 364–373. [CrossRef]
19. Lin, Y.H.; Lin, Z.H.; Chen, Q.T.; Lei, Y.P.; Fu, H.G. Laser in-situ synthesis of titanium matrix composite coating with TiB−Ti network-like structure reinforcement. *Trans. Nonferrous Met. Soc. China* **2019**, *29*, 1665–1676. [CrossRef]
20. Mielniczuk, B.; Hueckel, T.; Youssoufi, M.S.E. Laplace pressure evolution and four instabilities in evaporating two-grain liquid bridges. *Powder Technol.* **2015**, *283*, 137–151. [CrossRef]
21. Givenhchi, R.; Tan, Z.C. The effect of capillary force on airborne nanoparticle filtration. *J. Aerosol. Sci.* **2015**, *83*, 12–24. [CrossRef]
22. Feng, Q.; Ma, X.H.; Yan, Q.Z.; Ge, C.C. Preparation of soft-agglomerated nano-sized ceramic powders by sol–gel combustion process. *Mater. Sci. Eng. B* **2009**, *162*, 53–58. [CrossRef]
23. Feng, J.; Han, Y.F.; Han, X.C.; Wang, X.D.; Song, S.X.; Sun, B.D.; Chen, M.W.; Liu, P. Atomic insights into heterogeneous nucleation and growth kinetics of Al on TiB2 particles in undercooled Al-5Ti-1B melt. *J. Mater Sci. Technol.* **2023**, *156*, 72–78. [CrossRef]
24. Gorsse, S.; Miracle, D.B. Mechanical properties of Ti-6Al-4V/TiB composites with randomly oriented and aligned TiB reinforcements. *Acta Mater.* **2003**, *51*, 2427–2442. [CrossRef]
25. Lin, Y.H.; Lei, Y.P.; Li, X.Q.; Zhi, X.H.; Fu, H.G. A study of TiB_2/TiB gradient coating by laser cladding on titanium alloy. *Opt. Laser Eng.* **2016**, *82*, 48–55. [CrossRef]
26. Le, J.; Boehlert, C.J.; Huang, G.; Mao, J.; Lei, L.; Han, Y.; Lu, W. TiB whisker induced variant selection in titanium matrix composites. *Scr. Mater.* **2022**, *217*, 114772. [CrossRef]

27. Xu, S.Y.; Cai, Q.; Li, G.; Lu, X.F.; Zhu, X.L. Effect of scanning speed on microstructure and properties of TiC/Ni60 composite coatings on Ti6Al4V alloy by laser cladding. *Opt. Laser Technol.* **2022**, *154*, 108309. [CrossRef]
28. Zhang, L.L.; Zheng, Q.J.; Jiang, H.X.; Zhao, J.Z. Interfacial energy between Al melt and TiB_2 particles and efficiency of TiB_2 particles to nucleate α-Al. *Scr. Mater.* **2019**, *160*, 25–28. [CrossRef]

coatings

MDPI

Article

Effect of Mechanical Vibration on Microstructure and Properties of Laser Cladding WC-Reinforced Nickel-Based Alloy Coatings

Zhongtang Gao [1,*], Congcong Ren [1], Jinzhou Li [1], Zhiming Gao [1], Lifei Du [1], Zhuhui Qiao [2,3] and Chuanwei Zhang [1]

1 School of Mechanical Engineering, Xi'an University of Science and Technology, Xi'an 710054, China; conconren@163.com (C.R.); jzhouli@163.com (J.L.); gaozm@xust.edu.cn (Z.G.); dulifei@xust.edu.cn (L.D.); zhanggwei@xust.edu.cn (C.Z.)

2 State Key Laboratory of Solid Lubrication, Lanzhou Institute of Chemical Physics, Chinese Academy of Sciences, Lanzhou 730000, China; zhqiao@licp.cas.cn

3 Shandong Laboratory of Yantai Advanced Materials and Green Manufacture, Yantai 264006, China

* Correspondence: zhongtanggao@xust.edu.cn; Tel.: +86-29-85583157

Abstract: Ni-WC composite coatings on 35CrMoV alloy surface were successfully prepared by mechanical vibration field-assisted laser cladding technology. Optical microscopy (OM) and scanning electron microscopy (SEM) were used to study the microstructure of the composite coatings without vibration and at different vibration frequencies; the phase composition of the cladding layer was studied by X-ray diffraction (XRD); and an energy dispersive spectrometer (EDS) was used for elemental plane scanning analysis. The grain growth trend under different convection directions was simulated. The wear resistance and mechanical properties of the composite coating were analyzed by friction and wear testing machine, three-dimensional surface profiler, and microhardness tester. The vibration field generated by the self-improved shaking table device is used to assist laser cladding. The effect of mechanical vibration on the quality of the cladding layer was studied. The results show that compared with the coating without mechanical vibration, an appropriate increase in vibration frequency contributes to the refinement of the grains. The original coarse dendrite structure becomes a fine needle-like structure, and the fine grain size gradually decreases. The application of vibration can improve the effect of grain refinement. The vibration makes the grain size distribution more uniform and the microhardness fluctuation of the cladding layer decreases. The experimental results show that mechanical vibration can improve the microstructure uniformity of the coating by selecting suitable vibration parameters. The average friction coefficient and wear width are reduced, and the microhardness is also increased.

Keywords: mechanical vibration; laser cladding; WC; composite coating

1. Introduction

Laser cladding is a new laser processing technology with rapid melting and solidification. Through the rapid irradiation of its high-energy laser beam, the powder of the cladding layer is first melted into a molten pool in a short time and then rapidly solidified on the surface of the substrate so as to achieve a good metallurgical combination between the substrate and the cladding layer [1,2]. This is because the laser cladding technology can make the cheap plate and cladding powder form the metallurgical combination of high hardness, high wear resistance, corrosion resistance, and high-temperature oxidation resistance. At the same time, the thermal deformation of the matrix can be reduced, and the dilution rate of the cladding layer is low; therefore, it is widely used to improve the strengthening and repair of metal surfaces [3–5].

In recent years, Ni-based alloy powder has been widely used in the field of laser cladding because of its good wettability, abrasion resistance, corrosion resistance, impact resistance, fatigue resistance, and good self-lubrication under high-temperature conditions [6]. As a reinforcing particle in composite materials, WC has the advantages of high hardness and wear resistance and has good wettability between it and the Ni-based alloy powder. The uniform distribution of particles in the cladding layer can significantly improve the wear resistance of the cladding layer, which is an ideal hard phase material of the cladding layer [7–9].

In the laser cladding process, complex temperature fields will be generated during the rapid heating and cooling of the substrate surface. Under the influence of such factors, coarse columnar dendrites and uneven distribution of microstructure will often exist in the cladding layer, resulting in cracks or pores and other metallurgical defects [10–12]. The introduction of ultrasonic vibration in the laser cladding process has a significant effect on crack suppression and grain refinement. The introduction of ultrasonic vibration can reduce the porosity of the cladding layer, eliminate internal stress, and refine the microstructure, which plays an important role [13–15]. The ultrasonic vibration can increase the rate of liquid flow in the molten pool and make the elements evenly distributed, thus reducing the generation of various defects in the cladding layer. Increasing the flow speed of the liquid in the molten pool will also increase the cooling speed of the liquid itself, thus shortening the grain growth time in the cladding layer. In the process of crystal growth, the end and arm of the crystal are broken, and with the flow of liquid in the molten pool, the fine crystal structure formed is evenly distributed in the molten pool [16–18]. In addition, external electromagnetic fields have been introduced into laser cladding by some researchers at home and abroad, which has made some progress in reducing internal defects, reducing residual stress, refining and homogenizing microstructure, etc. [19–22]. The above studies mainly focus on the conditions of ultra-high frequency and high amplitude, and few kinds of literature report the influence of mechanical vibration on the laser cladding composite cladding layer under the conditions of low frequency and low amplitude. Therefore, this paper uses a mechanical vibration to assist the laser cladding process. The effect of mechanical vibration and convection on the quality of the laser cladding Ni-WC cladding layer of 35CrMoV alloy was studied.

2. Materials and Methods

2.1. Experimental Materials

In this experiment, 35CrMoV alloy plate is used as the base, and its composition is shown in Table 1. Before the experiment, the plates were cut to a predetermined size ($100 \times 60 \times 10$ mm^3). The surface of the substrate was polished with a 60-mesh grinding wheel and cleaned with absolute ethanol to remove the oil stains on the surface of the substrate. Nickel powders with an average diameter of 50 to 100 μm were used as the coating matrix material; WC powder with an average diameter of 100 to 150 μm was used as the reinforcing particle of the coating. The Ni-WC mixed powder used in this experiment is the commercial powder purchased, and the specific composition and content of the mixed powder are provided by the merchant. The mass fraction of WC powder in Ni-WC mixed powder is 60%, and its mixture composition is shown in Table 2. The SEM morphology of the two powders is shown in Figure 1.

Table 1. Chemical composition of 35CrMoV alloy plate (wt.%).

Elements	C	Si	Mn	Cr	Mo	V	P	S	Cu	Ni
35CrMoV	0.30–0.38	0.17–0.37	0.40–0.70	0.80–1.10	0.15–0.25	0.10–0.20	≤0.035	≤0.035	≤0.030	≤0.030

Table 2. Chemical composition of Ni-WC (wt.%).

Elements	Ni	W	C	Cr	Si	Fe
Ni-WC	Bal	57.3	2.45	2.36	1.11	0.12

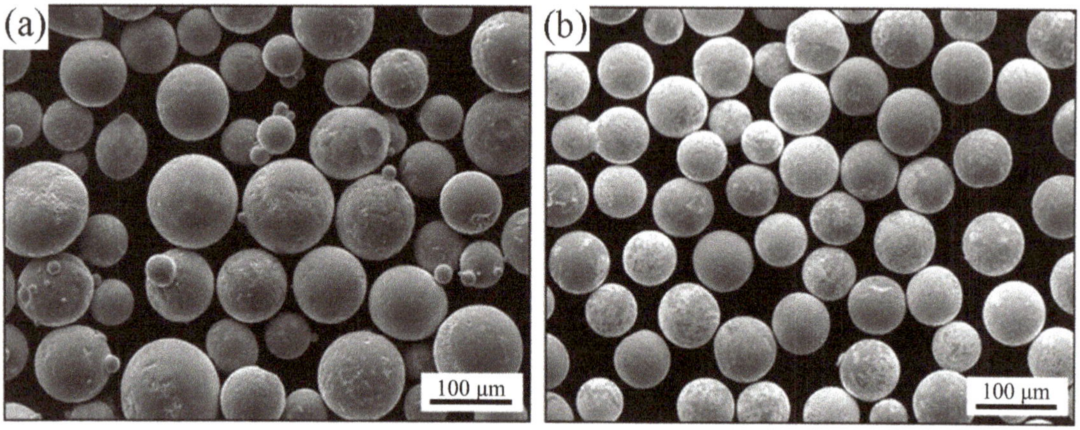

Figure 1. SEM morphology of power: (**a**) WC; (**b**) Ni.

2.2. Test Equipment, Parameters, and Test Process

Laser cladding coating was obtained using BS-OF-3000-15-4L universal laser heat treatment equipment. Before coating, the powder mixture was thoroughly mixed with a planetary ball mill for 2 h until it was well mixed and then dried in a drying box at 80 °C for 2 h. Through experimental analysis and verification, the method of coaxial powder feeding is adopted. The technological parameters of laser cladding are as follows: power 2000 W, spot size 15 mm × 3 mm, scanning speed 120 mm/min, feeding rotary speed 300 r/min, lap rate 50%, high-purity argon gas flow rate 15 L/min.

In this experiment, a self-improved shaking table device is adopted, as shown in Figure 2. During the experiment, the vibration device was placed on the laser cladding workbench. The vibration frequency can be adjusted by adjusting the current. Through the previous experimental comparison, different vibration frequencies were determined in the formal cladding, which were, successively 0 Hz, 50 Hz, 150 Hz, and 300 Hz, and the amplitude was set as 1 μm. The vibration was finally transmitted to the substrate in a direction perpendicular to the substrate. The clamp is used to fix the substrate on the vibration platform during cladding, and the vibration equipment shutdown delay is set for 60 s after laser cladding. After the experiment, the substrate was cut into samples with sizes of 15 × 10 × 10 mm^3 and 10 × 10 × 10 mm^3, respectively. The sample was ground and polished with different amounts of sandpaper and finally corroded with aqua regia for 4 s.

Figure 2. Laser cladding equipment and vibration equipment.

2.3. Sample Detect

The microstructure and structure of the coating were characterized by an optical microscope (OM). The microstructure and distribution of the coating were characterized by Nova Nano450 scanning electron microscopy (SEM) and the attached energy dispersive spectroscopy (EDS) analyzer. The phase changes of the coating surface were characterized by X-ray diffraction (XRD). The wear resistance of the sample surface was tested by the MFT-R4000 wear testing machine. The wear morphology of samples was characterized by Contour GT-K 3D optical profilometer. Vickers hardness measurements were performed on a through-thickness section normal to the deposition direction by means of a HMAS-D1000SZ microhardness tester. The load is 9.8 N, and the loading time is 15 s. The same horizontal plane of the cross-section of each sample was measured three times and then averaged.

3. Results and Discussion

3.1. Surface and Cross-Section Morphology of the Coating

Figure 3 shows the macroscopic and microscopic morphology of the cladding layer under different vibration frequencies in the previous experiment. In the preliminary experiment, laser cladding was carried out at different frequencies between 50 Hz and 300 Hz. It can be seen from the figure that the coating with good fusion quality can be successfully prepared on the surface of 35CrMoV alloy by using the mechanical vibration-assisted laser cladding technology. At the same time, the laser scanning track on the matrix surface can be clearly observed. It can be seen from the width of scanning lines that with the increase of vibration frequency, the width of the cladding layer also increases and is close to flat. When the vibration frequency increased from 50 Hz to 180 Hz, the cladding surface roughness gradually decreased. This is because the applied mechanical vibration frequency gradually increases to the resonance frequency of the substrate. When the vibration frequency increases from 185 Hz to 300 Hz, the surface roughness of the cladding layer gradually increases. This is because the frequency of the applied mechanical vibration slowly moves away from the resonant frequency of the substrate. It can be concluded that the appropriate vibration frequency can reach the resonance frequency of the substrate, which can better improve the quality of the cladding layer [14].

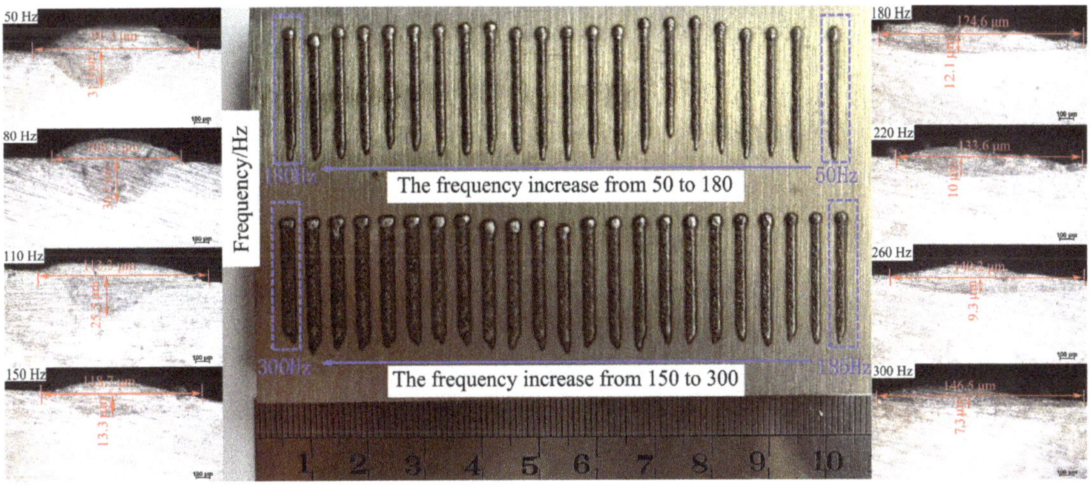

Figure 3. Morphology of cladding layer and typical cladding profile under different vibration frequencies.

In order to analyze the influence of mechanical vibration power on the coating dilution rate, single-channel laser cladding tracks at vibration frequencies of 50, 80, 110, 150, 180, 220,

260, and 300 Hz were observed by optical microscopy. The typical cross-sectional profiles are shown in Figure 3. The measurement results show that with the increase of vibration frequency, the width of the cladding layer increases from 91.3 μm to 146.5 μm, while the depth of the cladding layer decreases from 31.3 μm to 7.3 μm. When mechanical vibration is applied, part of the energy generated by vibration will be absorbed by the atoms or molecules in the molten pool, and the irregular movement of the atoms or molecules that have absorbed energy will become more violent, eventually leading to the improvement of the fluidity in the molten pool. When the vibration frequency is low, the mobility of the molten pool is relatively poor, which leads to poor energy mobility in the molten pool. Therefore, the dilution rate of the coating is relatively high. With the increase of vibration frequency, the fluidity of the molten pool increases relatively. This results in good energy mobility of the molten pool and a relatively low dilution rate of the coating. Through analysis based on previous data, we finally decided to use laser cladding coatings with vibration frequencies of 0, 50, 150, and 300 Hz for further research on the structure and properties of the composite coatings.

3.2. Optical Microscope Analysis of Coating

Figure 4 shows the optical micrographs of the coating surface under different mechanical vibration frequencies. As can be seen from Figure 4, the coating surface is mainly composed of a large number of secondary dendrites, rod-like crystals, equiaxed crystals, and carbides. It can be observed from Figure 4a that when mechanical vibration is not applied, a large number of developed secondary dendrites and carbide particles with relatively large particle sizes exist on the coating surface. When the vibration frequency is 50 Hz, as shown in Figure 4b, a large number of developed secondary dendrites are broken and many broken dendrites are formed. This is because in the process of solidification and growth of dendrite in the molten pool, due to external mechanical vibration, the energy generated by vibration is absorbed by the crystal, and when the crystal grows to a certain length, it will be broken. At the same time, the larger size of carbide particles will be broken after absorbing the external energy. When the vibration frequency is 150 Hz, as shown in Figure 4c, a large amount of secondary dendrite is eventually broken to form a large amount of fine crystal. When the vibration frequency reaches 300 Hz, as shown in Figure 4d, some equiaxed crystal structures appear on the coating surface.

In addition, the mechanical vibration accelerates the flow of liquid in the molten pool, which makes the temperature of the liquid tend to be uniform, leading to a decrease in the temperature gradient. The temperature gradient is calculated using the following formula [23]:

$$G = \frac{2\Pi K(T - T_0)^2}{\eta P} \tag{1}$$

where K is the thermal conductivity of the material, T is the liquid temperature of the alloy, T_0 is the preheating temperature of the substrate, η is the laser absorption coefficient, and P is the laser power. It can be seen from the formula that when the temperature gradient decreases under the action of mechanical vibration, the liquid metal temperature T decreases, and the initial substrate temperature T_0 remains unchanged, leading to accelerated cooling of the substrate heat dissipation interface. Therefore, the composition and temperature tend to be uniform, which reduces the conditions for secondary dendrite formation and promotes the formation of fine crystals when supercooled.

3.3. Microstructure and Phase Analysis of Coating

To further observe the morphology of the crystal, the microstructure of the coating was further analyzed by scanning electron microscopy. Figure 5 shows the microstructure of the coating surface under different vibration frequencies. It can be seen from Figure 5 that with the increase in vibration frequency, the microstructure refinement in the coating is very obvious. When no vibration is applied, as shown in Figure 5a, massive dendrites exist in the coating. This is because in the process of crystal growth without external conditions, the growth mode of the crystal is affected by its own. However, when mechanical vibration is applied, the crystal growth process is affected by the external field energy. As shown in Figure 5b, when the initially applied vibration is 50 Hz, the thick dendritic crystal will break when it grows to a certain length and form fine needle-like tissue due to the effect of vibration energy transfer during its growth. When the vibration frequency is 150 Hz, a large number of secondary dendrite arms will break. It can also be seen from the figure that some columnar crystals will also break under the condition of vibration, forming a large number of equiaxed crystals, and the grain refinement effect is the most obvious. When the vibration frequency reaches 300 Hz, as shown in Figure 5d, the coarse dendrites will be broken, but a few dendrites are not completely broken. Since mechanical vibrations are applied to speed up the flow of the molten pool, these broken crystals are distributed more evenly in the coating.

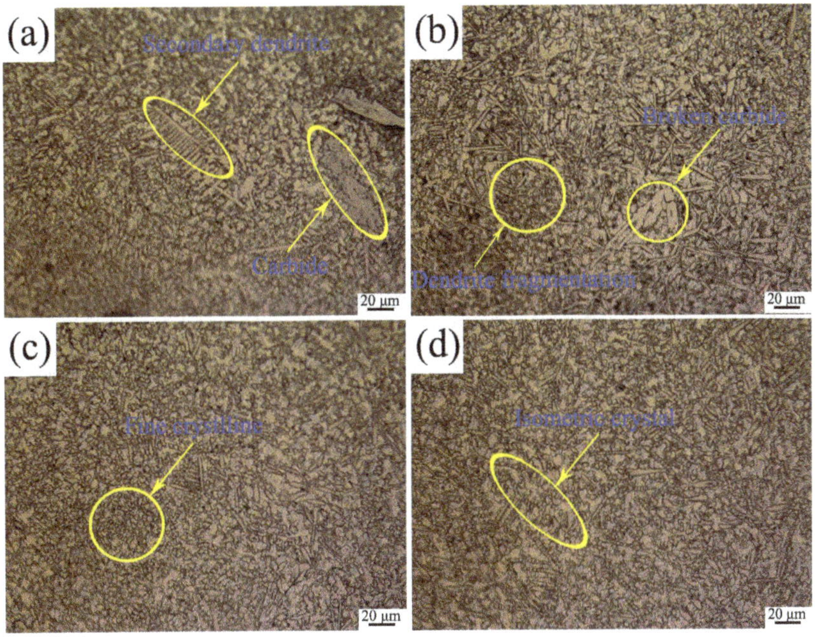

Figure 4. Optical micrographs of the coating surface at different vibration frequencies: (**a**) 0 Hz; (**b**) 50 Hz; (**c**) 150 Hz; (**d**) 300 Hz.

Figure 6 shows the evolution of carbide particle morphology in layers under different mechanical vibration frequencies. It can be seen from the figure that the degree of carbide fragmentation is different under different mechanical vibration frequencies. Figure 6a shows that no mechanical vibration is applied, and almost no carbide particles exist on the coating surface. There are some thick dendritic structures on its surface. This is because when mechanical vibration is not applied, the density of carbide particles in the molten pool is large, and a large number of carbide particles will sink to the bottom of the molten pool when the molten pool solidifies, which makes it difficult to find on the surface. However, when mechanical vibration is applied, because the direction of vibration is perpendicular to the processed surface of the substrate, the fluidity of the molten pool will be accelerated, so part of the carbide will solidify on the surface of the coating. When the vibration frequency is 50 Hz, as shown in Figure 6b, there are mainly some fishbone carbides on the surface of the coating. It can be seen from the figure that the boundary of the carbides is broken to a very small degree. When the vibration frequency is 150 Hz, as shown in Figure 6c, the carbide in the layer is broken to a great extent, and a large amount of broken triangular carbide is dissolved into the coating. When the vibration frequency reaches 300 Hz, as shown in Figure 6d, there are mainly some cylindrical carbides in the layer, and the fragmentation degree of carbides is relatively reduced. This may be because when the vibration frequency is 150 Hz, the vibration frequency reaches the resonance frequency of the matrix. This makes the carbide absorb most of the energy transferred by vibration at 150 Hz. Therefore, the degree of fragmentation is the largest.

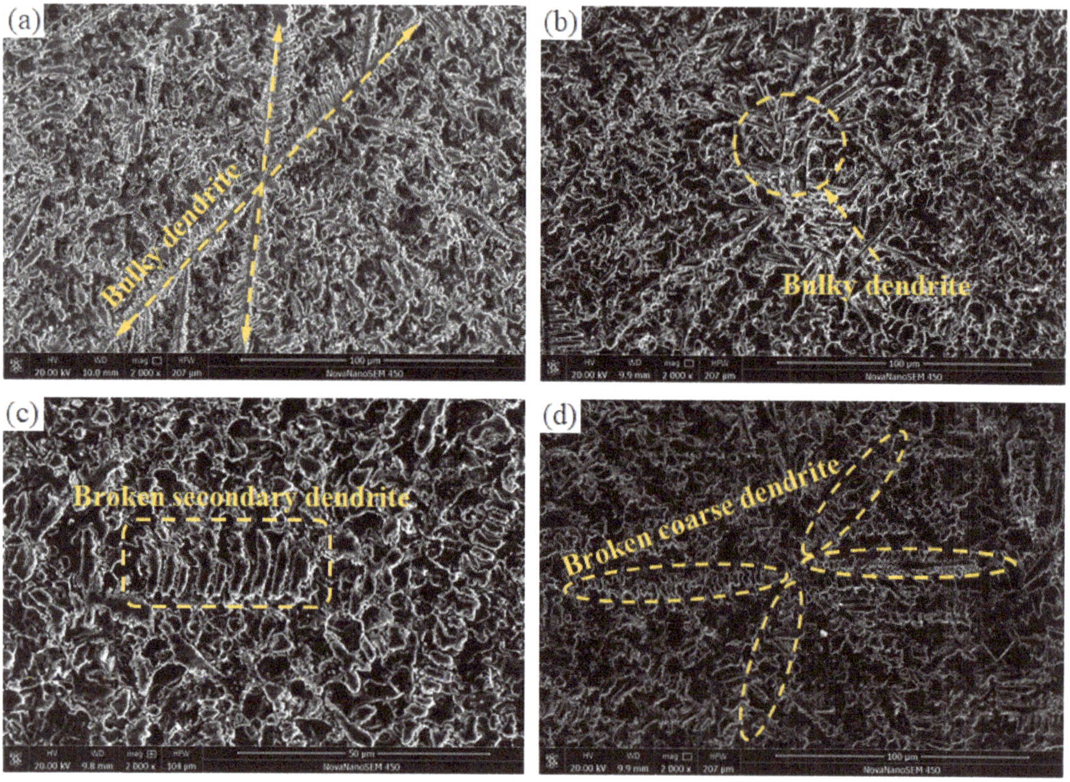

Figure 5. SEM diagram of crystal morphology evolution: (**a**) 0 Hz; (**b**) 50 Hz; (**c**) 150 Hz; (**d**) 300 Hz.

The X-ray diffraction spectra of WC-Ni composite coatings at different vibration frequencies are shown in Figure 7. It can be seen from the atlas that there is no obvious change in the cladding phase with and without vibration, and the main components of the cladding phase are $FeNi_3$, Mo_2C, W_2C, $(Mo, Fe, Mn)_2C$, and a small amount of WC phase. The formation of these phases indicates that there is a good metallurgical combination between the matrix and the cladding layer, and the newly generated hard alloy phases are evenly distributed in the cladding layer, which improves the mechanical properties of the cladding layer. The W_2C and Mo_2C produced in the cladding layer are the main reinforcement phases of the coating and Fe and Ni form $FeNi_3$ solid solution in the cladding layer. It can be seen from the figure that the diffraction peak of WC is very weak, indicating that WC particles are almost all dissolved in the Ni-based alloy solvent. Compared to that without mechanical vibration, the intensity of the diffraction peak of $FeNi_3$ solid solution decreases, indicating that mechanical vibration improves the solubility of $FeNi_3$ solid solution in the cladding layer. Diffraction peaks of W_2C, Mo_2C, and $(Mo, Fe, Mn)_2C$ are enhanced, which is conducive to the formation of a stable microstructure [24]. The results show that the mechanical vibration causes the distribution of elements in the cladding layer to be more uniform, the grain size to be finer, and the grain integrity to be improved.

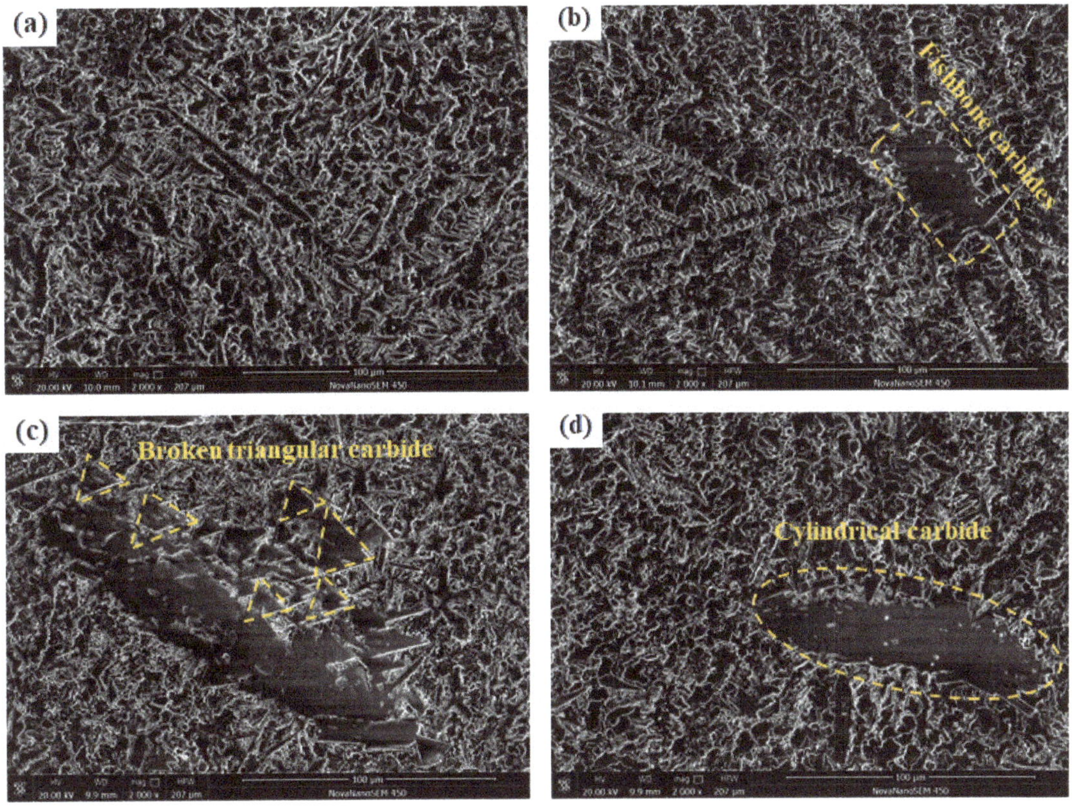

Figure 6. SEM diagram of carbide morphology evolution: (**a**) 0 Hz; (**b**) 50 Hz; (**c**) 150 Hz; (**d**) 300 Hz.

Figure 8 shows the schematic diagram of the laser cladding process assisted by mechanical vibration. The principle of mechanical vibration on microstructure refinement of composite coatings is revealed. When mechanical vibration is not applied, there are mainly some coarse dendrites in the coating. When mechanical vibration is applied in the process of laser cladding, the coarse crystals in the coating will be broken due to the convective effect of vibration (Figure 8a). The vibration promotes the fracture of the crystal, which, in turn, refines the grain size of the structure. During the solidification process of the molten pool, the convection effect caused by vibration produces a stirring effect on the molten pool, which is beneficial to the distribution of elements in the coating. At the same time, the effect of stirring greatly promotes the distribution of chemical components in the molten pool. Finally, the grain size is finer, and the distribution is more uniform. In Figure 8b, the change of crystal growth structure under the condition of external vibration is simulated. The angle between the convection direction and the transverse angle of the crystal are 0°, 30°, 45° and 60° from left to right, respectively. The morphology diagram of crystal step size at the same time, under the condition of vibration or its absence, is drawn by finite element analysis software. It is found that the applied mechanical vibration has a significant effect on the growth direction of the crystal. According to the theory of growth in backflow and no growth in oncoming flow, when vibration is applied, dendrites show strong right dendrite growth. This is because vibration can change the direction of convection in the molten pool, resulting in a different direction from that in the molten pool without vibration, thus changing the original growth path of dendrites. Therefore, mechanical vibration is an effective method to adjust dendrite morphology in the process of molten pool solidification.

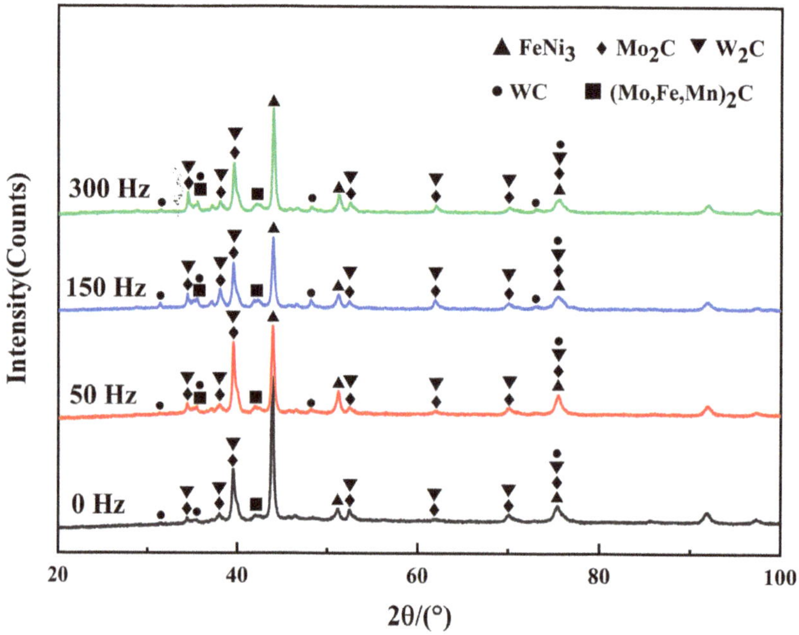

Figure 7. XRD patterns of the coating at different vibration frequencies.

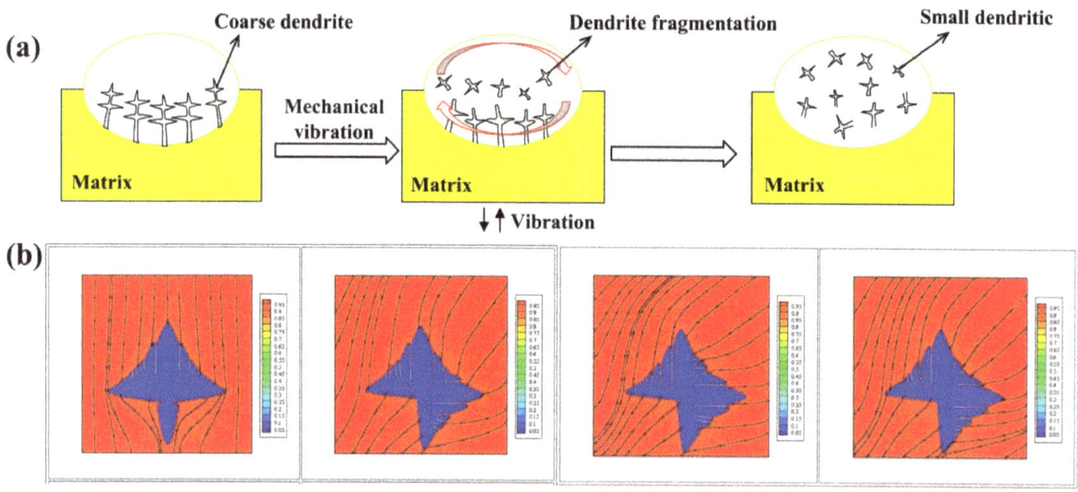

Figure 8. Schematic diagram and simulation of dendrite evolution under vibrational convection: (**a**) schematic diagram of dendrite evolution; (**b**) simulation results of dendrite evolution under different convection directions.

In order to further analyze the distribution of elements in the coating by vibration frequency, EDS surface scanning was performed on the coating surface at different frequencies, and the results were shown in Figure 9. The main elements distributed on the crystal are Cr, Si, and W. The Fe and Ni elements are mainly distributed in the intergranular region. This shows that SiC, WC, and Cr_3C_2 are mainly distributed on the crystal. Fe-Ni solid solution is mainly distributed in the intercrystalline region. When no vibration is applied, there is a certain degree of segregation between the Cr element on the crystal and the Fe element in the intercrystalline region. The degree of segregation of Cr on the crystal and Fe elements in the intercrystalline region is slightly improved when the vibration frequency is 50 Hz. The degree of segregation of the Cr on the crystal and the Fe element in the intercrystalline is further weakened when the vibration frequency is 150 Hz and 300 Hz. The results show that mechanical vibration can improve the segregation degree of Cr and intercrystalline Fe elements and promote the uniform distribution of Si and W elements on the crystal and Ni elements in the intercrystalline region; however, it does not change the atomic concentration of the elements. The reason for this phenomenon is that when the mechanical vibration is continuously applied to the liquid pool, the energy generated by the vibration can break the dendritic structure and the thick coating carbide into fine grains, thus reducing the segregation degree of the elements. As the Stokes deposition rate decreases, the fluidity of the molten pool increases, which eventually weakens the segregation of the microstructure and causes the distribution of elements to become even [25]. According to the grain growth theory, the grain size is determined by the ratio between the nucleation rate and growth rate, and the larger the ratio, the smaller the grain size [26].

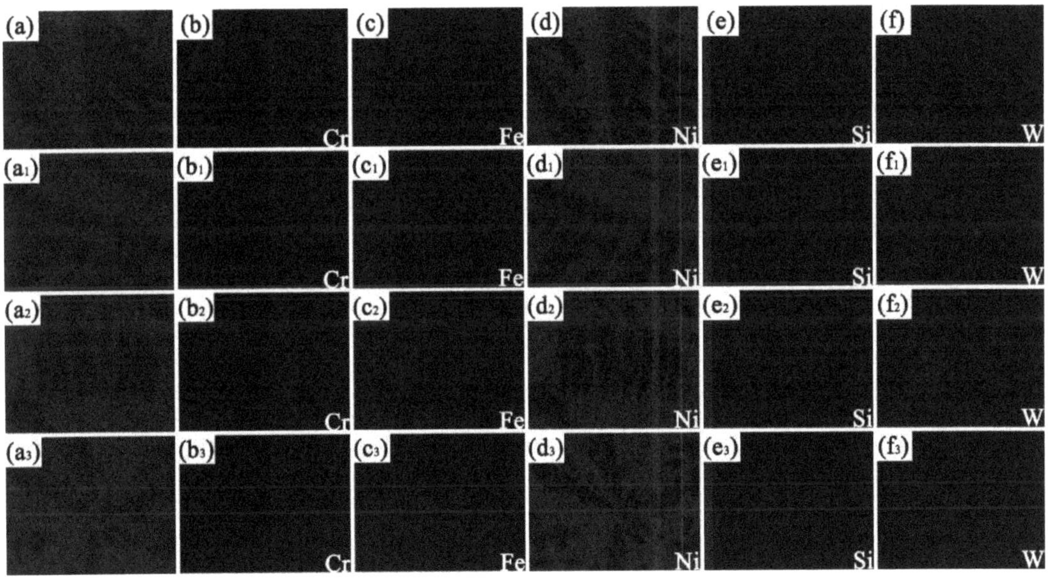

Figure 9. Scanning results of coating surface at different vibration frequencies: (**a**–**f**) 0 Hz; (**a$_1$**–**f$_1$**) 50 Hz; (**a$_2$**–**f$_2$**) 150 Hz; (**a$_3$**–**f$_3$**) 300 Hz.

3.4. Wear Resistance and Hardness of the Coating

Figure 10a shows the friction coefficient curves of the coating under different mechanical vibration frequencies. Under the same experimental conditions, the material with a small friction coefficient has better wear resistance [27]. It can be observed from Figure 10a that with the increase of wear test time, the friction coefficient of the composite coating without mechanical vibration gradually increases from 0.4 to 0.6. When mechanical vibration is applied, the effect of vibration significantly reduces the average friction coefficient to

0.359. When the initially applied vibration frequency is 50 Hz, the friction coefficient of the coating is significantly reduced, which is lower than that without applied vibration. At the same time, it can be seen that when the vibration frequency is 150 Hz, the time to reach the running-in period is reduced, the friction coefficient of the coating is also the smallest, and the fluctuation of the curve is relatively stable, and the wear resistance is also improved. When the vibration frequency reaches 300 Hz, the friction coefficient is higher than that when the vibration frequency is 150 Hz. Figure 10b shows the average friction coefficient curve of the coating. The average friction coefficient was the lowest when the vibration frequency was 150 Hz (0.359). The results show that when the vibration frequency exceeds a certain value, the wear resistance of the composite coating will be reduced.

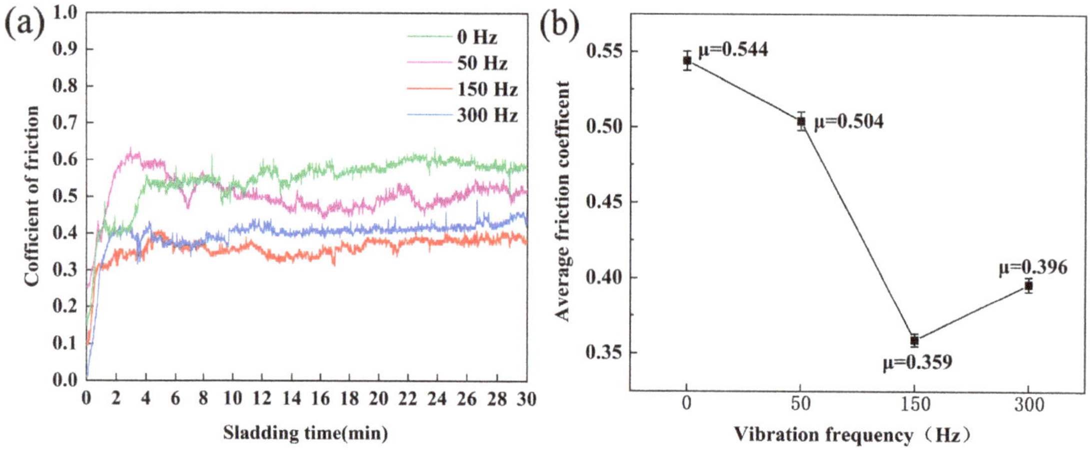

Figure 10. Friction coefficient curves of the coating at different vibration frequencies: (**a**) coefficient of friction; (**b**) average friction coefficient.

In order to further explore the wear resistance of the composite coating under different vibration frequencies, the 3D morphology of the wear marks was measured by a 3D optical profilometer. Figure 11 shows the three-dimensional morphology of the wear surface of Ni-WC composite coating under 5 N loads. By analyzing and comparing the width of the cross-section of the three-dimensional morphology of the composite coating under different vibration frequencies, the excellent degree of wear resistance of the composite coating is compared. As shown in Figure 11a, when no mechanical vibration is applied, the wear width of the wear surface reaches 172 μm. When the first applied vibration frequency is 50 Hz, as shown in Figure 11b, the wear width of the surface abrasion marks decreases to 161 μm. When the vibration frequency is 150 Hz, as shown in Figure 11c, the wear width of the surface abrasion marks decreases to 115 μm, and the wear width reaches the minimum at this time, indicating that the wear resistance of the composite coating reaches the best at the vibration frequency of about 150 Hz. As the vibration frequency continued to increase, when the vibration frequency reached 300 Hz, as shown in Figure 11d, the wear width of the surface increased to 130 μm. This further indicates that when the vibration frequency exceeds a certain value, the wear resistance of the coating will not increase but decrease.

Figure 12a shows the microhardness curve of the coating surface under different vibration frequencies. It can be inferred that the microhardness of the coating increases with the increase in vibration frequency. When the vibration frequency is 0, 50, 150, and 300 Hz, the average microhardness of the coating is 751.71, 790.45, 862.02, and 811.17 HV1 (Figure 12b). It can be seen from the fluctuation of the figure line. The average microhardness of the coating first increases and then decreases with the increase of vibration frequency. The microhardness of the matrix is about 341.4 HV1, and the microhardness of the composite coating is between 600 and 910 HV1. When the vibration frequency is 150 Hz, the curve fluctuation is relatively stable, and the average microhardness is the highest. At this time, the microhardness of the composite coating reaches 901.5 HV1, which is 2.6 times the hardness of the matrix. With the increase of vibration frequency, the average microhardness of the coating first increases and then decreases. It is speculated that the resonance frequency of the matrix may be reached during this process. When the resonance frequency of the matrix is reached, the energy absorbed by the molten pool is the maximum. The flow of the molten pool is greatly accelerated. The tissue is more evenly distributed. The uniformity of the tissue distribution has an important effect on the microhardness of the coating. The more homogeneous the tissue is, the higher the microhardness is. Compared with the matrix, the average microhardness of the composite coating is increased by 15% after the application of vibration. This fully indicates that the application of mechanical vibration can significantly improve the microhardness of the composite coating, indicating that the application of mechanical vibration is an effective method to enhance the mechanical properties of the laser cladding composite coating.

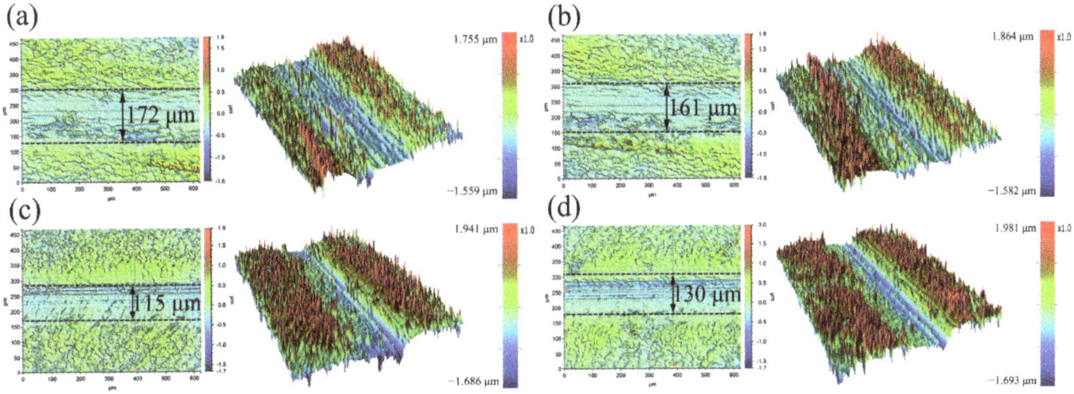

Figure 11. Three-dimensional morphology of coating wear at different vibration frequencies: (**a**) 0 Hz; (**b**) 50 Hz; (**c**) 150 Hz; (**d**) 300 Hz.

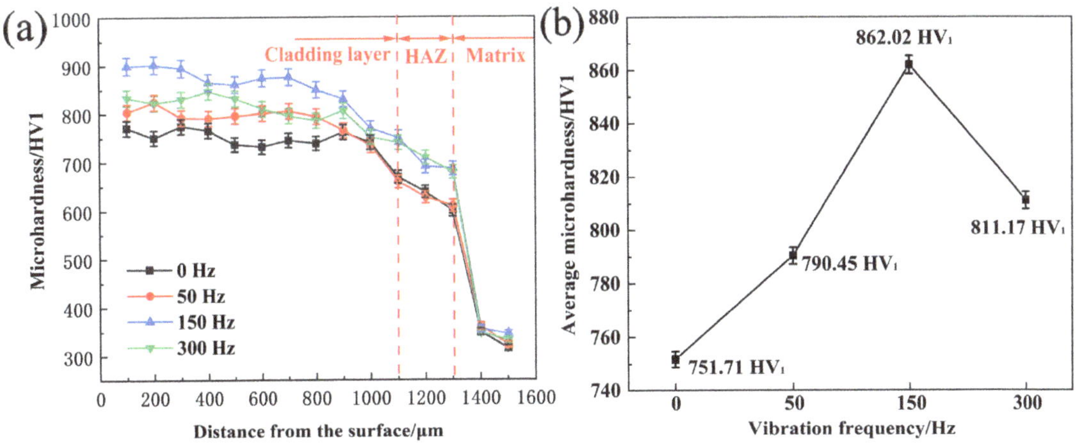

Figure 12. Microhardness curves of coating sections under different vibration frequencies: (**a**) microhardness; (**b**) average microhardness.

4. Conclusions

(1) WC-reinforced Ni-base composite coating was prepared on the surface of 35CrMoV alloy by applying a mechanical vibration field, and the dilution rate of the coating decreased with the increase of vibration frequency. With the increase in vibration frequency, the width of the molten pool increases and the depth decreases. Mechanical vibration promotes the fluidity of the molten pool;
(2) The application of mechanical vibration can cause the coarse grain in the coating to be broken, and the grain size can be reduced at the same time. The grain distribution in the layer is more uniform, and the element distribution is even. In the process of laser cladding, mechanical vibration can effectively improve the cladding effect and improve the quality of the coating;
(3) When the vibration frequency is 150 Hz, the microhardness of the composite coating reaches 901.5 HV1, which is 2.6 times that of the matrix. The average microhardness of the composite coating increased by 15% after external vibration;
(4) Under the 150 Hz vibration frequency, the average friction coefficient of the composite coating decreases from 0.544 to 0.359, which is 34% lower than that of the non-vibration composite coating. The wear width is shortened from 172 μm to 115 μm, the wear resistance is remarkable, and the mechanical properties are excellent.

Author Contributions: Conceptualization, Z.G. (Zhongtang Gao) and Z.G. (Zhiming Gao); data Analysis and writing, C.R.; simulation, L.D.; investigation, J.L.; supervision, Z.Q. and C.Z. All authors have read and agreed to the published version of the manuscript.

Funding: NSFC (51804251: 52104384); Science Fund of Shandong Laboratory of Yantai Advanced Materials and Green Manufacturing (AMGM2021F04); Planning project of Xi'an Science and Technology (21XJZZ0041); China Postdoctoral Science Foundation (2020M683675XB); NSF of Shaanxi Provincial Department of Education (21JK0761); Key Research and Development Program of Shaanxi (2023-YBGY-343).

Institutional Review Board Statement: Not applicable.

Informed Consent Statement: Not applicable.

Data Availability Statement: Not applicable.

Conflicts of Interest: The authors declare no conflict of interest.

References

1. Lv, J.; Zhou, J.; Zhang, T.; Meng, X.; Li, P.; Huang, S. Microstructure and Wear Properties of IN718/WC Composite Coating Fabricated by Ultrasonic Vibration-Assisted Laser Cladding. *Coatings* **2022**, *12*, 412. [CrossRef]
2. Zhuang, D.-D.; Du, B.; Zhang, S.-H.; Tao, W.-W.; Wang, Q.; Shen, H.-B. Effect and action mechanism of ultrasonic assistance on microstructure and mechanical performance of laser cladding 316L stainless steel coating. *Surf. Coat. Technol.* **2022**, *433*, 128122. [CrossRef]
3. Sun, F.; Cai, K.; Li, X.; Pang, M. Research on Laser Cladding Co-Based Alloy on the Surface of Vermicular Graphite Cast Iron. *Coatings* **2021**, *11*, 1241. [CrossRef]
4. Wang, K.; Du, D.; Liu, G.; Chang, B.; Hong, Y. Microstructure and properties of WC reinforced Ni-based composite coatings with Y2O3 addition on titanium alloy by laser cladding. *Sci. Technol. Weld. Join.* **2019**, *24*, 517–524. [CrossRef]
5. Weng, Z.; Wang, A.; Wu, X.; Wang, Y.; Yang, Z. Wear resistance of diode laser-clad Ni/WC composite coatings at different temperatures. *Surf. Coat. Technol.* **2016**, *304*, 283–292. [CrossRef]
6. Ma, Q.; Li, Y.; Wang, J.; Liu, K. Microstructure evolution and growth control of ceramic particles in wide-band laser clad Ni60/WC composite coatings. *Mater. Desing* **2016**, *92*, 897–905.
7. Guo, C.; Chen, J.; Zhou, J.; Zhao, J.; Wang, L.; Yu, Y.; Zhou, H. Effects of WC–Ni content on microstructure and wear resistance of laser cladding Ni-based alloys coating. *Surf. Coat. Technol.* **2012**, *206*, 2064–2071. [CrossRef]
8. Zhang, P.; Pang, Y.; Yu, M. Effects of WC Particle Types on the Microstructures and Properties of WC-Reinforced Ni60 Composite Coatings Produced by Laser Cladding. *Metals* **2019**, *9*, 583. [CrossRef]
9. St-Georges, L. Development and characterization of composite Ni–Cr + WC laser cladding. *Wear* **2007**, *263*, 562–566. [CrossRef]
10. Xie, S.; Li, R.; Yuan, T.; Chen, C.; Zhou, K.; Song, B.; Shi, Y. Laser cladding assisted by friction stir processing for preparation of deformed crack-free Ni-Cr-Fe coating with nanostructure. *Opt. Laser Technol.* **2018**, *99*, 374–381. [CrossRef]

11. Zhao, Y.; Wu, M.; Hou, J.; Chen, Y.; Zhang, C.; Cheng, J.; Li, R. Microstructure and high temperature properties of laser cladded WTaNbMo refractory high entropy alloy coating assisted with ultrasound vibration. *J. Alloy. Compd.* **2022**, *920*, 165888. [CrossRef]
12. VOcelík, V.; de Oliveira, U.; de Boer, M.; de Hosson, J. Thick Co-based coating on cast iron by side laser cladding: Analysis of processing conditions and coating properties. *Surf. Coat. Technol.* **2007**, *201*, 5875–5883. [CrossRef]
13. Xu, J.; Zhou, J.; Tan, W.; Huang, S.; Wang, S.; He, W. Ultrasonic vibration on wear property of laser cladding Fe-based coating. *Surf. Eng.* **2020**, *36*, 1261–1269. [CrossRef]
14. Li, C.; Zhang, Q.; Wang, F.; Deng, P.; Lu, Q.; Zhang, Y.; Li, S.; Ma, P.; Li, W.; Wang, Y. Microstructure and wear behaviors of WC-Ni coatings fabricated by laser cladding under high frequency micro-vibration. *Appl. Surf. Sci.* **2019**, *485*, 513–519. [CrossRef]
15. Li, C.; Li, S.; Zeng, M.; Sun, S.; Wang, F.; Wang, Y. Effect of high-frequency micro-vibration on microstructure and properties of laser cladding aluminum coatings. *Int. J. Adv. Manuf. Technol.* **2019**, *103*, 1633–1642. [CrossRef]
16. Wang, F.; Li, C.; Sun, S.; Zeng, M.; Liu, C.; Lu, Q.; Wang, Y. Al_2O_3/TiO_2-Ni-WC Composite Coatings Designed for Enhanced Wear Performance by Laser Cladding Under High-Frequency Micro-Vibration. *Jom* **2020**, *72*, 4060–4068. [CrossRef]
17. Gao, Z.; Ren, H.; Geng, H.; Yu, Y.; Gao, Z.; Zhang, C. Effect of CeO_2 on Microstructure and Wear Property of Laser Cladding Ni-Based Coatings Fabricated on 35CrMoV Steel. *J. Mater. Eng. Perform.* **2022**, *31*, 9534–9543. [CrossRef]
18. Liu, Q.; Zang, Y.; Song, Y.; Qi, F.; Zhai, Q. Influence of ultrasonic vibration on mechanical properties and microstructure of 1Cr18Ni9Ti stainless steel. *Mater. Design* **2007**, *28*, 1949–1952.
19. Wang, D.S.; Zhu, Y.Y.; Qu, G.; Zhou, X.H. Influential Effects of Follow-up Feed Pulse Current on a Laser Clad Ni-based Alloy Coating. *Lasers Eng.* **2022**, *52*, 151–167.
20. Zhai, L.; Wang, Q.; Zhang, J.; Ban, C. Effect of alternating current electric field on microstructure and properties of laser cladding Ni-Cr-B-Si coating. *Ceram. Int.* **2019**, *45*, 16873–16879. [CrossRef]
21. Zhang, T.; Li, P.; Zhou, J.; Wang, C.; Meng, X.; Huang, S. Microstructure evolution of laser cladding Inconel 718 assisted hybrid ultrasonic-electromagnetic field. *Mater. Lett.* **2021**, *289*, 1873–4979. [CrossRef]
22. Wang, Y.; Shi, H.; Hao, X.; Liu, H.; Zhang, X. Microstructure and Wear Resistance of Fe60 Laser Cladding Coating Assisted by Steady Magnetic Field-Mechanical Vibration Coupling Field. *Coatings* **2022**, *12*, 751. [CrossRef]
23. Zhou, S.; Xiong, Z.; Dai, X.; Liu, J.; Zhang, T.; Wang, C. Microstructure and oxidation resistance of cryomilled NiCrAlY coating by laser induction hybrid rapid cladding. *Surf. Coat. Technol.* **2014**, *258*, 943–949. [CrossRef]
24. Nurminen, J.; NaKKi, J.; Vuoristo, P. Microstructure and properties of hard and wear resistant MMC coatings deposited by laser cladding. *Int. J. Refract. Metals Hard Mater.* **2009**, *27*, 472–478. [CrossRef]
25. Gao, Z.; Geng, H.; Qiao, Z.; Sun, B.; Gao, Z.; Zhang, C. In situ TiBX/TiXNiY/TiC reinforced Ni60 composites by laser cladding and its effect on the tribological properties. *Ceram. Int.* **2023**, *49*, 6409–6418. [CrossRef]
26. Shen, Y.J.; Li, X.F.; Chen, X.Y. Effect of Ultrasonic Power on microstructure and properties of Laser cladding Fe-Cr-V Coatings. *Trans. China Weld. Inst.* **2017**, *38*, 74–78+132.
27. Gao, Z.; Zhang, S.; Ren, C.; Yu, Y.; Gao, Z.; Zhang, C. Effect of Rare-Earth La_2O_3 on Tribological Properties of Laser Cladding Nickel-Based Coatings on 35CrMoV Alloy Steel. *J. Mater. Eng. Perform.* **2023**. [CrossRef]

coatings

MDPI

Article

Effect of Scanning Strategies on Anisotropy of YCF104 Alloy Mechanical Properties by Laser Cladding

Yu Zhao [1,2], Wenkai Shi [1], Liaoyuan Chen [3], Wenzheng Wu [1,2,*] and Tianbiao Yu [3]

1 School of Mechanical and Aerospace Engineering, Jilin University, Changchun 130025, China
2 Chongqing Research Institute, Jilin University, Chongqing 401133, China
3 School of Mechanical Engineering and Automation, Northeastern University, Shenyang 110819, China
* Correspondence: wzwu@jlu.edu.cn

Abstract: YCF104 alloy specimens were fabricated by laser cladding on #45 steel following three different scanning strategies. The microstructure, phase composition, friction coefficient, microhardness, tensile strength, and compressive strength of the specimens formed by different scanning strategies were investigated. The results show that the uniformity of laser remelting of solidified layers improved the uniformity of microstructure and refined the grains. The strengthening effect of the Fe–Cr/Mo solid solution was primarily responsible for the high strength. YCF104 cladding specimens exhibit brittle failure, and with the change of metallurgical bonding form in the overlap area, the compressive strength (the maximum value is 3235 MPa) and the tensile strength (specimen via strategy 3 is 527.44 MPa), there appears significant anisotropy. With the improvement of the uniformity of temperature distribution, the friction coefficient as well as the microhardness decreased when the GCr15 was used as the friction pair. The mechanical anisotropy of the coating is related to both the microstructure and metallurgical bonding strength in the overlap area.

Keywords: laser cladding; YCF104 alloy; scanning strategies; mechanical properties anisotropy

1. Introduction

Compared to the mature machine tool remanufacturing industry abroad, the domestic remanufacturing industry is still in the primary stage. The lack of specifications and standards of remanufacturing processes restrict the development of the machine tool remanufacturing industry toward normalization, standardization, and industrialization. Laser cladding, as a promising surface modification technique, has attracted significant research interest. It is extensively studied for its advantages including fast speed, flexible clamping, no tool usage, no deformation, no slag inclusion, and low porosity and crack extension. Sexton et al. [1] reported that laser cladding technology has the potential to form pore-free and crack-free coatings. Li et al. [2] successfully obtained good metallurgical bonding between the 308L stainless steel (SS) coating and 316L SS substrate (with scrapped or erosive surface) without tools and fixtures. Laser cladding, as a new technology of the 21st century, has gradually become a key technology to promote the development of the remanufacturing industry. Most parts of machine tools are steel parts. In order to ensure the stability and service life of repaired parts, selection of the laser cladding process and the performance of remanufactured parts are very important [3]. Therefore, extensive research efforts have been devoted to the investigation of the laser cladding properties of Fe-based alloy powder [4,5].

Zhang et al. [6] found that the ceramics phase could improve the wear resistance of Fe-based alloy coating. Qu et al. [7] reported that VC-VB/Mo could significantly improve the hardness and wear resistance of the Fe-based alloy coating. Yang et al. [8] assumed that Mn has the potential to improve the non-magnetic properties of the Fe-60%WC composite coating. Yang et al. [9] investigated the effect of Nb contents on the performance of Fe-based alloy coating. Therefore, in order to improve the properties of Fe-based alloy and Fe-based composite coatings, the effects of the addition of some materials to Fe-based alloys has been studied. Tan et al. [10] investigated the effect of strengthening/reinforcement particles (Al_2O_3 and M_7C_3) on the microstructure and wear resistance of the matrix composite coatings. The results showed that thermite reactants could enhance the microhardness and wear resistance by increasing the amount of Al_2O_3 ceramic and M_7C_3 carbide. Zhang et al. [11] illustrated the relationship between pulse laser parameters and the size of the in situ-formed Ti–V carbide particle. The results indicated that the influence of pulse frequency on carbide particle size was not monotonous, while it exhibited a contrary effect on the hardness and corrosion resistance. By reducing the additional content of CeO_2 from 0.5 to 0.25 wt.%, Zhang et al. [12] improved the corrosion resistance of TiC–VC-reinforced Fe-based cladding layers from 3.5 times to 7.33 times. Furthermore, the microstructure that was generated in the laser cladding process was mainly lath martensite and retained austenite. Wang et al. [13] studied the effect of content of Mo on the wear resistance and microstructure of SS coatings. The results showed that 6.0 wt.% of Mo coating led to an excellent wear resistance (3.7 times) compared to free Mo, and the content of Mo played a positive effect on ferrite content. Zhao et al. [14] studied the effect of a ceramic particle (TiC, VC) on the Fe-based alloy coating. Generation of the composite particle (Ti, V)C during the laser cladding process is critical to improving the wear resistance and hardness of the cladding layer. According to the abovementioned studies, the reinforcement particles produced by the reaction between the additional elements and the Fe-based alloy matrix in the cladding process can lead to the enhancement of the properties of cladding layers.

Many systematic explorations and achievements have been made in the research of performance strengthening of Fe-based alloy [15]. However, the effect of process parameters on the performance of the coating is also very important, which can provide the theoretical basis and research direction for further research and application of Fe-based alloy. Zhou et al. [16] fabricated a Cu–Fe-based coating without cracks and pores using laser induction hybrid rapid cladding (LIHRC) technology. The LIHRC was able to provide good properties to the Fe-based alloy with impressive efficiency (four times that of individual laser cladding). S. Zhou and Dai [17] also studied the microstructure evolution of Fe-based/WC composite coating fabricated on steel surface by LIHRC. They reported that the microhardness of the composite coating was three times higher than that of the substrate. Zhou et al. [18] investigated the influence of a Fe-based amorphous matrix composite coating reinforced with various portions of 316L SS powders on the microstructure and mechanical properties. Li et al. [19] prepared a Co–Cr–Fe alloy cladding layer using a combination of laser cladding and friction stir processing, and illuminated the relevant mechanisms for the mechanical and wear properties. Wang et al. [20] investigated the effect of oil lubrication on the wear and contact fatigue damage resistance of Fe-based alloy fabricated by laser cladding. The results showed that the addition of suitable La_2O_3 content (1.2 wt.%) in the Fe-based alloy coating resulted in a good wear resistance under the oil condition. Karczewski et al. [21] successfully fabricated thin walls with a minimum wall thickness of 0.5 mm using the Fe–16Al alloy powder by laser engineered net shaping; however, the produced specimens exhibited the presence of a number of cracks. Fang et al. [22] found that the phase transformation obviously affected the stress evolution, and in turn, the martensitic transformation characteristic temperature was affected. Nonetheless, the study of the effect of the cladding process on properties and mechanism evolution still needs further exploration.

In this study, the correctness of the start–stop overlap model [23] was verified, and the effects of z-axis increment on the geometrical features and properties of the specimen fabricated by laser cladding using 316L SS [24] were studied. Moreover, Taguchi grey correlation analysis was used to realize the optimization of multiple parameters under multiple responses [25]. By adding ceramic phase in the Fe-based alloy, the wear resistance of the coating was improved [26]. The results indicated that the cladding strategies obviously affect the properties of the specimens. Distribution of the temperature field in the cladding process is closely related to the structure of the coating, which directly affects its performance. The selection of scanning strategy in the cladding process directly determines the temperature field distribution [27]. In order to reduce the laser-cladded layer's anisotropy, and thus improve the stability and reliability of the laser-cladded specimens, the different scanning strategies were designed and investigated. The relationship between the scanning strategies and the properties of the anisotropy of YCF104 alloy powder was analyzed. Moreover, the relationship between the properties of the anisotropy and the typical microstructure is studied.

2. Experimental

2.1. Experiment Materials

The specimens used in this study were fabricated by laser cladding on the #45 steel substrate using three different scanning strategies. The material of the specimen is YCF104 alloy (Fe-based alloy), and the powder consisted of circular particles of sizes between 53 and 150 μm. Zhang et al. [28] found that nitrogen can improve the hardness and corrosion resistance of the SS. Yadaiah et al. [29] found that the self-protective atmosphere of argon showed a positive effect on the weld bead dimensions and free surface profile formation. Therefore, in this study, the argon gas is used as sending powder gas and shielding gas. The selection of the value of gas flow rate should guarantee both the flow stability of powder feeding and the low oxidation rate of the cladding layer during the cladding process. The composition of #45 steel and YCF104 alloy is presented in Table 1. Drying of the powder is necessary prior to its use. On the one hand, it can prevent the existing water vapor in the cladding layer from forming pores and other defects responsible for reducing the quality of cladding parts. On the other hand, it could prevent the water vapor from reacting with elements in Fe-based alloys at high temperature, which then reduces the quality. The formed oxide dust and solid particles brought by water vapor could also be bonded to the protective mirror in the laser head, which could otherwise reduce the quality of the spot and affect the experimental results.

Table 1. Chemical composition (wt.%) of 45 steel and YCF104 steel powder (provided by suppliers).

	C	B	P,S	Si	Cr	Ni	Mn	Mo	Nb	Cu	W	Fe
RCF104	0.06	0.6	-	0.75	9.6	0.8	0.2	3	1	-	3	Bal.
45 steel	0.42–0.5		≤0.045	0.17–0.37	≤0.25	≤0.25	0.5–0.8	-	-	≤0.25	-	Bal.
mechanical properties of #45 steel												
Tensile strength	≥600 MPa		elongation	≥16%		Reduction of section			≥40%			
Yield strength	≥355 MPa		hardness	≤197 HB								

2.2. Process Design

According to the previous experimental studies, the process parameters selected for preparing specimens based on different scanning strategies in this study are laser power 450 W, scanning speed 0.55 mm s^{-1}, powder feeding rate 8.26 g min^{-1}, shielding gas flow rate 12 L min^{-1}, powder feeding gas flow rate 8 L min^{-1}, distance between bottom plane of nozzle and substrate surface 14–16 mm, and defocusing rate 0 mm. Distance between the centers of the adjacent clad tracks was 0.9 mm during transverse overlapping and the distance between the clad paths of the adjacent cladding layer was 0.35 mm during longitudinal overlap. The off-line software RobotArt (PQArt(×86)) was used to design the trajectories of different scanning strategies. Based on the horizontal and vertical overlapping distances obtained from previous experiments, the trajectories were designed and optimized according to the designed cuboid size (35 mm × 25 mm × 15 mm) as shown in Figure 1a–c, respectively. Preparation of the tensile pattern is presented in Figure 1c, and the cuboid size is 115 mm × 30 mm × 12 mm, as shown in Figure 2. The total length of the tensile specimen is 105 mm, the gauge is 30 mm in size with the section size 4 × 6 mm^2, the length of parallel segment for clamping is 25 mm, and the width is 20 mm. Furthermore, the scanning strategy 1 (Figure 1a) has the same trajectory in each layer, and the scanning strategy is consistent with the path 1 shown in Figure 2. For scanning strategy 2 (Figure 1b), the scanning directions of the adjacent two layers are perpendicular to each other, and the scanning strategy indicates that the paths 1 and 2 are alternate to each other as presented in Figure 2. Scanning strategy 3 (Figure 1c) involves paths 1, 2, 3 and 4 as presented in Figure 2. Path 1 is perpendicular to path 2, 3 is perpendicular to 4, and the angle between paths 2 and 3 is 45°.

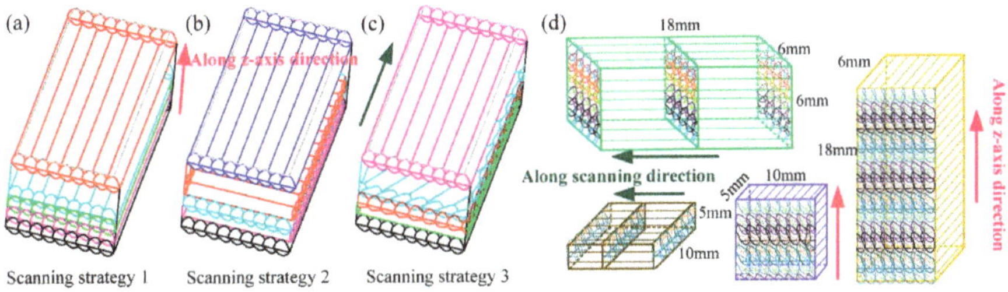

Figure 1. Compressive specimens design. (**a**) Scanning strategy 1; (**b**) Scanning strategy 2; (**c**) Scanning strategy 3; (**d**) Size of specimens.

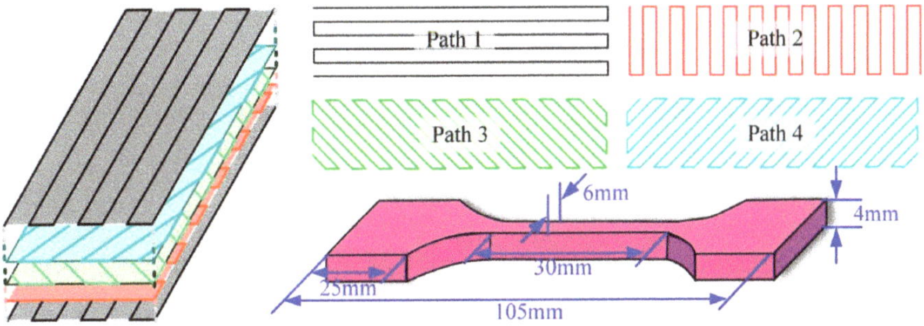

Figure 2. Tensile specimens design.

In order to study the effect of different scanning strategies on the compressive strength, microstructure, and wear resistance of cladded specimens, cuboid blocks completed by cladding were utilized (Figure 1d). The specimens were prepared parallel to the cladding direction and perpendicular to the cladding direction, respectively. The dimensions of the compressive specimens were 6 mm × 6 mm × 18 mm and those of the wear-resistant specimens were 10 mm × 10 mm × 5 mm. The surfaces of the compressive specimens were ground using a grinding wheel. The compressive strength and tensile strength of the specimens were tested using a universal testing machine ETM 305D (Shenzhen Xinlangpu Electronic Technology Co., Ltd, Shenzhen, China); during the tensile and compressive test, the loading rate was 2 mm min^{-1}. Effects of different scanning strategies on the compressive properties were analyzed and compared. The morphology of the crushed specimens was observed by scanning electron microscopy (SEM). The specimens used for wear resistance were sanded using sandpaper (240#, 400#, 600#, 800#, 1000#, 1500#, 2000#), and then polished by paste polishing. The wear resistance of polished specimens was tested using a reciprocating friction and wear tester MFT-4000 (Lanzhou Huahui Co., Ltd, Lanzhou, China). The experiment was carried out at room temperature in dry sliding friction; the load was 10 N, the wear speed was 200 mm min^{-1}, and the reciprocating distance was 5 mm. After grinding using sandpaper and polishing paste, the microhardness which was measured using a microhardness tester (EM500-2A, Shanghai Hengyi Co., Ltd, Shanghai, China) was compared under different overlapping modes, a 500 gf load was applied for 10 s, and the spacing between adjacent two points was 600 μm.

The microstructures were observed by confocal microscopy (LEXTOLS4100, Olympus, Tokyo, Japan), SEM, and energy dispersive spectroscopy (EDS). Moreover, the mechanism highlighting the effect of scanning strategies on the microstructure was analyzed. It was necessary to use the etching solution ($HCl:H_2O:FeCl_3$ = 20:10:1) to cause corrosion before observing the microstructure.

Figure 3 displays that the distribution of the laser energy density in laser cladding process is Gauss distribution [29]. Figure 3a exhibits the temperature distribution of scanning strategy 1's (Figure 1a) cross-section. Clearly, the temperature in the overlap zone is lower than that in the center of the molten pool when cladding is carried out with a single scanning strategy. The remelting area of the overlap zone is the smallest, which is prone to generate cracks and pores. The difference of temperature gradient in the overlapping zone and clad track central area is large, and thus significant microstructure heterogeneity will be observed. Figure 3b shows the temperature distribution of scanning strategy 2's (Figure 1b) cross-section. Compared to that in strategy 1, temperature inhomogeneity in strategy 2 gets significantly improved, but it still shows the main area for the appearance of the pores and cracks in the overlap area of adjacent two-layer clad tracks. Figure 3c exhibits the temperature distribution of scanning strategy 3's (Figure 1c) cross-section. More remelting happens and the temperature grade in overlap area is low. During cladding, the overlap area within the layer and the overlap area between the layers are often the areas where defects occur [27]. The areas where defects occur in each layer are concentrated in the overlap area, thus the direction of each layer gets changed to change the direction of defect generation, thereby weakening the defect [30].

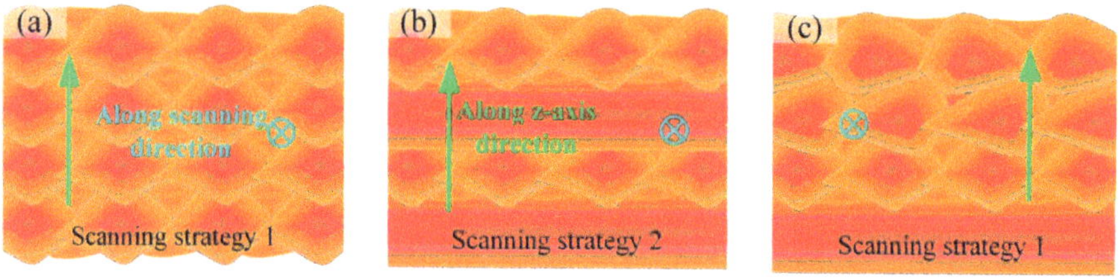

Figure 3. Distribution of the laser energy density. (**a**) Scanning strategy 1; (**b**) Scanning strategy 2; (**c**) Scanning strategy 3.

3. Results and Discussion

3.1. *Influence of Scanning Strategies on the Microstructure of Forming Parts*

Figure 4 shows the cross-section images obtained under different laser-cladded scanning strategies. The remelting area of scanning strategies 1, 2, and 3 increases in turn, and the six specimens prepared by three scanning strategies along the *z*-axis direction and parallel to the scanning direction show little cracks and small pores. Moreover, the metallurgical bonding quality in overlap area is good. Owing to the incompletely melted powder bonded on the clad track surface, defects (slag inclusion, porosity, cracks, etc.) tend to occur in the overlap zone [1].

Figure 4 demonstrates that the scanning strategy mainly changes the distribution of the heat-affected zone and the remelting mode of the solidified melted layer. Therefore, the defects in the coating are reduced by remelting the matrix metal with defects. Different scanning strategies change the temperature distribution in the melt layer, which results in changes in the temperature gradient and the heat dissipation ability of the melt layer, and then affects the structure of the melt layer and reduces the defects in the overlapping zone. In order to better understand this impact, different microstructures at the cross-sections were tested.

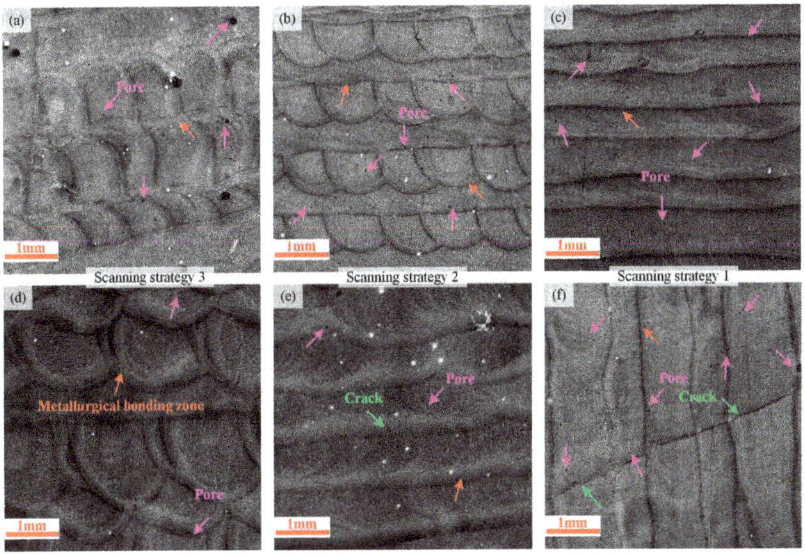

Figure 4. Cross-section morphology of specimens under different scanning strategies. Along the z-axis direction: (**a**) Scanning strategy 3; (**b**) Scanning strategy 2; (**c**) Scanning strategy 1. Parallel to the scanning direction: (**d**) Scanning strategy 3; (**e**) Scanning strategy 2; (**f**) Scanning strategy 1.

Figure 5 shows the microstructures of different areas in the cross-section along the z-axis direction of scanning strategy 3. According to the microstructure of the overlapping area for which the adjacent two-layer paths are distinguished along the direction of length and width of cladding the rectangular block (as shown in Figure 5a), and the microstructure of the overlapping area for which the adjacent two-layer paths are perpendicular to each other and have an angle of 45° with the rectangular edge (as shown in Figure 5b), the microstructure of the cross-section is found to be fine and uniform. Figure 5(a1,b1) exhibits magnified images of the microstructure of the molten pool. During the cladding process, the internal temperature of the molten pool is the highest, the temperature gradient is small, the heat dissipation is slow, and the grain growth time is long; thus, the main crystal is isometric in nature. Figure 5(a2) shows an enlarged image of the microstructure of the overlapping zone of the two clad tracks. With the increase of the temperature gradient at the overlap, columnar dendrite crystals appear. Remelting leads to continuous growth of columnar dendrite crystals, which eventually become dendrites. Friction, shear, and collision between dendrites break them up into fine dendrites, and thus in this zone, dendrite dominates. Figure 5(b2) exhibits the enlarged image of microstructure of the zone which is around the overlapping zone. During cladding, the temperature in this area is low, the temperature gradient is large, the grain growth time is short after nucleation, and the remelting effect is poor; thus, the secondary growth of grain is not obvious. Therefore, this zone mainly consists of columnar dendrite crystals. Figure 5(a3,b3) shows enlarged images of the microstructure of the overlapping zone between the two layers. This region has a large temperature gradient; therefore, the microstructure mainly consists of columnar dendrite and isometric crystals. There is little difference in the grain size of the entire section, and it has a high homogeneity.

Figure 6 shows the cross-section of the microstructures formed using scanning strategy 3 parallel to the scanning direction during the cladding process. The metallurgical bonding zone is the remelting mode of the other adjacent layers in the cladding layer. The microstructure in the molten pool still contains mainly isometric crystals, as shown in Figure 6(a1). Within the a2 area, there are mainly columnar dendrite crystals, as shown in Figure 6(a2). Formation of columnar dendrite crystals is effectively inhibited by isometric crystals near the metallurgical bonding zone. Within the a3 area, there are mainly dendrites and isometric crystals, as shown in Figure 6(a3). Complete remelting in the cladding area results in effective refining of the grains and improvement in the uniformity of the microstructure size.

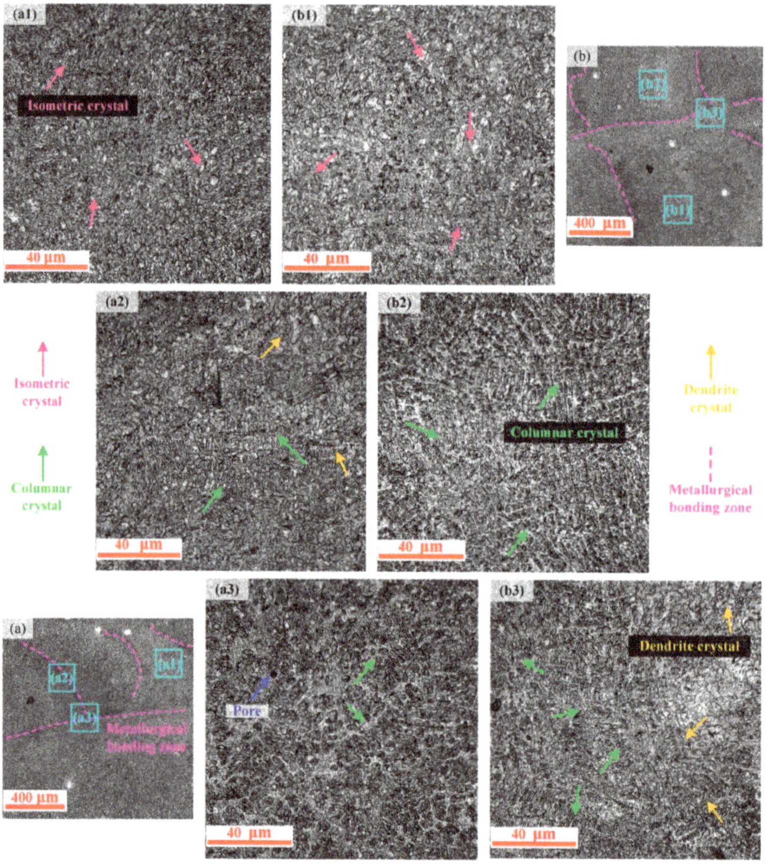

Figure 5. Microstructures of scanning strategy 3: (**a**,**b**) microstructures along the z-axis direction in scanning strategy 3; (**a1**–**a3**) enlarged images of an area of (**a**); (**b1**–**b3**) enlarged images of an area of (**b**).

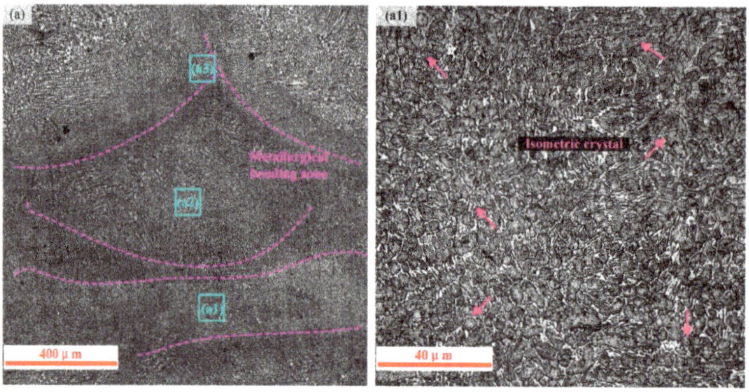

Figure 6. *Cont.*

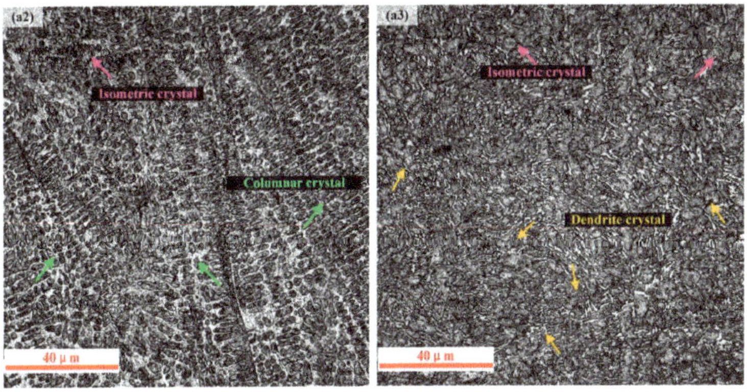

Figure 6. Microstructure obtained using strategy 3: (**a**) microstructure parallel to the scanning direction; (**a1–a3**) enlarged images of an area of (**a**).

In order to study the effect of scanning strategy on the microstructure of the clad layer, the microstructures obtained using scanning strategies 1 and 2 along the *z*-axis direction were observed. The measured results of the specimens are presented in Figure 7. The internal microstructure of the molten pool in scanning strategies 1 and 2 mainly consists of isometric crystals, as shown in Figure 7(a1,b3). Figure 7(a3) demonstrates that within the cross-section of scanning strategy 1, the thick dendrite crystals mainly occur in the overlapping zone. In the process of transverse overlapping and longitudinal overlapping, the overlapping area remelts at the edge of the clad track. The molten pool formed in the solidified layer is shallow and small, and thus the remelting quality is poor and the grain size is large in this area. Figure 7(b1,b2) reveals that dendrite and isometric crystals form the major structure. With the increase of remelting area in the overlapping region, the size of the grains decreases and the uniformity of the structure increases relative to scanning strategy 1. The metallurgical quality in this zone is good.

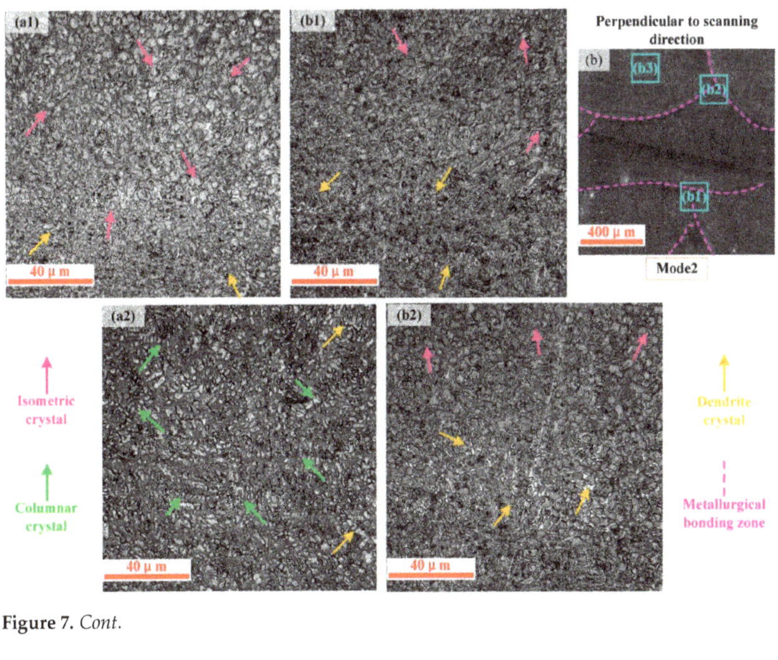

Figure 7. *Cont.*

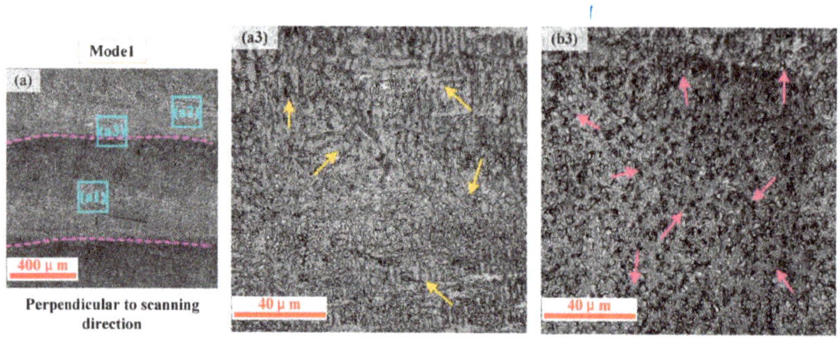

Figure 7. Microstructures of scanning strategies 1 and 2: (**a**) microstructure along the *z*-axis direction in scanning strategy 1; (**a1–a3**) enlarged image of an area of (**a**); (**b**) microstructure along the *z*-axis direction in scanning strategy 2; (**b1–b3**) enlarged image of an area of (**b**).

With the increase of the complexity of scanning strategy between layers, the remelting zone of the clad layer increases, the temperature gradient decreases, and the time difference of grain growth decreases, which reduces the grain size, improves the structural uniformity, and refines the microstructure. Comparative analysis of Figures 5 and 7 demonstrates that the difference of grain size in scanning strategies 1, 2, and 3 decreases in turn, while the homogeneity of the structure increases.

Figure 8 shows the microstructures of scanning strategy 1 and scanning strategy 2 in cross-sections parallel to the scanning direction. The microstructures in the molten pool consist of isometric crystals, and the uniformity of the structure in scanning strategy 2 is higher than that in scanning strategy 1. Regardless of the scanning strategy, columnar dendrite and dendrite crystals are mainly formed in metallurgical bonding zone and in its vicinity area, as shown in Figure 8(a1,b2).

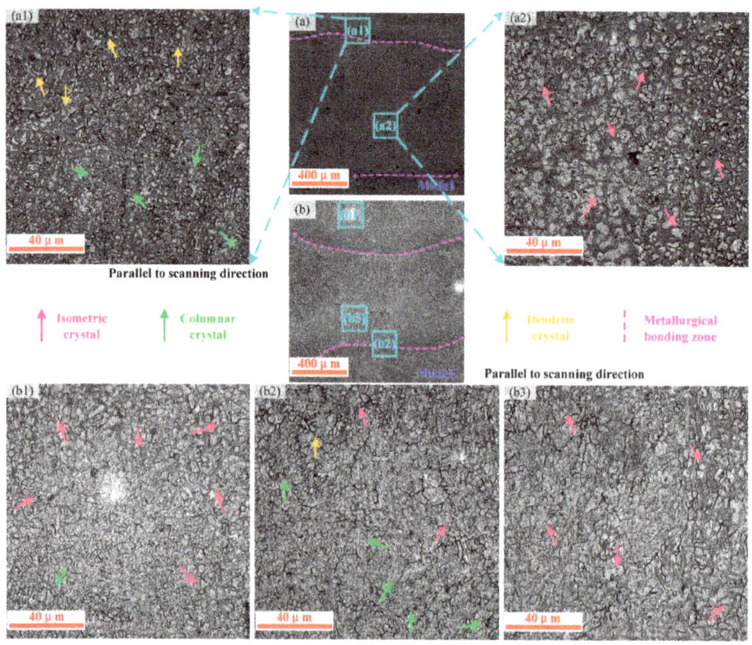

Figure 8. Microstructures of scanning strategies 1 and 2: (**a**) microstructure parallel to the scanning direction in scanning strategy 1; (**a1–a2**) enlarged image of an area of (**a**); (**b**) microstructure parallel to the scanning direction in scanning strategy 2; (**b1–b3**) enlarged image of an area of (**b**).

Through microstructural analysis under different scanning strategies, the internal structure can be effectively controlled by changing the scanning strategies between layers. The change of scanning strategy mainly leads to the change in the temperature distribution in the cladding process; thus, the structure can be effectively controlled. Owing to the low-temperature gradient, high temperature, and slow heat dissipation, dense isometric crystals are formed after nucleation for a longer growth time. In the overlapping area, the temperature gradient is large, secondary crystallization or even longer duration crystallization occurs during cladding, and the grain continues to grow or the structure remelts and regenerates and finally breaks into fine dendrites or isometric crystals. However, different scanning strategies generate different remelting modes; therefore, the structure of the overlapping zone is different. Defects are most likely to occur in the overlap zone, and thus controlling the structure of the overlapping zone is one of the key factors in controlling the properties of the coating.

3.2. Energy Dispersive Spectroscopy Analysis

The SEM image and corresponding EDS spectrum of the specimen fabricated via scanning strategy 3 are shown in Figures 9 and 10. The EDS results reveal that the grain boundary is rich in Cr, Ni, Mo, and Nb (as shown in Figure 9d), whereas the grain interior is rich in Si, Fe, and W (as shown in Figure 9e). The grain boundary consists of lamellar phase, as shown in Figure 9c. Figure 10 shows the microstructure of the metallurgical bonding zone of overlapping regions. The grain interior does not contain Nb and W, and the content of elements Fe and Cr are basically the same, as shown in Figures 9e and 10e. Figure 10d demonstrates that the entire rectangular area of spectrum 1 is dominated by Fe, and a significant amount of Cr is also observed. It can be concluded that the structure along the grain boundary is some displacement solid solution about Fe-Cr/Si/Ni/Mo. Figure 10f exhibits the existence of significant amounts of Fe, Cr, and Mo. The Mo enriched in the metallurgical bonding zone formed the solid solution Fe-Cr/Mo/Si/Ni/Nb/W. High Cr content in the laser cladding layer reduced the tenacity of the Fe-based alloy. The solid solution strengthening effect resulted in the lattice distortion of ferrite and then improved the hardness and strength of the coatings.

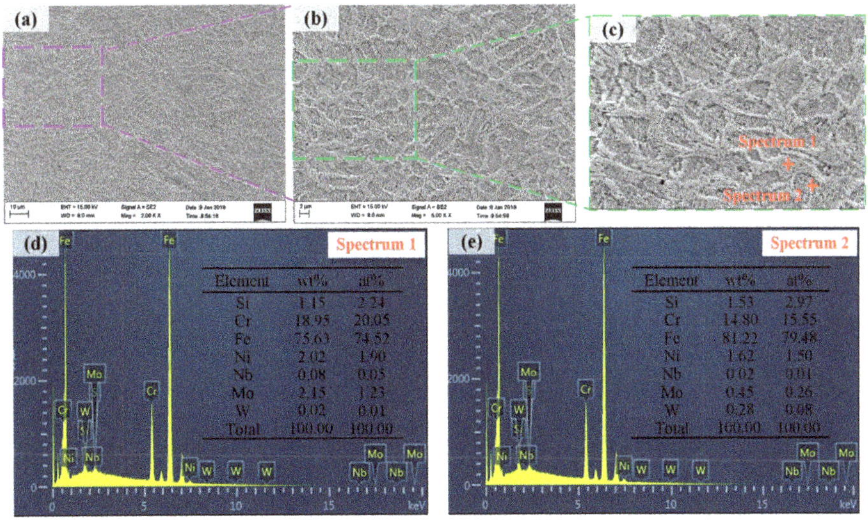

Figure 9. The SEM image and corresponding EDS spectra obtained from scanning strategy 3: (**a**–**c**) microstructure in scanning strategy 3; (**d**) Spectrum 1; (**e**) Spectrum 2.

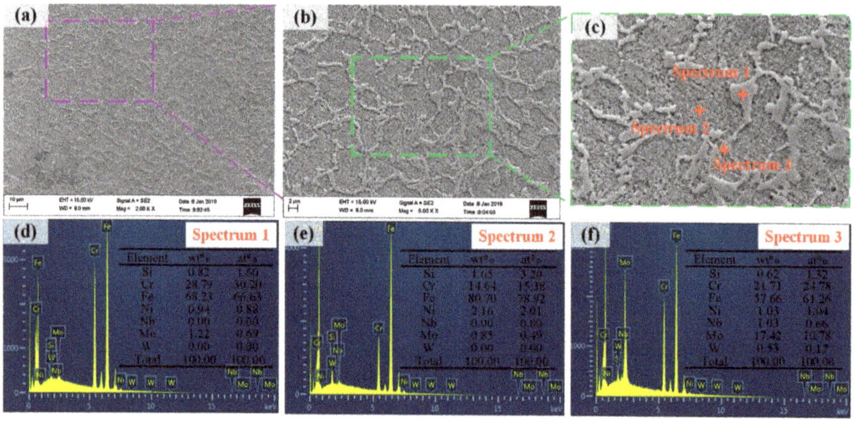

Figure 10. SEM image and corresponding EDS spectra obtained from scanning strategy 3: (**a–c**) microstructure in scanning strategy 3; (**d**) Spectrum 1; (**e**) Spectrum 2; (**f**) Spectrum 3.

3.3. Analysis of Friction Coefficient and Microhardness

The friction coefficient with wear time was measured using a reciprocating friction and wear tester under different scanning strategies. The experimental results of friction and wear on the cross-section of the specimen parallel to the scanning direction and along the z-axis direction are shown in Figure 11.

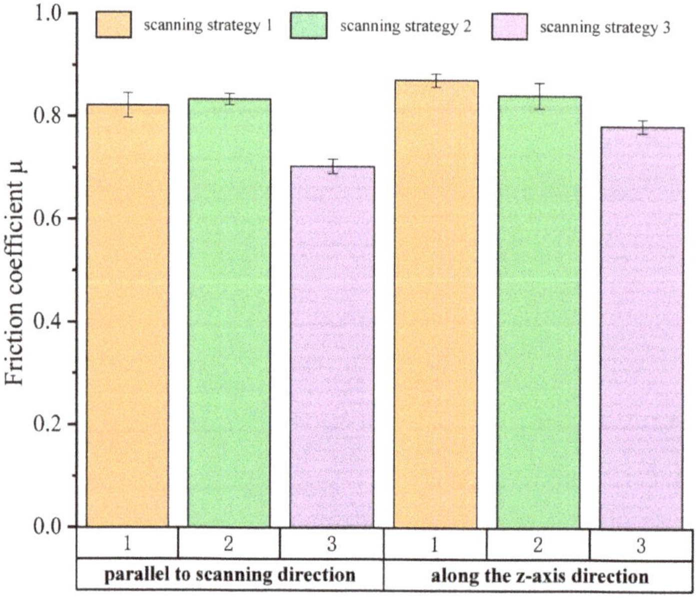

Figure 11. Friction coefficient under different scanning strategies.

The friction coefficient of the specimen in the 'horizontal direction' (the specimen along the scanning direction) is lower than that in the 'vertical direction' (the specimen along the z-axis direction). The temperature gradient at the bottom of the molten pool in the 'vertical direction' is larger than that in the 'horizontal direction'. In contrast, the temperature gradient at the top of the molten layer in the 'horizontal direction' is larger than that in the 'vertical direction'. Therefore, in the overlapping area, the direction of the growth of the microstructure is different. In the 'horizontal direction', the direction of the temperature gradient is basically the same, the uniformity of the temperature distribution is high, and the direction of the growth of the structure in the overlap area is high, as shown in Figure 5(b2), Figure 6(a2), Figure 7(a2) and Figure 8(a1).

Scanning strategy 3 shows the smallest friction coefficient in both 'horizontal direction' and 'vertical direction' compared to scanning strategy 1 and scanning strategy 2. In scanning strategy 3, the uniformity of structures in the two directions is high, as shown in Figures 5 and 6. This result benefits from the uniform distribution of remelting zone, a small difference of temperature gradient, and uniform distribution of temperature field during the cladding process. In the cladding process, there is no change in the powder composition, and the laser power, scanning speed, and powder feeding rate, etc., remain the same; that is, the energy input is the same. Different scanning strategies can change the temperature input mode, thus affecting the temperature distribution, further controlling the microstructure, and eventually affecting the wear resistance of the clad layer. The dense microstructure leads to an increase in the number of boundaries that need to be destroyed in the wear process, and the wear resistance increases.

Figure 12 shows the cross-section wear morphology along the z-axis and scanning direction of specimens under different scanning strategies. Figure 12a–f exhibits that the wear surface mainly consists of plow-groove wear scars, and the wear scar width of scanning strategy 1, scanning strategy 2, and scanning strategy 3 decreases in turn. This is consistent with the result of the scratch section profile obtained in Figure 13. Slight adhesion wear is observed on the wear scar surface. The homogeneity of the structure in the 'horizontal direction' is higher than that in the 'vertical direction', and thus the flaking in the 'vertical direction' is more serious than that in the 'horizontal direction', as shown in Figure 12(a1–f1). This is mainly attributed to the fact that the structural homogeneity of the 'horizontal direction' is higher than that of the 'vertical direction', which can be confirmed by Figure 13; scanning strategy 3 (parallel to scanning direction) exhibits the shallowest abrasion mark. Although the internal composition does not change, the number of grains increases and refines. With the increase of remelting area produced by cladding paths and the improvement of uniformity of temperature distribution, the number of grains increases, and then the wear resistance increases.

Figures 14 and 15 show the microhardness profiles along the z-axis direction and scanning direction, respectively. The values of average microhardness of different scanning strategies are between 633.0 and 729.9 $HV_{0.5}$. Different scanning strategies have an obvious effect on the microhardness value of the cladding layer. The microhardness of scanning strategies 1, 2, and 3 decreased in turn. The microhardness of scanning strategy 3 was the minimum, with the value of about 633.0 $HV_{0.5}$ (the specimen was parallel to the scanning direction), and the microhardness of scanning strategy 1 was the maximum, the value being about 729.9 $HV_{0.5}$ (the specimen was parallel to the scanning direction).

In the cladding process, scanning strategy 3 shows the best remelting uniformity for the cladded layer, and the effect of reducing defects is obvious. Uniformity of the inner microstructure of the cladding layer is high, and the surface wear resistance is the best. However, the microhardness of the cross-section gets obviously reduced. The scanning strategy providing good wear resistance and uniform structure may not show the best microhardness for the fabricated specimens. Therefore, the scanning strategy in the fabrication process of the YCF104 material should be selected reasonably according to the requirements in the application process.

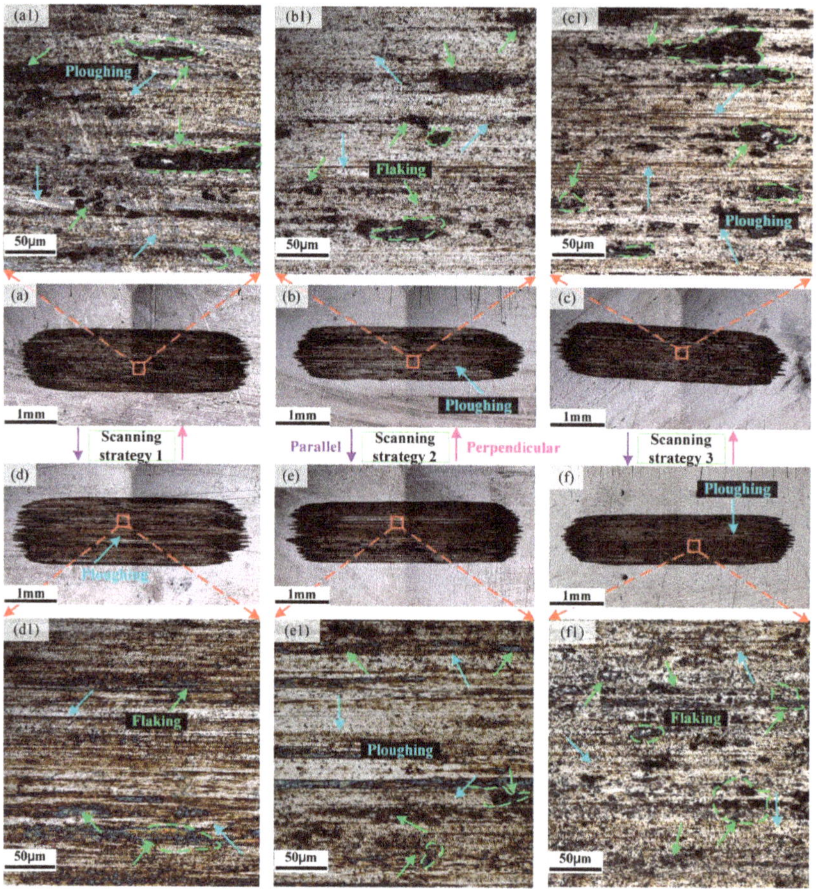

Figure 12. Wear morphology. Perpendicular to the scanning direction: (**a**,**a1**) scanning strategy 1; (**b**,**b1**) scanning strategy 2; (**c**,**c1**) scanning strategy 3. Parallel to the scanning direction: (**d**,**d1**) scanning strategy 1; (**e**,**e1**) scanning strategy 2; (**f**,**f1**) scanning strategy 3.

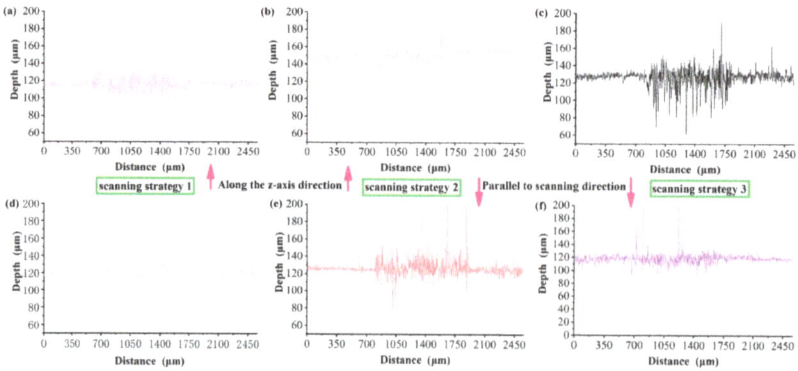

Figure 13. Sectional profile of scratch under different scanning strategies. Along the *z*-axis direction: (**a**) scanning strategy 1; (**b**) scanning strategy 2; (**c**) scanning strategy 3; Parallel to scanning direction: (**d**) scanning strategy 1; (**e**) scanning strategy 2; (**f**) scanning strategy 3.

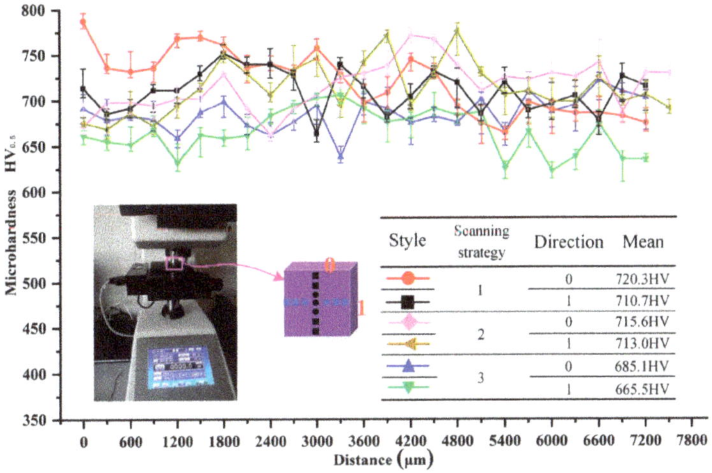

Figure 14. Microhardness of layers along *z*-axis direction.

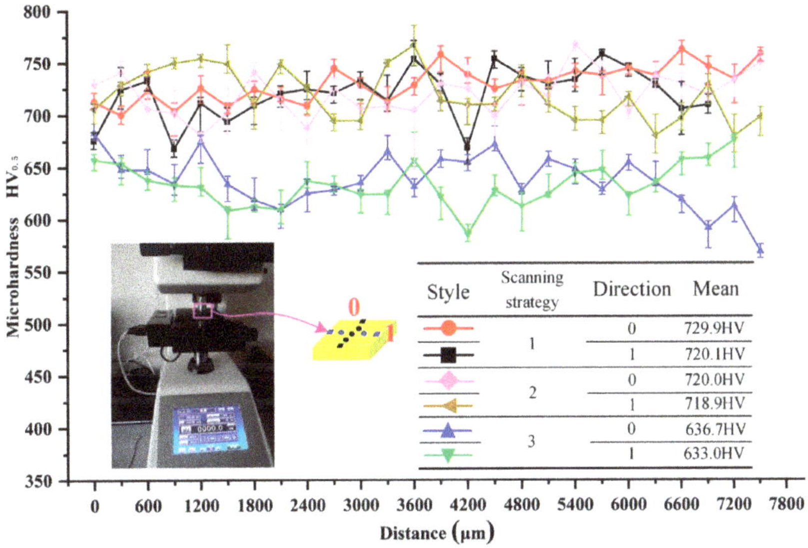

Figure 15. Microhardness of layers along scanning direction.

3.4. Strength Measurement and Analysis

For a comprehensive understanding of the influence of material properties of YCF104 and scanning strategy on the compressive strength of laser cladding specimens, compression experiments were carried out on the specimens. The corresponding experimental results are presented in Figure 16. There is no obvious straight-line part in the curve, and the yield stage is also not observed in the curve. Moreover, the specimen collapsed directly when the deformation was very small. Noteworthy, the compressive strength of YCF104 melt layer was found to be very high. The minimum compressive strength of the specimens was measured to be 2640 MPa (scanning strategy 2 along *z*-axis direction). The surface of the compression pad was damaged during compression due to the extreme hardness of the specimen. The figure exhibits that the overlap direction of the specimen prepared under scanning strategy 2 (specimen prepared along *z*-axis direction) is perpendicular to the load direction. Compared to the internal part of the molten pool, the microstructure of the overlap zone is thicker. During compression, the compaction of the microstructure of the overlap zone is easier due to the heterogeneity of the microstructure. Moreover, the vertical overlapping also prevents the splitting of the overlapping region along the X or Y direction during compression. The compressive strength of specimens fabricated under strategy 3 along z-axis and scanning direction is 3235 and 3204 MPa, respectively. The microstructure of the melt layer prepared via strategy 3 is more uniform and compact along the z-axis and scanning direction. Moreover, it shows high resistance toward external compressive stress during compression. When the strain is less than 0.08, the stress increases rapidly, and then it increases slowly with the strain until the specimen gets crushed. Although the extension direction of the overlap area of scanning strategy 3 (specimen prepared along z-axis direction) is perpendicular to the direction of load, the grain size of the overlap zone is refined and the quality of metallurgical bonding in the overlap zone is improved. Furthermore, the microstructure heterogeneity of the molten pool and the overlapping zone is reduced. Therefore, the compressive strength is higher than that for the specimen prepared via scanning strategy 2 following the same method. Microstructural analysis of the specimen fabricated via scanning strategy 3 (specimen prepared along scanning direction) indicates that the angle between overlap direction and load direction is 90°, 0°, and 45°. It will not reduce the deformation resistance unless the scanning direction is at a certain angle to the load direction. This is mainly attributed to the fact that scanning strategy 3 can effectively refine the microstructure of the metallurgical bond and improve its performance. Moreover, the direction of the temperature gradient in the melting layer is also disordered, which increases the anisotropy of the growth direction of the microstructure in the melting layer. The stress–strain curve and the morphology of the specimen after crushing indicate that YCF104 is a brittle material. The fracture morphology after crushing is shown in Figure 17.

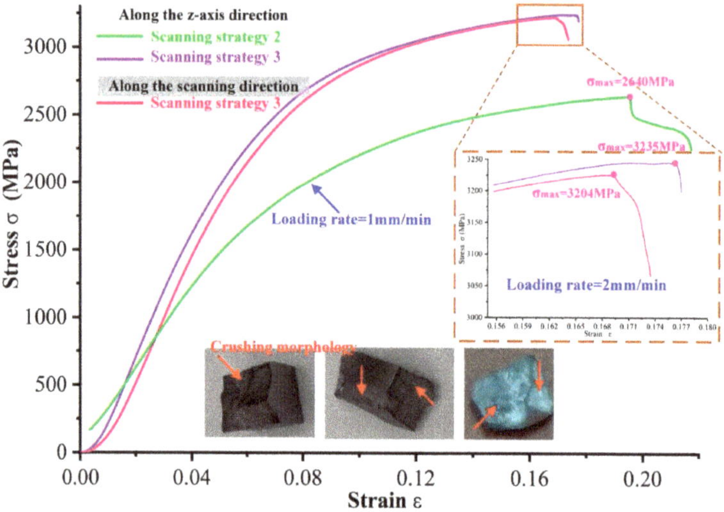

Figure 16. Compressive strength of scanning strategy 2 and 3 specimens.

Fracture morphology reflects the characteristics of part failure to a certain extent. The brittle fracture should be avoided for engineering applications. During compression, all the specimens are directly crushed, and the fractured surfaces show the appearance of many steps. During the process of the propagation of cleavage cracks, the cleavage steps converge to form a river-like pattern, as presented in Figure 17a,e. Under the action of external normal stress, the fracture of the interatomic bond causes brittle transgranular fracturing along a specific crystal plane, as shown in Figure 17(b–d,f,g). Brittle fracturing occurs in YCF104 alloy during compression. However, YCF104 alloy has a high compressive strength when crushed, which can meet the requirements of users based on their specific project.

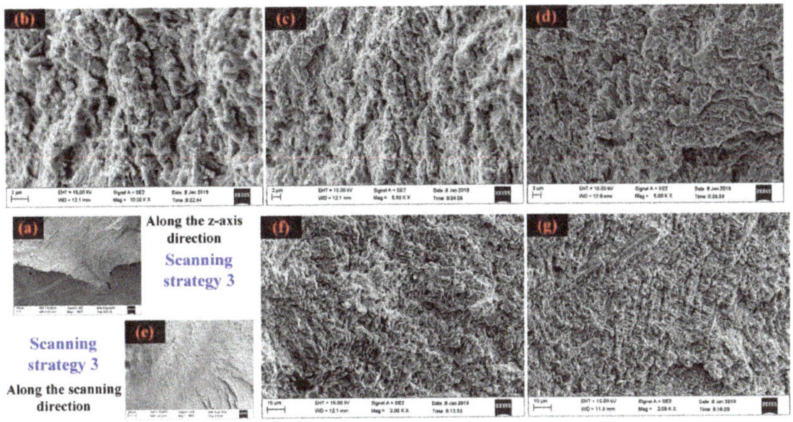

Figure 17. Crushing morphology of scanning strategy 3 specimens. (**a**) Specimen along the *z*-axis direction; (**b–d**) Enlarge images of (**a**); (**e**) Specimen parallel to scanning direction; (**f,g**) Enlarge images of (**e**).

Figure 18 shows the tensile strength of the YCF104 specimen fabricated using scanning strategy 3. The tensile test was conducted to measure the tensile strength of YCF104 alloy powder. The metallurgical bonding direction and overlapping direction of the specimen under scanning strategy 3 are uniform and systematic. A uniform remelting mode improves the quality of metallurgical bonding and reduces internal defects. When the stress is lower than σ_e, it is proportional to the strain of the specimen, and the original shape can be restored when the stress disappears. In this case, the specimen is in the stage of elastic deformation, and the elastic limit is σ_e = 337.5 MPa. When the stress exceeds σ_e, the plastic deformation of the specimen occurs, and the stress disappears. Only part of the deformation can be restored, and the yield limit is σ_s = 358.71 MPa. When the stress of the specimen exceeds σ_s, it increases continuously with the increase of stress until it reaches the strength limit of the specimen and is broken, i.e., at this stage the tensile strength of the specimen is σ_b = 527.44 MPa. According to the fracture morphology based on compressive test and the tensile process, the specimen prepared using YCF104 powder should be brittle material.

The obvious tension–compression asymmetry in laser-formed parts was observed in Figures 16 and 18, which agrees with what was previously reported in the literature [31]. The difference in loading direction of tensile and compression tests resulted in the different work hardening [32]. During the tension process, the alloy exhibited a limited work-hardening behavior and failed at a strain of 0.141. The cracking of grain boundaries of the overlap zones caused the drop in the flow stress [32,33]. With the crack's extension, the specimen is a failure. However, during the compression test, the alloy showed a more substantial and sustained work-hardening behavior than the tension specimen and failed at a strain of 0.174 (scanning strategies 3) and 0.217 (scanning strategies 2). In this paper, owing to the high microhardness of the laser-cladded specimens (ranges from 633–729.9 $HV_{0.5}$ (Figure 15)) and big brittleness, the specimen exhibited bigger compressive strength in the compression process. The differences in microstructure characteristic resulted in an uneven stress distribution in the specimen, causing the overlap zone to simultaneously suffer shear and press stress. Therefore, the compressive strength is larger. During the compressive process, the failure form of the specimen is crushed, not jumping up (plastic deformation), and the fracture morphology is an inclined plane [32,34].

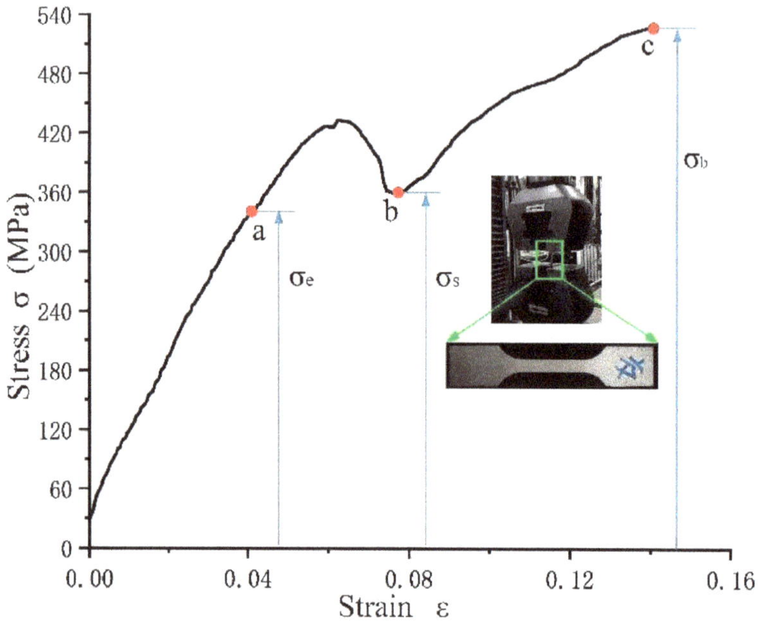

Figure 18. Tensile strength of scanning strategy 3 specimen.

4. Conclusions

The YCF104 (Fe-based) alloy specimens were successfully fabricated by laser cladding on #45 steel using three different scanning strategies. The effect of scanning strategies on the microstructure and mechanical properties of YCF104 alloy layer were systematically investigated. The following results were obtained:

(1) With the increase of remelting area produced by cladding paths and the improvement of uniformity of temperature distribution, the microstructure becomes more uniform and the grains become more refined. Moreover, the compressive strength and wear resistance are improved significantly, and the porosity and cracks are reduced.

(2) The Fe–Cr/Mo solid solution strengthening effect led to improvement in the strength of the layers. The grain sizes of specimens fabricated under scanning strategies 1, 2, and 3 decreased in turn, and the homogeneity of the structure increased. The microstructure in the cross-section parallel to the scanning direction was more uniform than along the *z*-axis direction.

(3) Scanning strategies were closely related to the microstructure. The dense microstructure led to the increase in the number of boundaries that needed to be destroyed in the wear process, reducing the size of the abrasion mark. Compared to *z*-axis direction, the friction coefficient parallel to the scanning direction (scanning strategies 1, 2, and 3) reduced by 5.7%, 0.1%, and 10.1%, respectively. However, the microhardness of scanning strategies 1, 2, and 3 decreased in turn, and the minimum microhardness of YCF104 was 633.0 $HV_{0.5}$ among specimens prepared under different scanning strategies.

(4) Laser-cladded YCF104 material belongs to the brittle materials, and with the increase of the uniformity of laser remelting, the compressive strength improved, reaching 3235 MPa. Under the action of external normal stress, the fracture of the interatomic bond caused brittle transgranular fracture along a specific crystal plane, and the cleavage fracture is the main failure of the YCF104 alloy. The tensile strength of the specimen under scanning strategy 3 is 527.44 MPa, the elastic limit is 337.5 MPa, and the yield limit is 358.71 MPa.

Author Contributions: Y.Z.: experimentation, data curation, methodology, resources, and writing the original draft. W.S.: data collection, and manuscript revision. L.C.: experimentation, and data curation, W.W.: supervision, conceptualization, and methodology. T.Y.: conceptualization, methodology, and resources. All authors have read and agreed to the published version of the manuscript.

Funding: This work was supported by the Special Financial Aid to China Postdoctoral Science Foundation (No.2022TQ0116), the China Postdoctoral Science Foundation (No. 2022M721315), the General Program of Chongqing Natural Science Foundation (No. CSTB2022NSCQ-MSX1623), and the Ministry of Industry and Information Technology of China (No. 201675514).

Institutional Review Board Statement: The article follows the guidelines of the Committee on Publication Ethics (COPE) and involves no studies on human or animal subjects.

Informed Consent Statement: Not applicable.

Data Availability Statement: All data generated or analyzed during this study are included in the present article.

Conflicts of Interest: The authors declare no competing interests.

References

1. Sexton, L.; Lavin, S.; Byrne, G.; Kennedy, A. Laser cladding of aerospace materials. *J. Mater. Process. Technol.* **2022**, *122*, 63–68. [CrossRef]
2. Li, K.; Li, D.; Liu, D.; Pei, G.; Sun, L. Microstructure evolution and mechanical properties of multiple-layer laser cladding coating of 308 L stainless steel. *Appl. Surf. Sci.* **2015**, *340*, 143–150. [CrossRef]
3. Liu, Y.; Ding, Y.; Yang, L.; Sun, R.; Zhang, T.; Yang, X. Research and progress of laser cladding on engineering alloys: A review. *J. Manuf. Process.* **2021**, *66*, 341–363. [CrossRef]
4. Yin, Y.; Pan, C.; Zhang, R.; Zhao, C.; Qu, Y. The effect of Ti addition on the microstructure and properties of high chromium iron-based coatings. *J. Alloys Compd.* **2018**, *765*, 782–790. [CrossRef]
5. Shanmugam, K.; Lakshminarayanan, A.K.; Balasubramanian, V. Tensile and Impact Properties of Shielded Metal Arc Welded AISI 409M Ferritic Stainless Steel Joints. *J. Mater. Sci. Technol.* **2009**, *25*, 181–186.
6. Zhang, M.; Luo, S.X.; Liu, S.S.; Wang, X.H. Effect of Molybdenum on the Wear Properties of (Ti,Mo)C-TiB_2-Mo_2B Particles Reinforced Fe-Based Laser Cladding Composite Coatings. *J. Tribol.* **2018**, *140*, 051603. [CrossRef]
7. Qu, K.L.; Wang, X.H.; Wang, Z.K.; Niu, W.Y. Effect of Mo on the VC–VB particles reinforced Fe-based composite coatings. *Mater. Sci. Technol.* **2017**, *33*, 333–339. [CrossRef]
8. Yang, J.; Miao, X.; Wang, X.; Yang, F. Influence of Mn additions on the microstructure and magnetic properties of FeNiCr/60% WC composite coating produced by laser cladding. *Int. J. Refract. Met. Hard Mater.* **2014**, *46*, 58–64. [CrossRef]
9. Yang, J.; Huang, J.; Fan, D.; Chen, S. Microstructure and wear properties of Fe-6 wt.% Cr-0.55wt.% C-Xwt.%Nb laser cladding coating and the mechanism analysis. *Mater. Des.* **2015**, *88*, 1031–1041. [CrossRef]
10. Tan, H.; Luo, Z.; Li, Y.; Yan, F.; Duan, R.; Huang, Y. Effect of strengthening particles on the dry sliding wear behavior of Al_2O_3-M_7C_3/Fe metal matrix composite coatings produced by laser cladding. *Wear* **2015**, *324–325*, 36–44. [CrossRef]
11. Zhang, H.; Chong, K.; Zhao, W.; Sun, Z. Effects of pulse parameters on in-situ Ti-V carbides size and properties of Fe-based laser cladding layers. *Surf. Coat. Technol.* **2018**, *344*, 163–169. [CrossRef]
12. Zhang, H.; Zou, Y.; Zou, Z.; Wu, D. Microstructures and properties of low-chromium high corrosion-resistant TiC-VC reinforced Fe-based laser cladding layer. *J. Alloys Compd.* **2015**, *622*, 62–68. [CrossRef]
13. Wang, K.; Chang, B.; Chen, J.; Fu, H.; Lin, Y.; Lei, Y. Effect of Molybdenum on the Microstructures and Properties of Stainless Steel Coatings by Laser Cladding. *Appl. Sci.* **2017**, *7*, 1065. [CrossRef]
14. Zhao, G.L.; Zou, Y.; Zou, Z.D.; Zhang, H. Research on in situ synthesised (Ti,V)C/Fe composite coating by laser cladding. *Mater. Sci. Technol.* **2015**, *31*, 1329–1334. [CrossRef]
15. Wang, Q.T.; Zeng, X.B.; Chen, C.R.; Lian, G.F.; Huang, X. An Integrated Method for multi-objective optimization of multi-pass Fe50/TiC laser cladding on AISI 1045 steel based on grey relational analysis and principal component analysis. *Coatings* **2020**, *10*, 151. [CrossRef]
16. Zhou, S.; Zhang, T.; Xiong, Z.; Dai, X.; Wu, C.; Shao, Z. Investigation of Cu-Fe-based coating produced on copper alloy substrate by laser induction hybrid rapid cladding. *Opt. Laser Technol.* **2014**, *59*, 131–136. [CrossRef]
17. Zhou, S.; Dai, X. Microstructure evolution of Fe-based WC composite coating prepared by laser induction hybrid rapid cladding. *Appl. Surf. Sci.* **2010**, *256*, 7395–7399. [CrossRef]
18. Zhou, H.; Zhang, C.; Wang, W.; Yasir, M.; Liu, L. Microstructure and mechanical properties of fe-based amorphous composite coatings reinforced by stainless steel powders. *J. Mater. Sci. Technol.* **2015**, *31*, 43–47. [CrossRef]
19. Li, R.; Yuan, T.; Qiu, Z.; Zhou, K.; Li, J. Nanostructured Co-Cr-Fe alloy surface layer fabricated by combination of laser clad and friction stir processing. *Surf. Coat. Technol.* **2014**, *258*, 415–425. [CrossRef]

20. Wang, W.J.; Fu, Z.K.; Cao, X.; Guo, J.; Liu, Q.Y.; Zhu, M.H. The role of lanthanum oxide on wear and contact fatigue damage resistance of laser cladding Fe-based alloy coating under oil lubrication condition. *Tribol. Int.* **2016**, *94*, 470–478. [CrossRef]
21. Karczewski, K.; Pe, M.; Zie, M.; Pola, M. Fe-Al thin walls manufactured by Laser Engineered Net Shaping. *J. Alloys Compd.* **2017**, *696*, 1105–1112. [CrossRef]
22. Fang, J.X.; Dong, S.Y.; Wang, Y.J.; Xu, B.S.; Zhang, Z.H.; Xia, D.; He, P. The effects of solid-state phase transformation upon stress evolution in laser metal powder deposition. *Mater. Des.* **2015**, *87*, 807–814. [CrossRef]
23. Zhao, Y.; Yu, T.; Wang, Z.; Chen, H. Calculation and verification of Start/Stop optimum overlapping rate on metal DLF technology. *Int. J. Adv. Manuf. Technol.* **2018**, *99*, 437–452. [CrossRef]
24. Yu, T.; Sun, J.; Qu, W.; Zhao, Y.; Yang, L. Influences of z-axis increment and analyses of defects of AISI 316L stainless steel hollow thin-walled cylinder. *Int. J. Adv. Manuf. Technol.* **2018**, *97*, 2203–2220. [CrossRef]
25. Yu, T.; Yang, L.; Zhao, Y.; Sun, J.; Li, B. Experimental research and multi-response multi-parameter optimization of laser cladding Fe313. *Opt. Laser Technol.* **2018**, *108*, 321–332. [CrossRef]
26. Yang, L.; Yu, T.; Li, M.; Zhao, Y.; Sun, J. Microstructure and wear resistance of in-situ synthesized Ti (C, N) ceramic reinforced Fe-based coating by laser cladding. *Ceram. Int.* **2018**, *44*, 22538–22548. [CrossRef]
27. Zhao, Y.; Yu, T.; Sun, J.; Xi, W.; Bi, X. Effect of laser cladding on forming qualities of YCF101 alloy powder in the different lap joint modes. *Int. J. Adv. Manuf. Technol.* **2018**, *96*, 1991–2001. [CrossRef]
28. Zhang, Z.; Jing, H.; Xu, L.; Han, Y.; Zhao, L.; Zhou, C. Effects of nitrogen in shielding gas on microstructure evolution and localized corrosion behavior of duplex stainless steel welding joint. *Appl. Surf. Sci.* **2017**, *404*, 110–128. [CrossRef]
29. Yadaiah, N.; Bag, S.; Paul, C.P.; Kukreja, L.M. Influence of self-protective atmosphere in fiber laser welding of austenitic stainless steel. *Int. J. Adv. Manuf. Technol.* **2016**, *86*, 853–870. [CrossRef]
30. Birger, E.M.; Moskvitin, G.V.; Polyakov, A.N.; Arkhipov, V.E. Industrial laser cladding: Current state and future. *Weld. Int.* **2011**, *25*, 234–243. [CrossRef]
31. Dadbakhsh, S.; Hao, L.; Sewell, N. Effect of selective laser melting layout on the quality of stainless steel parts. *Rapid Prototyp. J.* **2012**, *18*, 241–249. [CrossRef]
32. Joseph, J.; Stanford, N.; Hodgson, P.; Fabijanic, D.M. Tension/compression asymmetry in additive manufactured face centered cubic high entropy alloy. *Scr. Mater.* **2017**, *129*, 30–34. [CrossRef]
33. Zhao, X.; Lin, X.; Chen, J.; Xue, L.; Huang, W. The effect of hot isostatic pressing on crack healing, microstructure, mechanical properties of Rene88DT superalloy prepared by laser solid forming. *Mater. Sci. Eng. A* **2009**, *504*, 129–134. [CrossRef]
34. Guduru, R.K.; Darling, K.A.; Kishore, R.; Scattergood, R.O.; Koch, C.C.; Murty, K.L. Evaluation of mechanical properties using shear-punch testing. *Mater. Sci. Eng. A* **2005**, *395*, 307–314. [CrossRef]

coatings

Article

Influence of the Relative Position of Powder–Gas Jet and Laser Beam on the Surface Properties of Inconel 625 Coatings Produced by Extreme High-Speed Laser Material Deposition (EHLA)

Matthias Brucki [1,2,*], Tobias Schmickler [1], Andres Gasser [1] and Constantin Leon Häfner [1,2]

1 Fraunhofer ILT—Institute for Laser Technology, Steinbachstraße 15, D-52074 Aachen, Germany
2 Chair for Laser Technology, RWTH Aachen University LLT, Steinbachstraße 15, D-52074 Aachen, Germany
* Correspondence: matthias.brucki@ilt.fraunhofer.de; Tel.: +49-241-890-6314; Fax: +49-241-8906-121-10

Abstract: Laser material deposition (LMD) is a widely used coating process in industry. However, to increase its economic appeal, higher process speeds are required. The solution to this challenge is an innovative modification known as extreme high-speed laser material deposition (EHLA). EHLA allows for an impressive increase in process speed from 2 m/min for conventional LMD to 500 m/min. With the ability to adjust process parameters, EHLA can generate tailor-made surface properties, expanding its potential application beyond current industrial uses. In this novel study, we explore the effects of relative positioning between tools (laser beam and powder–gas jet) and substrate on the surface properties of EHLA coatings. By laterally and axially offsetting the tools, the proportional energy coupling of the laser radiation into the powder–gas jet and substrate can be modified. Altering the position of the powder–gas jet can also affect the weld pool flow or number of particle attachments, thereby affecting surface properties. This approach allows for the adjustment of surface roughness over a wide range—from smooth, quasi-laser-polished surfaces to rough surfaces covered with particle adhesions.

Keywords: extreme high-speed laser material deposition; EHLA; laser material deposition; LMD; surface properties; roughness; relative position; axial offset; lateral offset; direct energy deposition; DED; laser cladding

1. Introduction

The coating of components, e.g., against wear or corrosion, has become an indispensable part of industry and technology. In addition to the reduction of costs through the construction of components from low-cost base materials with high-end coatings for the extension of service life, the functionality of components is often determined by their geometrical surface properties.

Laser material deposition (LMD), also known as direct energy deposition (DED), is a well-established coating process used for maintenance, wear, and corrosion protection in areas such as tools, engines, or mechanical engineering [1]. In this coating process (Figure 1a), a laser beam (LB) focused on the surface generates a melt pool at the edge of the component, into which the metallic filler material (powder) is introduced and liquefies. By moving the component relative to the laser beam and powder–gas jet (PGJ), the weld pool, including the filler material, leaves the laser spot and solidifies, creating a metallurgical and thus firmly bonded layer. The disadvantage of this process is the process speed of typically 0.2 m/min to 2 m/min [2], which is too low for many applications.

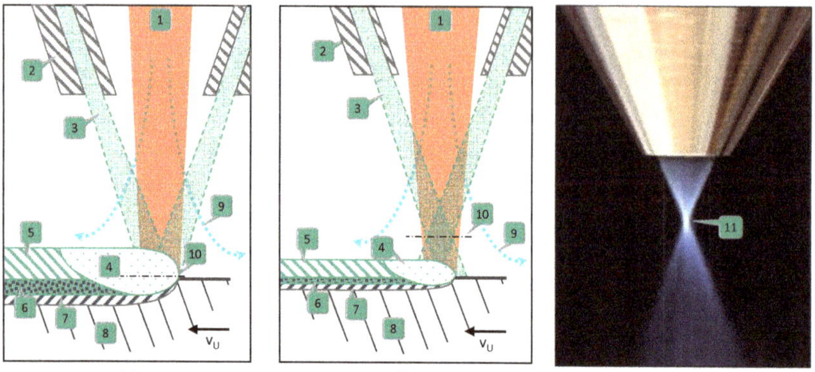

(**a**) (**b**) (**c**)

Figure 1. Schematic representation of the principles of (**a**): conventional LMD; (**b**): EHLA; (**c**): picture of the tip of the powder feed nozzle and the powder–gas jet for the EHLA process [3]. 1: laser beam; 2: powder nozzle; 3: powder–gas jet; 4: weld pool; 5: coating layer; 6: dilution zone; 7: heat-affected zone; 8: substrate; 9: shielding gas; 10: level of powder–gas jet focus; 11: focus area of the powder–gas jet; v_U: peripheral speed.

To increase the process speed and thus productivity and economic efficiency, LMD was developed for extreme high-speed laser material deposition (German acronym: EHLA, Figure 1b). Here, the powder focus and preferably also the laser focus are located above the component surface, so that the powder is already strongly heated—preferably even completely melted—above the component surface. In this way, powder material reaches the weld pool as liquid droplets [2]. Since the particles are already melted on the path to the surface, the time needed to melt the particles in the weld pool is eliminated [2], and the process speed can be increased to up to 500 m per minute [4].

An important aspect of the component functionality is the geometric shape of the surface, which defines the surface condition. The shape of the surface results from the deviations of the actual surface from the geometric (ideal, technical drawing) surface due to manufacturing and random factors [5]. The actual surface can be divided into coarse and fine shapes, with waviness (structures with relatively long wavelengths) and roughness (structures with relatively small wavelengths) attributed to the fine shape surface [6]. Coatings manufactured by high-speed LMD reach a surface roughness of approx. Rz = 50–100 μm, Ra = 11.3–20.5 μm [7–9]. In comparison to this, coatings manufactured by conventional LMD reach a surface roughness of approx. Ra = 13.8–31.7 μm (Sa = 15–105 μm) [10–12]. The material and coating thickness differ in the mentioned studies. Since different applications demand different surface shape requirements and postprocessing such as grinding or turning is cost-intensive, targeted processing of surface conditions is desirable. This article presents an approach to shaping the surface condition by axial and lateral offset of the laser beam, powder–gas jet, and component surface with the EHLA process. Thus, surfaces ranging from rough to smooth and from porous to dense can be generated.

2. Materials and Methods

2.1. Materials

2.1.1. Filler Material

For this study, the nickel-based superalloy Inconel 625 (German material number 2.4856) is used as filler material in the form of Oerlikon Metco's MetcoClad 625F powder. Its chemical composition is given in Table 1.

Table 1. Chemical composition of the filler material [max. mass%].

Ni	Cr	Mo	Nb	Fe	rest
58.0–63.0	20.0–23.0	8.0–10.0	3.0–5.0	≤5.0	<2.0

In Figure 2, a scanning electron microscope (SEM, Leo 1455 EP, Carl Zeiss AG, Jena, Germany) shows images of the powder. The particles have a predominantly nearly spherical shape and, in some cases, small satellites. Both are characteristic of gas-atomized powders such as MetcoClad 625F. Moreover, 53.2% of the particles have a sphericity of SPHT = 0.9, and the average sphericity is SPHT = 0.892. Sphericity indicates how closely the shape of a particle corresponds to a sphere, where SPHT = 1 is equivalent to a perfect sphere.

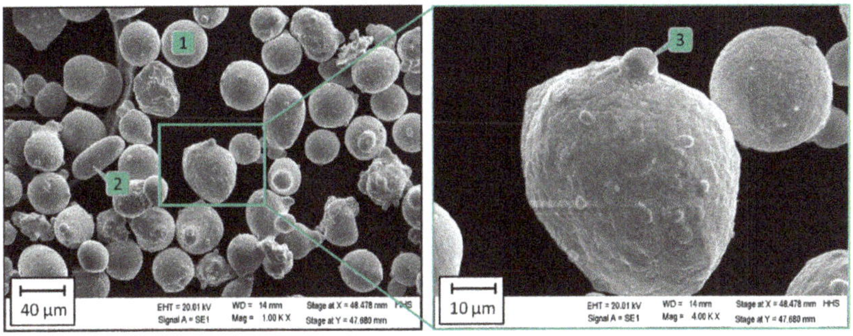

Figure 2. Scanning electron microscope (SEM) images of the powder. 1: nearly spherical particle; 2: aspherical particle; 3: satellite.

The nominal particle size distribution is specified by the manufacturer as −53 + 20 µm. The measured particle size distribution is shown in Figure 3. In terms of volume, 99.7%, and in terms of number, 98.5% of the particles have a particle diameter x_{area} in the range of the manufacturer's specifications. The fine fraction (x_{area} < 20 µm) is 0% by volume and 1.4% by number. The coarse fraction (x_{area} > 53%) is 0.3% by volume and 0.1% by number.

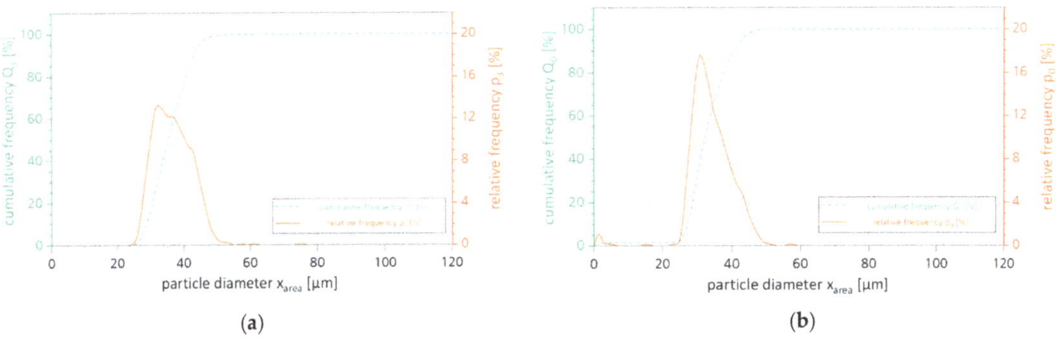

Figure 3. Particle size distribution: cumulative Q and relative p frequency against the particle diameter x_{area}. (**a**): volume-weighted; (**b**): number-weighted.

2.1.2. Substrate

Cold-drawn, seamless precision steel tubes with an outside diameter of 100 mm, a wall thickness of 8 mm, and a length of 400 mm are used as substrates. The base material is the unalloyed structural steel S355J0 + AR (German material number 1.0553, chemical composition: Table 2), which is mainly used in mechanical engineering [13]. Before the experiments, the substrates are turned at a feed rate of 0.2 mm per revolution and cleaned with ethanol immediately before the coating process.

Table 2. Chemical composition of the substrate material [max. mass%].

C	Si	Mn	P	S
≤0.22	≤0.55	≤1.60	≤0.025	≤0.04

2.2. *Experimental Setup and Tools*

2.2.1. Experimental Setup

The experiments are carried out at a facility specially designed for the EHLA process by Hornet Laser Cladding BV, Netherlands. It contains a powder conveying system with a powder feed nozzle, a laser beam source including optics, as well as a work cell including rotary handling, as shown schematically in Figure 4.

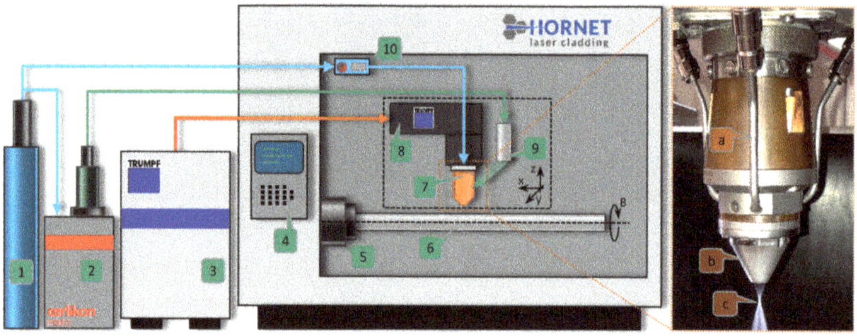

Figure 4. Schematic representation of the EHLA machine. 1: conveying and shielding gas supply (argon); 2: powder feeder; 3: laser beam source; 4: CNC control panel; 5: lathe; 6: substrate; 7: powder feed nozzle; 8: optics; 9: powder distributor; 10: gas mass flow meter; a: powder supply pipe; b: nozzle tip; c: powder–gas jet (focus area).

The Oerlikon Metco powder feeder Twin-150 (dosing disc type: NL 5.0/1.0) conveys the powder as a powder–gas mixture (conveying gas: argon) via a powder distributor (by HD Sonderoptiken, Übach-Palenberg, Germany), which divides the powder–gas mixture into three equal streams, to the nozzle. This coaxial powder feed nozzle, HighNo 40 from HD Sonderoptiken, is specially designed for the EHLA process and is positioned at a standard distance of 10 mm from the component surface to the nozzle tip. Inside the nozzle, the powder–gas mixture is fed through the nozzle coaxially to the laser beam and focused as a cone over the substrate, where it is melted by the laser beam. The laser beam as well as the shielding gas (argon, for oxygen shielding of the process area and for protection of the nozzle) are guided through the nozzle centrally along the z-axis. The adjustment of the shielding gas volume flow is carried out with the gas mass flow meter of the Redy Compact 2 series from Vögtlin Instruments GmbH, Muttenz, Switzerland.

As a laser beam source, the disc laser TruDisk4002 from Trumpf GmbH, Ditzingen, Germany, with a wavelength of 1030 nm, is used. The nominal output power in continuous wave operation is 80 W up to a maximum of 4000 W, with an output power stability of ±1%. The laser radiation is coupled into the processing optics via a fiberoptic cable LLK-D 06 from Trumpf GmbH, Ditzingen, Germany, with a diameter of 600 µm and a numerical aperture of 0.125.

The BEO D70 collimating optics from Trumpf GmbH enables automated positioning of the laser focal level in the z-direction, thus the axial offset of the laser beam with respect to standard EHLA settings can be realized.

Together with the powder feed nozzle, the optics form a unit that is automatically moved in the x- and z-directions relative to the substrate surface during the coating process. For the axial offset of the powder–gas jet, this unit is shifted in the z-direction, whereby the laser focus position is corrected by the collimator. For the lateral offset between the laser beam and the powder–gas jet, the nozzle is adjusted accordingly via (x-y) set screws.

2.2.2. Laser Beam Caustic

To determine the laser beam caustics, the FocusMonitor FM+ (Primes GmbH, Pfungstadt, Germany) with the detector DFY-PS+ (photodiode) and the measuring tip NIR high div (pinhole diameter 20 μm; sensitivity 1560 cts/(MW/cm^2)), as well as the software LaserDiagnosticsSoftware (version 1.4.3) from Primes, is used. The laser beam is measured at a power of P_L = 1800 W and for the axial offsets of the laser beam of Δ_{LB} = −20 mm, Δ_{LB} = 0 mm and Δ_{LB} = +20 mm. For each measurement, the intensity distribution is recorded on 31 planes in a range of about ±25 mm around the laser focus position (Figure 5).

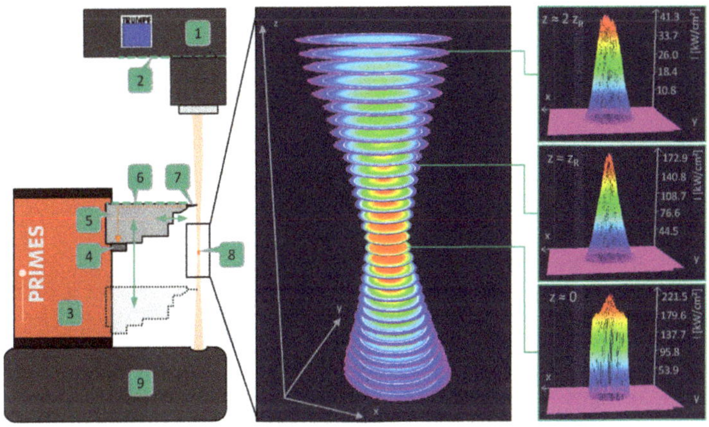

Figure 5. Analysis of laser beam caustic. left: Schematic representation of the caustic measurement device; middle: laser power density distribution at 31 levels around the focal area; right: laser power density distribution near the focal plane, the Rayleigh length, and twice the Rayleigh length 1: optics; 2: reference edge; 3: FocusMonitor; 4: detector; 5: vertical and horizontal slide; 6: reference edge (FocusMonitor); 7: measuring tip; 8: laser focus range; 9: beam trap.

The results of the caustic measurement are summarized in Table 3. At standard settings (no axial offset of the laser beam), the laser focus is on the substrate surface and has a diameter of about 1.147 mm. With an axial offset of the laser focus position by means of the collimator, the diameter at the beam waist and the Rayleigh length become smaller from negative offset to positive offset, and the divergence angle becomes larger. The diffraction measure M^2 and the beam parameter product BPP remain almost constant.

Table 3. Laser beam parameters.

Axial Offset of the Laser Beam Δ_{LB} [mm]	Radius at Beam Waist w_0 [μm]	Rayleigh Length z_R [mm]	Divergence Angle Θ [mrad]	Diffraction Measure M^2	Beam Parameter Product BPP [mm × mrad]
−20	596	13.54	88.05	80.07	26.25
0	573	12.53	91.5	80.03	26.24
+20	548	11.48	9557	79.97	26.22

In the area of focus position, the intensity distribution has a top-hat-like profile. This means that the intensity within the beam radius is relatively large and drops abruptly at the beam radius. However, since the intensity distribution within the beam radius is not homogeneously distributed but rather an intensity peak of about 30% in terms of power at the beam radius is present in the center, the distribution deviates slightly from an ideal top-hat distribution.

Along the optical axis, the power density distribution changes with increasing distance to the focus position into a Gaussian-like pattern (Figures A1 and 5).

2.2.3. Powder–Gas Jet Analysis

For powder–gas jet analysis, the powder jet monitor developed and patented at Fraunhofer ILT (Pat. No.: DE 10 2011 009 345 B3) is used (Figure 6). Thereby, the PGJ is illuminated by a line laser on different planes and viewed by a high-speed camera directed through the powder nozzle, so that the particle density distribution per plane can be determined. For the measurement of the PGJ (powder mass flow rate $\dot{m}$ = 9.6 g/min, conveying gas volume flow rare $\dot{V}_F$ = 6 L/min, protective gas volume flow rare $\dot{V}_S$ = 10 L/min), 1000 images per plane are taken, and superimposed on 14 planes (distances of the planes: see Table 4). From these, the powder density distribution as well as the diameters per plane can be determined, and the spatial position of the PGJ focus with respect to the nozzle tip as well as the propagation of the PGJ can be derived. The results are summarized in Table 4.

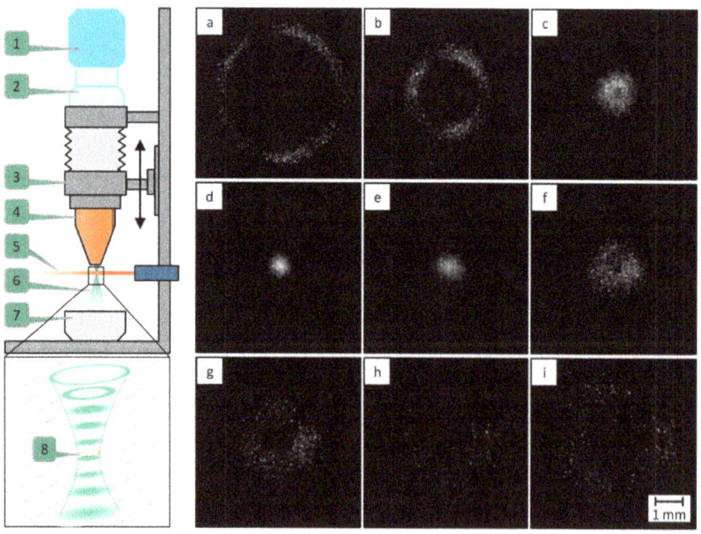

Figure 6. Powder–gas jet analysis. Left: schematic representation of the powder jet monitor; right: 1000 superimposed images, each of powder–gas jet measurements taken from (**a**–**i**) at a distance of 4, 6, 8, 9, 10, 11, 12, 13, 14 mm from the nozzle tip; 1: high-speed camera; 2: focusing optics; 3: nozzle mount and linear axle; 4: powder feed nozzle; 5: illumination laser; 6: powder–gas jet; 7: powder collection container; 8: calculated powder–gas jet focus level.

The smallest measured diameter is on the measuring plane at 9 mm, but the calculated focus position is located at a distance of 9.06 mm with respect to the nozzle tip, which corresponds to a distance of the focus to the substrate surface at the standard setting (no axial offset of the PGJ) of about 0.94 mm. At this position, the PGJ has a calculated focus diameter of 1.162 mm. Above the focus, the PGJ has the shape of a hollow cone, so the particle density distribution is annular. Towards the focus position, the powder–gas jet narrows to the area of the PGJ focus and widens below the focus position to a jet with a particle density distribution that is no longer annular.

Table 4. Powder–gas jet parameters.

	Measuring Plane (Distance to Nozzle Tip) [mm]	Corresponding Δ_{PGJ} [mm], Where the Measuring Plane Would Be on the Substrate Surface	Diameter [mm]
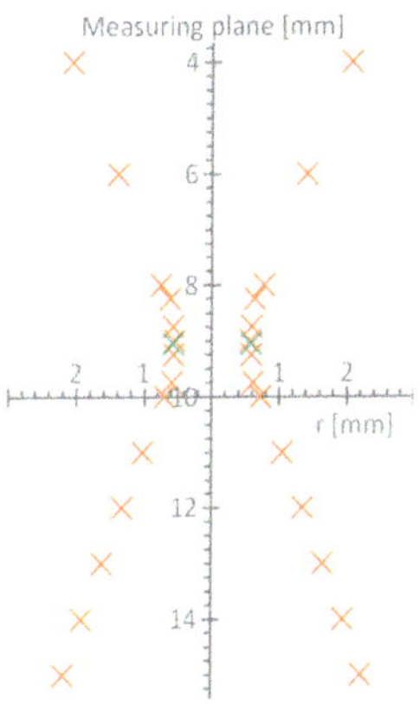	4	−6	4.124
	6	−4	2.795
	8	−2	1.528
	8.5	−1.5	1.259
	8.75	−1.25	1.167
	9	−1	1.126
	9.25	−0.75	1.149
	9.5	−0.5	1.203
	10	0	1.434
	11	+1	2.070
	12	+2	2.678
	13	+3	3.279
	14	+4	3.882
	15	+5	4.409
Calculated focal plane	9.06	−0.94	1.162

2.3. Methods

2.3.1. Experimental Approach

The investigation of the influence of the relative positioning of the laser beam, powder–gas jet, and substrate on the surface properties is carried out on the one hand on 12 mm-wide coatings and on the other hand on single tracks. To generate the coatings, several single tracks are placed with an overlap that results from an axial feed rate of the substrate relative to the processing head of $v_f = 0.2$ mm/rev. For the single tracks (without overlap), helices with four turns and a pitch of 2 mm ($v_f = 2$ mm/rev) are applied. The selected settings of the process parameters laser power P_L, powder mass flow rate $\dot{m}$, peripheral speed v_U of the substrate surface relative to the processing head, feed rate v_f, conveying gas volume flow rate $\dot{V}_F$, and protective gas volume flow $\dot{V}_S$ rate can be seen in Table 5. With these settings, the lateral and axial offsets of LB and PGJ are varied in succession.

Table 5. Parameter settings.

Process Parameter	Unit	Setting
Laser power P_L	W	1800
Powder mass flow rate $\dot{m}$	g/min	9.6
Peripheral speed v_U	m/min	50
Feed rate v_f	mm/rev	0.2 (coating) 2 (single track)
Conveying gas volume flow rare $\dot{V}_F$	L/min	6
Protective gas volume flow rare $\dot{V}_S$	L/min	10
Axial offset of the laser beam Δ_{LB}	mm	−20–+20
Axial offset of the powder–gas jet Δ_{PGJ}	mm	−1–+4
Lateral offset of the powder–gas jet φ_{PGJ} (angle)	°	0–315
r_{PGJ} (distance)	mm	0.6

Figure 7 shows schematically the lateral and axial offsets of the LB and PGJ. The axial offset of the laser beam Δ_{LB} indicates the distance in the z-direction (axial distance) of the laser focus from the substrate surface. At Δ_{LB} = 0 mm, the LB focus is on the substrate surface. A positive axial offset means that the laser focus is above the substrate surface, i.e., it is moved towards the powder feed nozzle, whereas a negative axial offset means that the laser focus is below the substrate surface. During preliminary investigations, it was possible to determine the range from $\Delta_{LB} = -20$ mm to $\Delta_{LB} = +20$ mm in which coatings or parts of them are completely metallurgically bonded to the substrate. The axial offset of the laser focus position affects, on the one hand, the interaction between the laser beam and the powder–gas jet, so that both the interaction time and distance, as well as the experienced intensity distribution of each powder particle, and thus the absorbed energy, are changed. On the other hand, the laser spot on the substrate surface changes both in size and intensity distribution, which in turn affects the heating of the substrate and the weld pool.

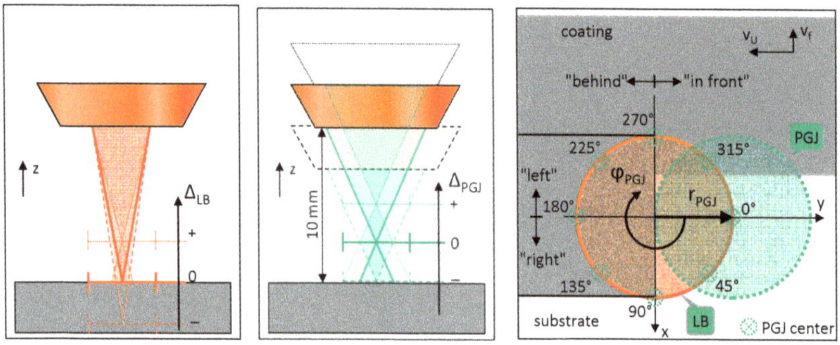

(**a**) (**b**) (**c**)

Figure 7. Schematic representation of the parameter settings of the offsets: (**a**) axial offset of the laser beam; (**b**) axial offset of the powder–gas jet (by adjusting the distance between nozzle tip and component surface); (**c**) lateral offset of the powder–gas jet relative to the laser beam.

The axial offset of the powder–gas jet Δ_{PGJ} is set via the nozzle tip distance to the substrate surface, whereby the laser focus position is corrected in each case. By default (Δ_{PGJ} = 0 mm), a nozzle distance of 10 mm is selected, which corresponds to a distance of the PGJ focus to the substrate surface of 0.94 mm. A positive axial offset of the PGJ means that the distance between the focus and substrate surface is increased, and vice versa. The axial offset is varied in 1 mm increments, with the limit set at $\Delta_{PGJ} = -1$ mm for negative offset. This corresponds practically to conventional LMD, so that a further negative offset is not considered. The upper limit in the positive range is chosen at $\Delta_{PGJ} = +4$ mm since preliminary investigations have shown that at larger offsets, it is no longer possible to ensure working coatings with the parameter settings used here. The axial offset Δ_{PGJ} affects the interaction distance and time between the laser beam and the powder particles, as well as the intensity distribution experienced by the particles. Furthermore, the PGJ spot on the substrate surface is influenced in terms of size and particle distribution.

In the case of a lateral offset between the LB and the PGJ, the PGJ is displaced laterally with respect to the optical axis of the LB. The offset is given by the polar coordinates φ_{PGJ} (angle with respect to v_U) and r_{PGJ} (distance of the axis of the PGJ to the optical axis of the LB). For the investigations concerning the lateral offset, r_{PGJ} = 0.6 mm is chosen. A larger displacement of the PGJ is not possible due to the nozzle geometry, since the laser beam could collide with the inner nozzle wall and damage it. The angle φ_{PGJ} is varied from $\varphi_{PGJ} = 0°$ to $\varphi_{PGJ} = 315°$ in 45° steps. Due to the lateral offset, the powder particles no longer pass through the laser beam radially symmetrically. As a result, the interaction distance and time of the particles within the PGJ differ. This affects the experienced intensity distribution as well as the heating of the individual particles of the PGJ and the substrate.

Furthermore, the lateral offset can influence the overspray so that, e.g., the number of particle adhesions can be increased. On the other hand, a kind of laser polishing effect is expected from the remelting of already applied filler material without the supply of further filler material. For the later description of the results, locations on the substrate surface with respect to the laser spot are labeled "behind", "in front", "on the left", and "on the right", as in Figure 7c.

2.3.2. Analysis of Surface Condition

The surface condition of the coatings is analyzed and evaluated with respect to the parameters "mean arithmetic height" Sa, "maximum height" Sz, and "developed transition area ratio" Sdr according to DIN EN ISO 25178-2. Furthermore, the number of particle adhesions per mm^2 is determined. Sa is calculated from the arithmetic mean of the absolute ordinate values within a defined range [14]. Sz is calculated from the sum of the largest profile peak and the largest slump height within a defined area [14]. Sdr describes the ratio of the increase in the transition area of a scale-limited surface within the definition range over the defined area [14]. Thus, Sdr indicates the percentage of the additional area of the definition area that is due to texture compared to the absolutely flat definition area [15].

To determine the surface parameters, the surfaces of the coatings are imaged using the Zygo Nexview NX2 white light interferometer (WLI) from Ametek (Berwyn, PA, USA). For this purpose, a 6.31 mm × 6.31 mm measuring field is first imaged for each coating using a 1.4× objective. Subsequently, images are taken at three different points in this measurement field with a 5.5× objective (Figure 8). These WLI files are prepared with the MX software (Zygo Corporation, Middlefield, CT, USA) for subsequent evaluation (for the preparation settings, see Table A1). From the 5.5× WLI images, the characteristics Sa, Sz, and Sdr are determined with MX.

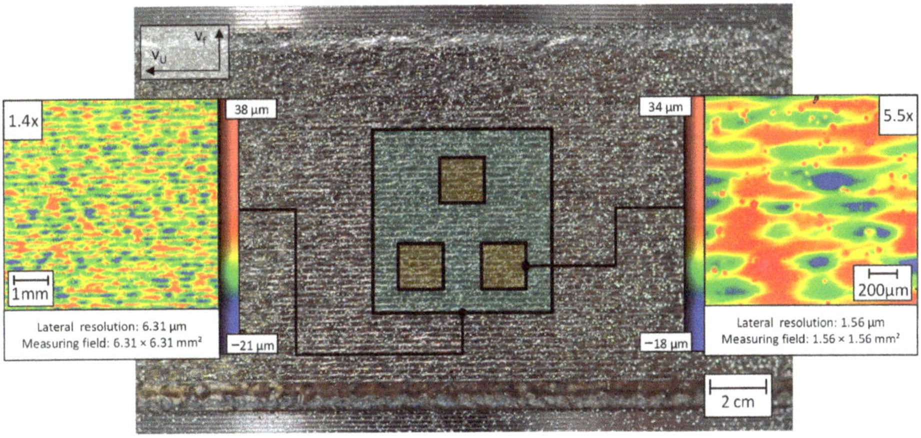

Figure 8. Schematic representation of the WLI imaging strategy. **Middle**: EHLA—coating, **left**: WLI—height profile image with 1.4× magnification, **right**: WLI—height profile image with 5.5× magnification.

To determine the number of particle adhesions on coatings, the 1.4× WLI images are prepared according to Table A1. With the software Gwyddion and the integrated function "Automated Threshold Using Otsu's Method on Heights", the particles are detected (orange dotes in Figure 9a,b), and their number can be determined (Figure 9a). This method reaches its limits with a large particle density on the surface if individual particles can no longer be resolved separately from each other. If individual particles are deposited close together, these particles can also form agglomerates, which are interpreted and counted as a single particle (Figure 9b, 2). In these cases, the determined particle number decreases. Another limit exists for surfaces with predominantly long-wave (>>grain fraction) structures and comparatively few particle adhesions. Filtering reduces the heights of particles so that the filtered surfaces are very flat and the detection of individual particles with Otsu's method is no longer possible (Figure 9c).

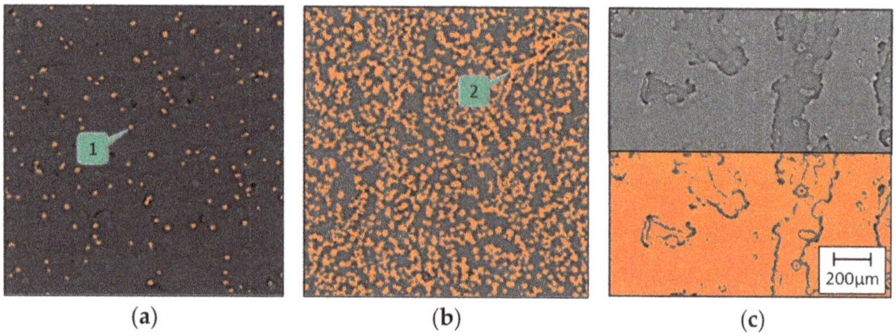

Figure 9. Detection of the particle adhesions on the coatings: (**a**) at standard position; (**b**) at Δ_{PGJ} = 2 mm; (**c**) at Δ_{LB} = 20 mm (top: without Otsu's method, bottom: with Otsu's method); 1: detected particle; 2: agglomerate of particles.

In addition, a distribution in the x-direction (Figure 10) of adhering particles, agglomerates, as well as any structures of the track edges is determined. Since a helix with 4 turns and a pitch of 2 mm was generated for each parameter setting, and the distance between two neighboring single tracks is sometimes so small that the particles cannot be assigned to the corresponding single track, the outer areas with a size of 2 mm × 2 mm of the outer single tracks are considered in each case and then merged (Figure 10). For the determination of the distribution, the 1.4× WIM images are prepared according to Table A1. By means of Gwyddion and the integrated function "Automated Threshold Using Otsu's Method on Heights", the structures are detected, and their distribution is given out in 0.05 mm intervals.

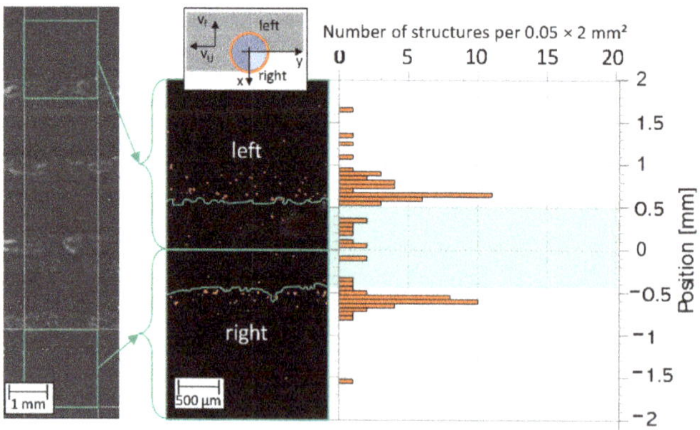

Figure 10. Schematic representation of the determination of the distribution of adhering structures for single tracks. Left: WLI file; middle: merged areas of the outer tracks, including marking of the detected adhesions and of the track edges; right: distribution of adhering structures (standard settings: Δ_{LB} = 0 mm, Δ_{PGJ} = 0 mm, r_{PGJ} = 0 mm).

Photron's FastCam SA5 (Photron, Tokyo, Japan) is used to observe the weld pool (Figure 11). The camera is aligned perpendicular to the peripheral speed of the substrate v_U (α = 90°) and at an angle of β = 45° to the horizontal on the weld pool. The weld pool is illuminated by the Cavilux HF illumination laser (wavelength: 808 ± 10 nm; power: 500 W ± 10%) from Cavitar Ltd. A bandpass filter (CWL = 810 nm, FWHM = 10 nm) is placed in front of the camera objective to reject the processing laser. The camera is operated at a frame rate of 40,000 fps.

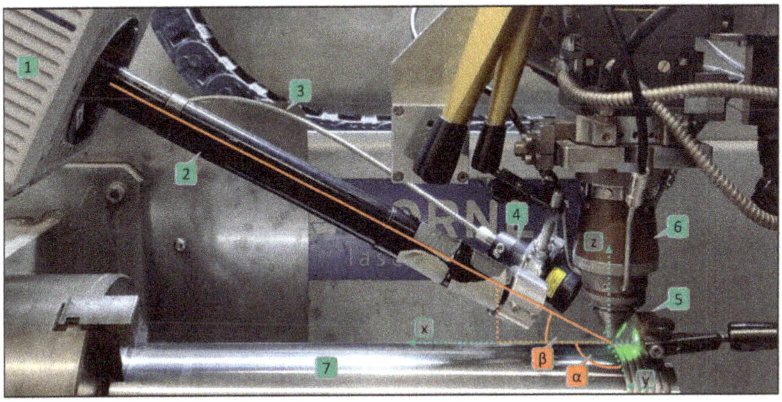

Figure 11. Setup of the high-speed imaging: 1: high-speed camera; 2: objective of the high-speed camera; 3: fiber of the illumination laser; 4: objective of the illumination laser; 5: mirror; 6: powder feeding nozzle; 7: substrate; α: angle with respect to scanning direction; β: angle with respect to the horizontal.

3. Results and Discussion

3.1. *Axial Offset of the Laser Beam* Δ_{LB}

To evaluate the surface condition, the height profile in the WLI images (Figure 12) shall be considered first. Based on the surface structures, the coatings created in a range of -15 mm $\leq \Delta_{LB} \leq 12.5$ mm have comparable coating surfaces at standard settings (Δ_{LB} = 0 mm). The microstructure is fine and has a small height difference. The surface structure changes significantly with a negative offset from $\Delta_{LB} = -17.5$ mm or with a positive offset from Δ_{LB} = 15 mm. A coarse macrostructure is formed, and defined melt islands are visible. This applies to both the lateral extent and the axial height of the structures (note the height scales). In these areas, large structures spreading over many single tracks can be recognized, which do not show any orientation.

The same can be seen in the determined parameters Sa, Sz, and Sdr (Figure 13). In the range of -15 mm $\leq \Delta_{LB} \leq 12.5$ mm, the mean arithmetic height is comparable with values of about Sa ≈ 6.3 μm. Outside this range, the value increases to Sa = 53.9 μm at $\Delta_{LB} = -20$ mm and to Sa = 68.8 μm at Δ_{LB} = 20 mm. The mean roughness takes values between Sz = 110.4 μm and Sz = 188 μm in the range of -15 mm $\leq \Delta_{LB} \leq$ 12.5 mm and more than doubles outside this range to Sz = 483.7 μm at $\Delta_{LB} = -20$ mm and to Sz = 443.7 μm at Δ_{LB} = 20 mm due to the large structure heights. Since the real surface area is increased by the resulting structures, the developed area ratio also shows this trend, so that it is Sdr ≈ 3% inside the range of -15 mm $\leq \Delta_{LB} \leq$ 12.5 mm and increases to Sdr ≈ 42.3% at $\Delta_{LB} = -20$ mm and to Sdr ≈ 68.4% at Δ_{LB} = 20 mm outside this range. The particle number could only be determined from $\Delta_{LB} = -17.5$ mm to Δ_{LB} = 15 mm using the method described above. In the range from $\Delta_{LB} = -15$ mm to Δ_{LB} = 12.5 mm, the coating surfaces have a number of adhering particles of 59 to 72 per mm^2, whereas at $\Delta_{LB} = -17.5$ mm and at Δ_{LB} = 15 mm, the number increases to 93 and to 106 per mm^2, respectively.

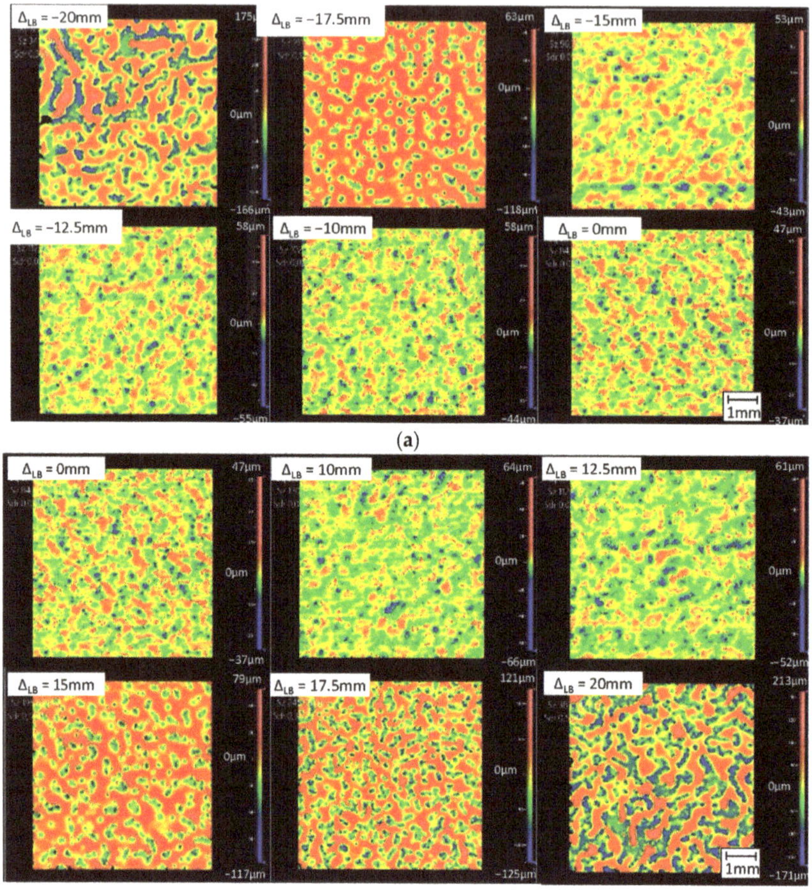

(**b**)

Figure 12. WLI images of the coatings at Δ_{LB}: (**a**) negative offsets (**b**) positive offsets.

The Rayleigh length z_R of the laser beam could be the reason why the surface properties of the coatings remain almost constant over a wide range of the axial offset of the laser beam and change significantly only from $\Delta_{LB} = -17.5$ mm to $\Delta_{LB} = 15$ mm. The Rayleigh length indicates the axial distance from the beam waist at which the laser beam expands to $\sqrt{(2)}$-fold [16], i.e., the laser spot area doubles. Within the Rayleigh length, intensity changes due to the caustic can be neglected (for Gaussian beams) [1]. The range from $z = -z_R$ to $z = z_R$ with respect to the beam waist is called the focus length [16]. For the laser beam used here, an approximately constant energy distribution and input into both the substrate and the PGJ is also expected in this range, so that the surface conditions are expected to be constant over the range of the focus length.

For negative offset, effects on the surface condition are first noticeable at $\Delta_{LB} = -17.5$ mm, whereas for positive offset, they are already noticeable at $\Delta_{LB} = 15$ mm. This asymmetry could be related to the change in Rayleigh length that occurs when the laser beam focus is shifted by the collimator. A negative shift increases the Rayleigh length, and a positive shift decreases it. Thus, with a positive offset, the substrate surface emerges earlier from the range of the focus length, so that differences in the energy input into the substrate and thus differences in the weld pool formation can occur earlier here than with a negative offset of the laser beam.

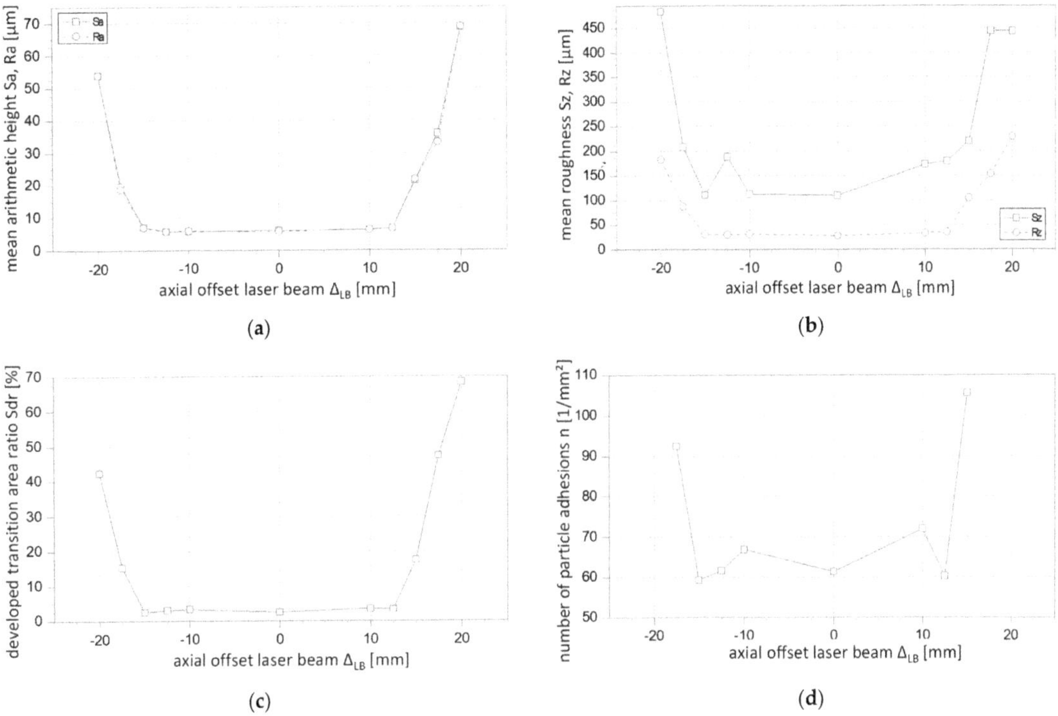

Figure 13. Results for axial offset of the laser beam. (**a**) Mean arithmetic height vs. Δ_{LB}; (**b**) mean roughness vs. Δ_{LB}; (**c**) developed transition area ratio vs. Δ_{LB}; (**d**) number of particle adhesions vs. Δ_{LB}.

Thus, it can be said that the EHLA process with the parameter settings of Table 5 is stable at least over the range of the focal length ($\pm z_R$). A shift beyond $\pm z_R$ can lead to an insufficient power density on the substrate surface to generate or permanently sustain a weld pool. This is shown by the single tracks (Figure 14). With the standard setting, a clear, almost uniform boundary between the deposited track and the substrate can be seen, whereas the edges of the single tracks become increasingly frayed with negative axial offset. Furthermore, up to $\Delta_{LB} = -15$ mm, the material is deposited continuously in the core areas of the single tracks. From $\Delta_{LB} = -17.5$ mm, however, an increasing number of defects can be seen in the core areas of the single tracks, so that no longer continuous tracks are produced. The behavior of the positive offset is equivalent to that of the negative offset.

The high-speed images (Figure 15a) show that at $\Delta_{LB} = -15$ mm, there is still a continuous, permanent weld pool. At $\Delta_{LB} = -17.5$ mm, the weld pool is no longer continuous, which could be caused by the reduced energy density on the substrate surface. So, instead of a single melt line, several individual structures are formed on the substrate surface. During the next overlapping run, the material already deposited is re-melted, and due to surface tension, the newly deposited material preferably forms a molten bond with the remelted material of the existing structures. This can be seen in Figure 15b. In the top image, the area of the already coated substrate marked by the orange rectangle is still in front of the process area of the single track now to be applied. Here, a recess with relatively small crater walls can be seen (green arrow). In the bottom image, the marked area has moved to the left, behind the process area. During the process, additional material has accumulated on the crater wall, making it more pronounced, and this structure thus extends over another single track.

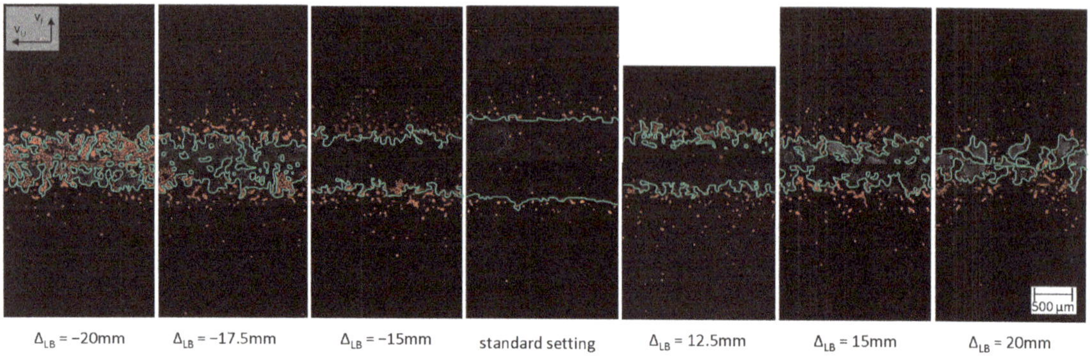

Figure 14. Shapes (green line) and particle adhesions (orange dots) of the single tracks at different axial offsets Δ_{LB}.

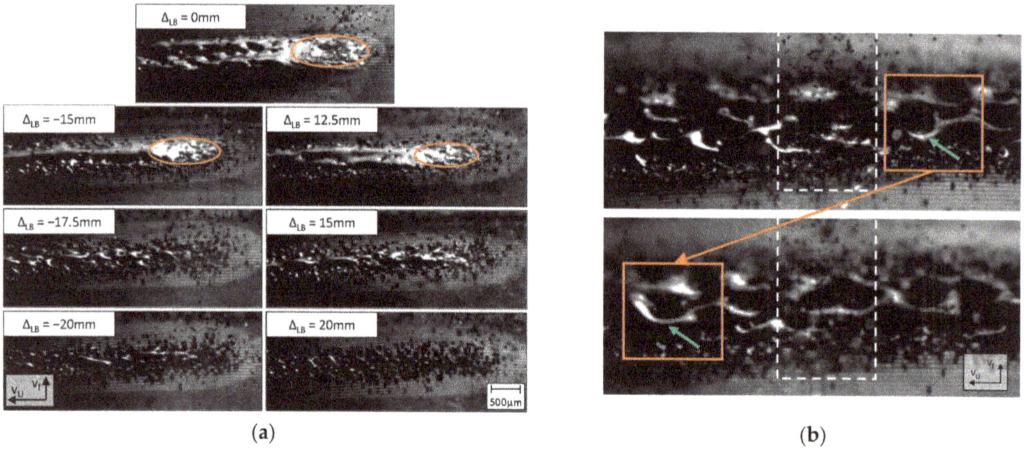

Figure 15. High-speed images of the weld pool at different offsets Δ_{LB}: (**a**) change in the extent of the weld pool due to different axial offsets Δ_{LB} (orange ellipse marks weld pool); (**b**) creation of large, overlapping structures by bonding new material with existing structures (Δ_{LB} = −17.5 mm: white dashed lines mark the PGJ and the process area; orange rectangles mark the same area in front of and behind the weld pool; green arrows mark the enlarging structure).

The same is observed with a positive axial offset. In this case, non-continuous core areas occur in single tracks with strongly frayed edges from Δ_{LB} = 15 mm. In the high-speed images, processes comparable to those described above can be seen. The fact that no more continuous weld pools are generated above a certain axial offset suggests that the power density on the substrate surface is no longer sufficient to apply enough energy per area and time to reach the melting temperature of the substrate material.

3.2. *Axial Offset of the Powder–Gas Jet* Δ_{PGJ}

Figure 16 shows the WLI images of the coatings produced with varied axial offsets of the PGJ. At Δ_{PGJ} = −1 mm, the focus is approximately on the substrate surface. At this axial offset, the structure of the surface is comparable to that at the standard setting (Δ_{PGJ} = 0 mm). Thus, at Δ_{PGJ} = −1 mm, the characteristic values are Sa = 6.3 µm, Sdr = 2.7%, and a particle number of 54 per mm^2, comparable to those at Δ_{PGJ} = 0 mm, with values of Sa = 6.1 µm, Sdr = 2.4% and a particle number of 61 per mm^2. The average roughness depth is larger for Δ_{PGJ} = −1 mm with Sz = 169 µm than for Δ_{PGJ} = 0 mm with Sz = 110.4 µm.

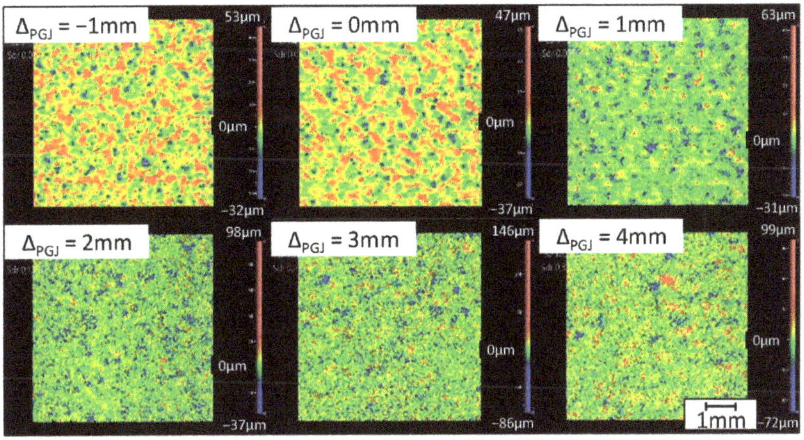

Figure 16. WLI images of the coatings at different Δ_{PGJ}, scanning direction from left to right.

With a positive axial offset, the WLI images show that finer structures increasingly appear on the surfaces. These fine structures cause the roughness values of Sa = 16.3 µm and Sz = 174.5 µm at Δ_{PGJ} = 4 mm (Figure 17). In addition, these structures enlarge the real surface area, causing the developed transition area ratio to increase to Sdr = 54.4% at Δ_{PGJ} = 4 mm. The investigations regarding particle adhesions suggest that these fine structures are caused by adherent particles. The number of particle adhesions is already 185 per mm^2 at Δ_{PGJ} = 1 mm. From Δ_{PGJ} = 2 mm, the particle density on the coating surface is so high that individual particles can often no longer be resolved separately using the method described above, and the number of particles on the coating therefore increases only slightly to 213 per mm^2. This effect is amplified with a further positive offset Δ_{PGJ}. In addition, since the particles are often deposited so close together or even one above the other, they merge more and more into agglomerates. Thus, the determined number of particles decreases.

The fact that the PGJ focus in the present investigations is shifted in a range that lies within the Rayleigh length of the laser beam leads to the assumption that the energy input into the powder particles is still sufficient to melt the majority of the particles on their way to the substrate. However, due to positive Δ_{PGJ}, the PGJ spot on the surface becomes larger, so that an increased number of molten particles outside the weld pool hit the surface. This overspray can cause molten particles that impinge, perhaps on a sufficiently preheated substrate or coating surface, to fuse with it [17]. This is well visible on the single tracks (Figure 18). With greater Δ_{PGJ}, the particle adhesion increases on the single track and on both sides next to it. No significant differences in the form of the track edges can be seen over the investigated offset range, which indicates a continuously existing weld pool, so a sufficiently high energy input in the substrate and thus homogeneously enclosed coatings are to be assumed.

Since the overspray increases with the positive offset, the width of the particle distribution increases on the one hand, and the total number of particles deposited on the surface per run increases on the other. This means that more particles are deposited on an already coated area by the following run, but also that the number of following runs in which particles are deposited at this area increases with the offset Δ_{PGJ}, which is the reason for the strong increase in particle adhesion and fine structures and thus in the parameters Sa, Sz, and Sdr.

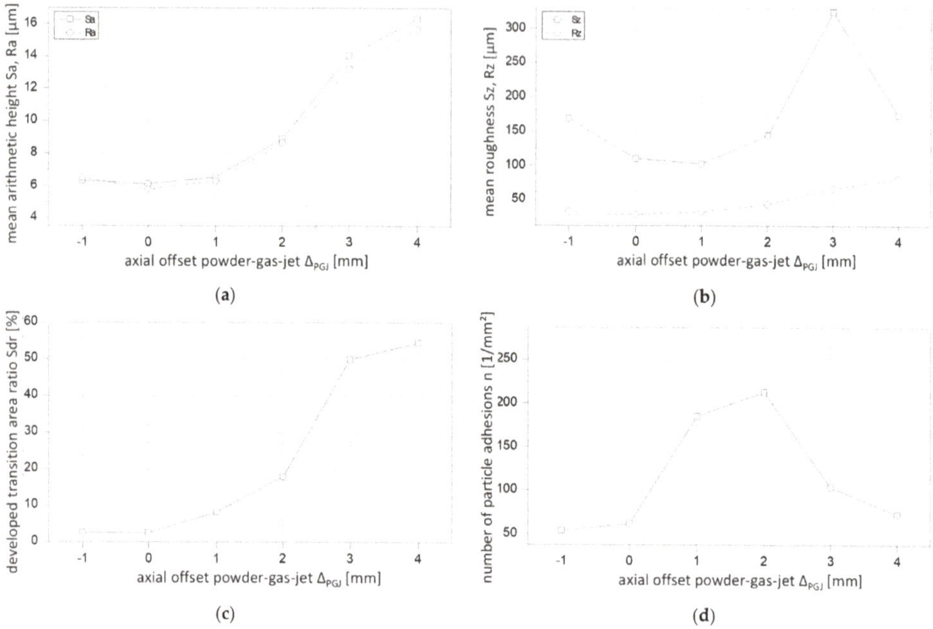

Figure 17. Results for axial offset of the powder–gas jet. (**a**) Mean arithmetic height vs. Δ_{PGJ}; (**b**) mean roughness vs. Δ_{PGJ}; (**c**) developed transition area ratio vs. Δ_{PGJ}; (**d**) number of particle adhesions vs. Δ_{PGJ}.

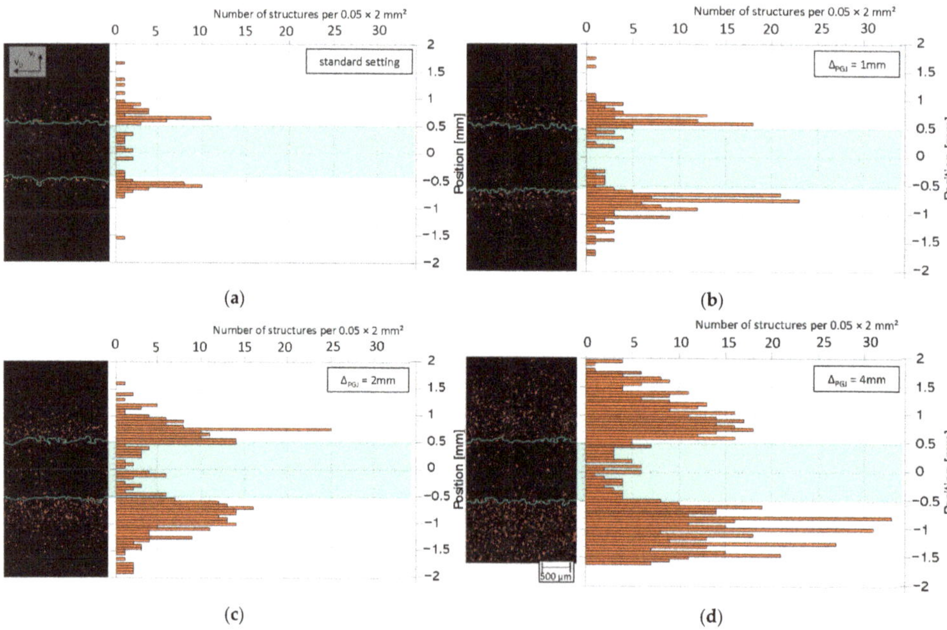

Figure 18. Particle distribution on single tracks at different axial offsets Δ_{PGJ}: (**a**) Δ_{PGJ} = 0 mm; (**b**) Δ_{PGJ} = 1 mm; (**c**) Δ_{PGJ} = 2 mm; (**d**) Δ_{PGJ} = 4 mm. The detected particles are marked with orange dots and the edge of the single tracks with green lines.

3.3. *Lateral Offset of the Powder–Gas Jet φ_{PGJ}*

Since the lateral offsets lead to very different surface textures (Figure 19), which are caused by different mechanisms, the lateral offsets investigated are evaluated individually. For this purpose, the WLI images of the coatings and the single tracks, the determined parameters, as well as the high-speed images, are considered.

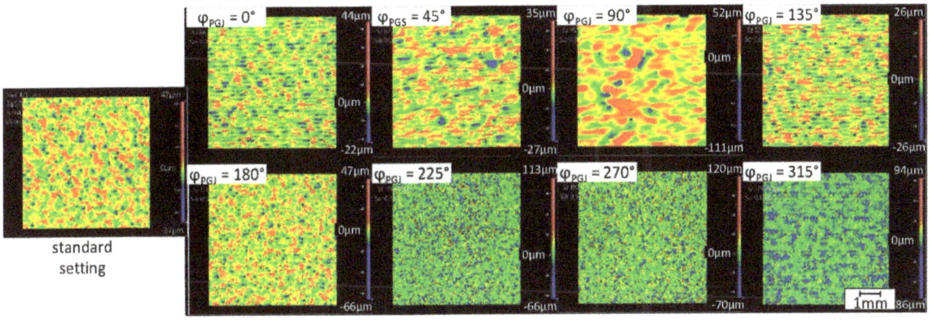

Figure 19. WLI images of the coatings for different φ_{PGJ}, scanning direction from left to right.

With an offset of $\varphi_{PGJ} = 0°$, compared to the standard setting ($r_{PGJ} = 0$ mm), coatings with smaller overlapping structures result, which have an orientation in the scanning direction (Figure 19). With $\varphi_{PGJ} = 0°$, the PGJ spot lies in front of the laser spot (see Figure 7). Thus, less powder enters the weld pool directly; rather, the powder is first deposited in front of the weld pool and then absorbed by the weld pool when moving over it. Since the powder is not fed directly into the weld pool, fewer waves are induced in it than with the standard setting (Figure 20).

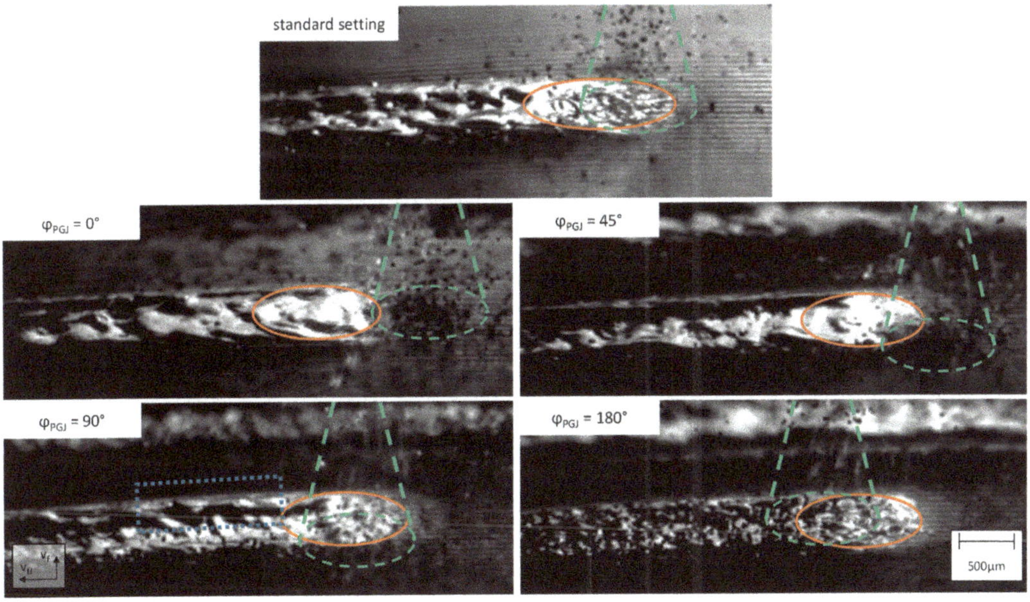

Figure 20. High-speed images of the weld pool at different offsets φPGJ. The powder–gas jet is marked with green dashed lines, the weld pool with an orange ellipse, and the raised track edge with a blue dotted square.

With the standard setting, structures develop out of the waves when the weld pool solidifies, which is less prominent with $\varphi_{PGJ} = 0°$ due to the smoother weld pool. So, the structures are less pronounced laterally and merely expand in the scanning direction. Due to the following weld pool, almost no further particles are deposited behind it (Figure 21a), and the surface is smoothed by the laser beam. In addition, fewer particles adhere on the left side next to the single track because the particles hitting the substrate's surface there do not pass the laser beam anymore and thus are not heated up enough to adhere. Thus, all investigated parameters at $\varphi_{PGJ} = 0°$ are reduced to Sa = 5.2 µm, Sz = 82.3 µm, Sdr = 1.8%, and the number of particle adhesions to 43 per mm^2 (Figure 22).

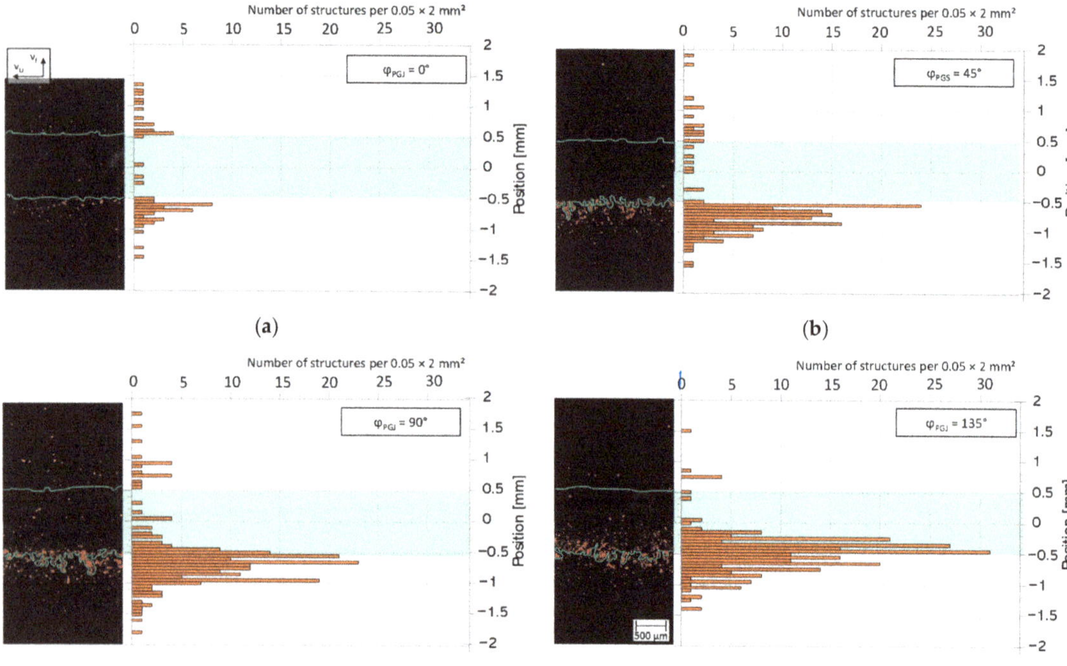

(a) (b) (c) (d)

Figure 21. Particle distribution on single tracks at different axial offsets φ_{PGJ}: (**a**) $\varphi_{PGJ} = 0°$; (**b**) $\varphi_{PGJ} = 45°$; (**c**) $\varphi_{PGJ} = 90°$; (**d**) $\varphi_{PGJ} = 135°$. The detected particles are marked with orange dots, and the edges of the single tracks are marked with green lines.

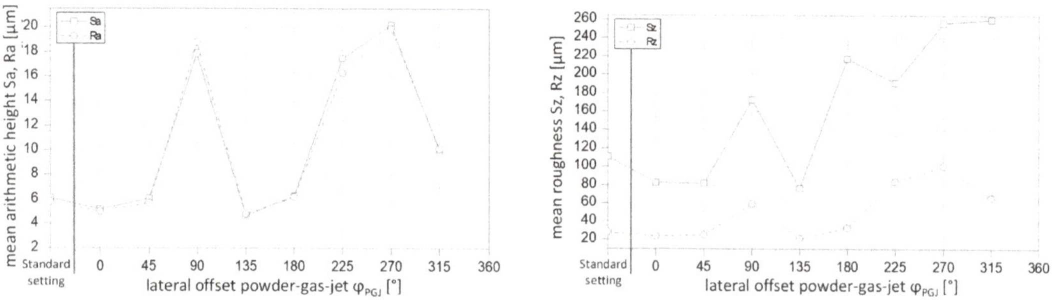

(a) (b)

Figure 22. *Cont.*

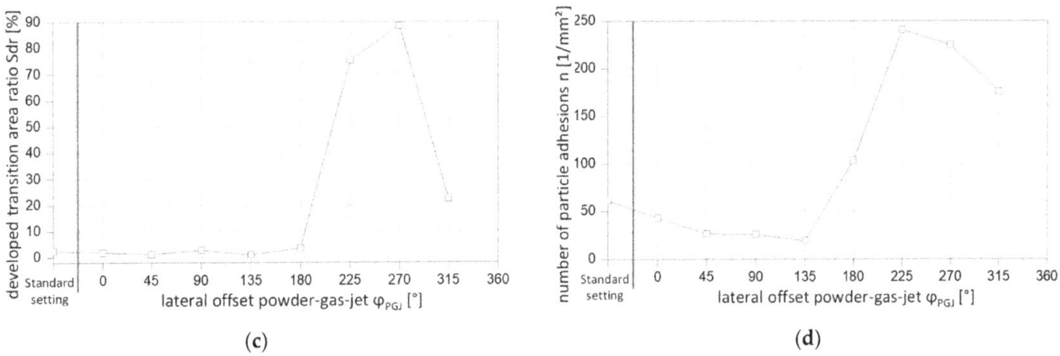

Figure 22. Results for lateral offset of the powder–gas jet. (**a**) Mean arithmetic height vs. φ_{PGJ}; (**b**) mean roughness vs. φ_{PGJ}; (**c**) developed transition area ratio vs. φ_{PGJ}; (**d**) number of particle adhesions vs. φ_{PGJ}.

At $\varphi_{PGJ} = 45°$, structures with an orientation in the direction of v_U appear again, whereby these structures are larger and partially spread over several tracks (Figure 19). The PGJ is in front of the laser beam, whereby the PGJ also lies to the right of the laser beam. Due to this relative positioning, the number of particles adhering to the single tracks is relatively small due to the following laser beam as described for $\varphi_{PGJ} = 0°$ (Figure 21b). Additionally, since the PGJ is located to the right of the laser beam, many particles adhere to the right side, and fewer melted particles reach the area to the left of the track. However, the particle adhesions on the right side do not contribute directly to the surface condition since these are re-melted during the next run. The low particle number on the left side leads to a reduction in the particle number to 26 per mm^2 on the generated coating. Because fewer particles are deposited directly into the weld pool, which is similar to $\varphi_{PGJ} = 0°$, the weld pool is relatively smooth (Figure 20). But at $\varphi_{PGJ} = 45°$, the particles are mainly fed to the weld pool from the side, so that they induce a wave and thus a material transport from the right to the left side of the weld pool. This leads to an accumulation of material on the left side of single tracks (Figure 20). During the coating process, the material is thus transported in the direction of the previous track and can be deposited there. Since the weld pool extends relatively far behind the laser beam and even further behind the PGJ, transported material can partially flow back from the previous track to the center of the now deposited track (Figure 20). This material flow could result in structures at the left edge of a single track, which could also spread over several track offsets due to the material returning to the weld pool. The combination of few particle attachments and the formation of (comparatively long-wavelength) structures due to a modified weld pool flow, leads to values of Sa = 6.1 µm, Sz = 81.6 µm, and Sdr = 1.2% at $\varphi_{PGJ} = 45°$ (Figure 22).

With a further shift to $\varphi_{PGJ} = 90°$, the resulting overlapping structures on the surface become even larger, so that they now extend over several single tracks (Figure 19). These structures again have an orientation in the scanning direction. Since the PGJ is located to the right of the laser beam with respect to the scanning direction, powder is significantly added to the weld pool on this side. As a result, many particles are deposited on this side, too (Figure 21c). Furthermore, adding the powder from the side forms a wave in the weld pool, which forces the material to the opposite, left side of the weld pool (Figure 20). On the one hand, this material transport makes the left edge higher, but on the other hand, it also creates unevenness, which leads to large and deep structures. Due to these structures, the roughness values increase to Sa = 17.9 µm and Sz = 172.3 µm. Since also here relatively few particles adhere to the left side of the track, the particle number of 26 per mm^2 remains unchanged compared to $\varphi_{PGJ} = 45°$. The developed transition area ratio is significantly increased by particle adhesions and less by large, long-wavelength structures, as they occur here, so this value increases to only Sdr = 2.7%.

At $\varphi_{PGJ} = 135°$, smaller structures with an orientation in the scanning direction appear again, which are comparable to those at $\varphi_{PGJ} = 45°$. Since very few particles adhere to the left edge as well as to the left of the single track (Figure 22d), the number of particles on the entire surface assumes the smallest value of 19 per mm^2 observed for all offsets. The other parameters also assume the smallest values of Sa = 4.6 µm, Sz = 75.9 µm, and Sdr = 0.9%. Thus, the roughness can be reduced by an offset of $\varphi_{PGJ} = 135°$ with respect to the standard setting.

At $\varphi_{PGJ} = 180°$, the PGJ is located behind the laser beam with respect to the scanning direction. Since the weld pool also extends to an area behind the laser spot, the powder particles impinge on an area in which the weld pool is still present or is cooling down. This results in a very turbulent weld pool with many particles adhering to the surface (Figure 20). These particles are not remelted by further runs, so an increased occurrence of particle adhesion on the single track (Figure 23a) and therefore also on the coating surfaces (103 per mm^2) is observed. The structures on the coatings are therefore, again, smaller and without orientation, comparable to the standard setting. Thus, the developed transition area ratio also increases to Sdr = 3.4%. In addition, the roughness values increase to Sa = 6.2 µm and Sz = 217.2 µm.

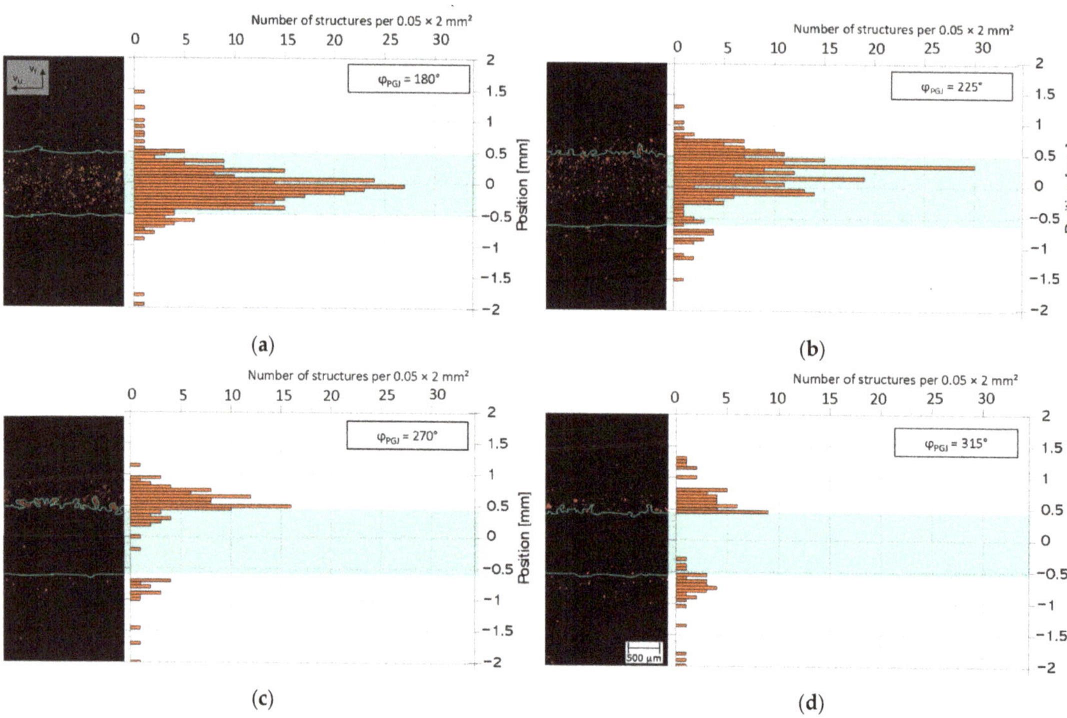

Figure 23. Particle distribution on single tracks at different axial offsets φ_{PGJ}: (**a**) $\varphi_{PGJ} = 180°$; (**b**) $\varphi_{PGJ} = 225°$; (**c**) $\varphi_{PGJ} = 270°$; (**d**) $\varphi_{PGJ} = 315°$. The detected particles are marked with orange dots, and the edge of the single tracks marked are with green lines.

At $\varphi_{PGJ} = 225°$, the PGJ is located behind the laser beam and to its left with respect to the scanning direction. For this reason, the particle distribution shifts to the left when looking at the single tracks, so that more particles adhere to the left edge area of the track and to any tracks that are already present but even on the single track itself (Figure 23b). Since the particles on the left edge area are not re-melted during the following run and further particles adhere during the following runs, a particle number on the coatings of 240 per mm^2 is determined. However, since the particle density on the surface is so high, particles can form agglomerates, and the surface shows fine structures without orientation. Due to these fine structures, the developed transition area ratio increases rapidly to Sdr = 75.1%. In addition, a further increase in roughness values to Sa = 17.5 μm and Sz = 191.1 μm can be seen.

With a further rotation of the lateral offset to φ_{PGJ} = 270°, the PGJ is at the same level but to the left of the laser beam, so that few particles impinge on the right side and the weld pool dominates there. So, the single tracks on the right have few particle adhesions (Figure 23c). The particle distribution of the single tracks thus shifts further to the left, and an increased particle adhesion can be observed both at the left edge of the track and in the area to the left of the track (i.e., on previous tracks). For this reason, the particle density on the coatings increases to such an extent that an increased formation of agglomerates takes place. Therefore, the determined number of particle adhesions on the coating surface decreases again to 225 per mm^2. The other parameters, however, increase again slightly due to the fine structures produced, so that with regard to the lateral offsets investigated, the largest values could be determined with Sa = 20.2 μm, Sz = 256.3 μm, and Sdr = 88.5% at φ_{PGJ} = 270°.

With φ_{PGJ} = 315°, the PGJ is located to the left of the laser beam as with φ_{PGJ} = 225°, but this time the PGJ is placed in front of the laser beam. As a result, although the particle distribution to the left of the single track at φ_{PGJ} = 315° is comparable to that at φ_{PGJ} = 225°, fewer particles are deposited on the edge region of the track this time because the following weld pool absorbs these particles (Figure 23d). Since the number of particles adhering to the single tracks is now so small, the particles adhering to the coatings are virtually only caused by the particles adhering to the left of the single tracks. Thus, the number of particles on the coatings decreases to 175 per mm^2. At the same time, the other parameters also decrease to Sa = 10 μm, Sz = 260.1 μm, and Sdr = 22.2%.

In summary, it can be said that the lateral offset φ PGJ influences the interaction between the particle adhesions on and in the area next to the track as well as the material transport in the weld pool. On the one hand, the lateral offset of the PGJ to the left or right side causes increased particle deposition on the respective side, and on the other hand, the lateral addition of powder into the weld pool forces material transport to the opposite side. If the PGJ is placed in front of the laser beam, a smooth weld pool is created, and the surface is smoothed by the laser beam. Therefore, hardly any particles are deposited on the single tracks. If the PGJ is located behind the laser beam, more particles are deposited on the single tracks. The surface properties are influenced by the synthesis of all these effects.

4. Conclusions

Coatings manufactured by high-speed LMD reach a surface roughness of approx. Rz = 50–100 μm, Ra = 11.3–20.5 μm, [7–9]. In the present investigations, it has been shown that the surface condition of coatings can be adjusted in a broader range from Sa = 4.6–68.8 μm (Ra = 4.6–67 μm) and Sz = 75–483 μm (Rz = 18–232 μm) by shifting the relative position of the powder–gas jet, laser beam, and component surface. A specific near-net shape of the coated component in terms of mean arithmetic height, mean roughness, developed transition area ratio, and number of particle adhesions can be realized with the relative positioning techniques presented in the article. Since different applications demand different surface shape requirements and postprocessing is cost-intensive, targeted processing of surface conditions is desirable. The near-net shape with the aim of achieving a smooth surface can be adjusted in the EHLA process to reduce the postprocessing work, e.g., by turning or grinding for the manufacturing of hydraulic cylinders, brake discs, and journal bearings. Increased surface roughness and the number of particle adhesions lead to a larger surface of the coating, which can be specifically adjusted in the EHLA process, e.g., for the manufacturing of heat exchangers or electrodes for electrolysers [18].

In the investigations, a common and well-established material combination of Inconel 625 as a filler material and construction steel S355 as a base material was used.

The measured quality feature is the surface condition. Application-targeted and specific metallographic properties should be investigated accordingly.

Author Contributions: Conceptualization, M.B.; methodology, M.B. and T.S.; validation, M.B., A.G. and C.L.H.; formal analysis, M.B.; investigation, M.B. and T.S.; data curation, T.S.; writing—original draft preparation, M.B., T.S. and A.G.; writing—review and editing, M.B. and T.S.; visualization, M.B. and T.S.; supervision, A.G. and C.L.H.; project administration, M.B. All authors have read and agreed to the published version of the manuscript.

Funding: This research received no external funding.

Institutional Review Board Statement: Not applicable.

Informed Consent Statement: Not applicable.

Data Availability Statement: Not provided.

Conflicts of Interest: The authors declare no conflict of interest.

Appendix A

Table A1. Settings at MX.

Parameter	Form Remove	Filter	Data Fill
Sa, Sz, Sdr	Cylinder	–	Fill All Voids
Number of particles on coatings	Cylinder	Filter Type: Spline Filter: Band Pass Type: Gaussian Spline Fixed Period: 20–53 µm	Fill All Voids
Distribution of adhering structures on single tracks	Cylinder	Filter Type: Spline Filter: Band Pass Type: Gaussian Spline Fixed Period: 20–53 µm	Fill All Voids

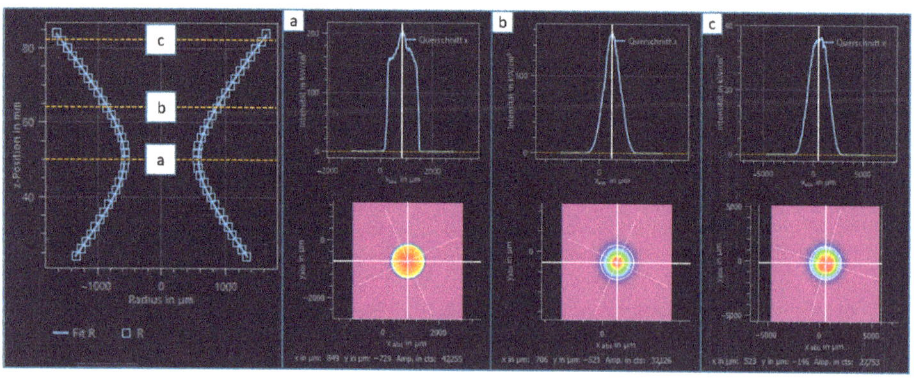

Figure A1. Intensity distribution at focal position (**a**), at the Rayleigh length (**b**) and at the double Rayleigh length (**c**).

References

1. Poprawe, R. *Lasertechnik für Die Fertigung: Grundlagen, Perspektiven und Beispiele für den Innovativen Ingenieur*; Springer: Berlin/Heidelberg, Germany, 2005.
2. Küppers, W.; Backes, G.; Kittel, J. Extremes Hochgeschwindigkeitslaserauftragsschweißverfahren (Offenlegungsschrift). DE Patent 102011100456A1, 4 May 2011.

3. Gasser, A.; Backes, G. Systemtechnik für das Pulverbasierte Laserauftragschweissen. Themenbroschüre Systemtechnik für das Pulverbasierte Laserauftragschweißen. 2009. Available online: https://www.ilt.fraunhofer.de/de/mediathek/prospekte/themenbroschuere-systemtechnik.html (accessed on 23 April 2023).
4. Schopphoven, T.; Gasser, A. Fraunhofer-Institut für Lasertechnik ILT; Press Release: Umweltfreundliche Alternative zum Verbotenen Hartverchromen Mit Chrom(VI). Available online: https://www.ilt.fraunhofer.de/de/presse/pressemitteilungen/pm2017/pressemitteilung-10-08-2017.html (accessed on 23 April 2023).
5. Weingraber, H.v.; Abou-Aly, M. *Handbuch Technische Oberflächen: Typologie, Messung und Gebrauchsverhalten*; Springer: Berlin/Heidelberg, Germany, 2013.
6. Schorr, D.-I.D. Rauheitsmessung zur Oberflächencharakterisierung—Aber Richtig. WOTech GbR. 6 September 2021. Available online: https://www.wotech-technical-media.de/womag/ausgabe/2018/04/24_schorr_rauheit_04j2018/24_schorr_rauheit_04j2018.php (accessed on 23 April 2023).
7. Lindner, T.; Liborius, H.; Töberling, G.; Vogt, S.; Preuß, B.; Rymer, L.-M.; Schubert, A.; Lampke, T. High-Speed Laser Metal Deposition of CrFeCoNi and AlCrFeCoNi HEA Coatings with Narrow Intermixing Zone and Their Machining by Turning and Diamond Smoothing. *Coatings* **2022**, *12*, 879. [CrossRef]
8. Xu, X.; Lu, H.; Qiu, J.; Luo, K.; Su, Y.; Xing, F.U.; Lu, J. High-speed-rate direct energy deposition of Fe-based stainless steel: Process optimization, microstructural features, corrosion and wear resistance. *J. Manuf. Process.* **2022**, *75*, 243–258. [CrossRef]
9. Bobzin, K.; Wietheger, W.; Knoch, M.A.; Sommer, J.; Brucki, M.; Schleifenbaum, J.H. Constantin Häfner: Combination of a laser deposition welded corrosion protection coating and a thermally sprayed wear protection coating. *DVS-Verl. Verl. Schweißen Verwandte Verfahren.* **2020**, *13*, 114–121.
10. Pauly Stahlhandel, Stahl Werkstoff-Nr.: 1.0553 Datenblatt. Available online: https://pauly-stahlhandel.com/de/werkstoff-nr/1.0553 (accessed on 10 August 2021).
11. Mariani, F.E.; Ribeiro, K.S.B.; Lombardi, A.N.; Casteletti, L.C.; Coelho, R.T. Effect of Laser Polishing Post-Processing Technique on the Roughness and Wear Resistance of Inconel 625 Deposited by Laser Cladding on AISI 304L Stainless Steel. *J. Mater. Eng. Perform.* **2021**, *30*, 6713–6721. [CrossRef]
12. Cus, F.; Zuperl, U.; Irgolic, T. Effects of laser cladding parameters on microstructure properties and surface roughness of graded material. *Proc. Manuf. Syst.* **2015**, *10*, 77.
13. Przestacki, D.; Majchrowski, R.; Marciniak-Podsadna, L. Experimental research of surface roughness and surface texture after laser cladding. *Appl. Surf. Sci.* **2016**, *388*, 420–423. [CrossRef]
14. *DIN EN ISO 25178-2*; Geometrische Produktspezifikation (GPS)—Oberflächenbeschaffenheit: Flächenhaft—Teil 2: Begriffe und Oberflächen-Kenngrößen (ISO 25178-2:2012). Deutsche Fassung EN ISO 25178-2:2012; DIN Deutsches Institut für Normung e. V.: Berlin/Heidelberg, Germany, 2012.
15. Keyence Deutschland GmbH, Parameter der Oberflächenrauheit: Hybrid. Sdr (Entwickeltes Grenzflächenverhältnis). Available online: https://www.keyence.de/ss/products/microscope/roughness/surface/tab03_b.jsp (accessed on 8 September 2021).
16. Eichler, H.J.; Eichler, J. *Laser: Bauformen, Strahlführung, Anwendungen*, 8th ed.; Springer: Berlin/Heidelberg, Germany, 2015.
17. Schopphoven, T. Experimentelle und modelltheoretische Untersuchungen zum Extremen Hochgeschwindigkeits-Laserauftragschweißen. Ph.D. Thesis, Maschinenwesen, Rheinisch-Westfälische Technische Hochschule, Aachen, Germany, 2019.
18. Bernäcker, C.; Loos, S. Fraunhofer IFAM: Electrolysis Technology—Development and Qualification of Stack Components. Available online: https://www.ifam.fraunhofer.de/content/dam/ifam/en/documents/dd/Infobl%C3%A4tter/electrolysis_technology_fraunhofer_ifam_dresden.pdf (accessed on 23 April 2023).

coatings

MDPI

Article

Effect of Zn on Phase Evolution and Shear Resistance of Stainless Steel and Aluminum Alloy Interface by Laser Cladding

Keyan Wang [1], Xianqing Yin [1,*], Chengxin Li [1] and Kaiping Du [2]

[1] State Key Laboratory for Mechanical Behavior of Materials, School of Materials Science and Engineering, Xi'an Jiaotong University, Xi'an 710049, China; wangkeyantc@stu.xjtu.edu.cn (K.W.); licx@xjtu.edu.cn (C.L.)
[2] BGRIMM Technology Group, Beijing 100160, China; dukaiping@bgrimm.com
* Correspondence: xqyin@mail.xjtu.edu.cn

Abstract: The connection between aluminum and iron alloys is of immense significance in the pursuit of lightweight industrial products. However, the Fe-Al interface's inherent weakness restricts its widespread application. This study investigates the impact of Zn at the interface of Al-Fe laser cladding on the phase and mechanical properties of the interface. Specifically, we examine the influence of the applied Zn powder layer and alloying Zn layer on the morphology of the Fe-based cladding layer. The inclusion of Zn enhances the spreadability of the Fe-based cladding layer. Additionally, we elucidate the effect of Zn on the composition and phase of the Fe-Al laser cladding interface. Notably, the affinity between Zn and the η phase surpasses that of the θ phase, and an increased Zn content significantly thickens the η phase. Shear tests reveal that the failure mode of shear fracture encompasses both brittle and ductile fractures. Density functional theory (DFT) calculations indicate that Zn has a limited effect on the strength of the η phase but reduces the enthalpy of formation of the η phase. Our findings demonstrate that the alloyed Zn layer initially facilitates the formation of a continuous and uniform η layer, while an increased Zn content enhances and stabilizes the shear strength of the interface.

Keywords: laser cladding; dissimilar joining; interface; intermetallic compounds

1. Introduction

Aluminum alloys have compelling potential as a substitute for traditional steel but require surface strengthening to broaden their applications due to their lower hardness. Laser cladding offers an efficient and versatile technology that can create a high-hardness metal cladding layer on the surface of aluminum alloy, improving its wear resistance and thus making it more suitable for various uses such as automotive lightweighting and electric vehicle battery manufacturing [1–3]. Furthermore, laser cladding can address the limitations of traditional methods such as embedding cast iron cylinder liners or thermal-spraying wear-resistant coatings on aluminum alloy surfaces. The use of martensitic stainless steel for the inner wall of the cylinder of the cast aluminum engine, prepared through rotatable inner hole laser cladding technology, offers a promising approach for improving the wear resistance of aluminum alloys [4]. This method has the advantages of low cost, high efficiency, and good wear resistance, making it an essential application for the automotive industry.

However, the fusion welding process involving steel and aluminum is challenging due to the creation of fragile intermetallic compounds (IMCs) at the joint interphase. The Fe-Al phase diagram shows that the metallurgical reactions in the weld pool lead to the creation of complicated IMCs in the joint area, which can adversely affect joint performance. The laser cladding interface presents another complication where rapid solidification leads to chaotic characterization, further derailing the interface, with the η phase and θ phase being the most harmful due to their lower formation enthalpy [5]. According to Šlapáková et al. [6], the growth and development of these intermetallics are primarily governed by diffusion, with the kinetics of growth of the intermetallic layer following a parabolic law. To enhance the ductility continuity of the Fe-Al binary IMC layer at the joint, the introduction of alloying elements becomes necessary. Zhou et al. [7] confirmed that the average tensile–shear strength of a Si-added joint was higher than that of a Mn-added joint, owing to an improvement in the metallurgical reaction. Furthermore, the strength of both joints surpassed that of a no-added joint.

Zn serves as a vital element in joining iron and aluminum through the formation of a ternary IMC with Fe and Al. This presence of a ternary IMC can have a positive impact on the joint's properties by improving its mechanical performance, corrosion resistance, and bonding strength. Zn is commonly used as a coating material on steel surfaces to improve corrosion resistance, which is known as galvanization. According to Yan et al. [8], the addition of Zn effectively improved the interface bonding behavior. Zn is a low-melting-point element, and the IMCs formed with Al and Fe have a low melting point, which reduces the interface metallurgical reaction temperature, resulting in a process similar to instant liquid phase diffusion welding. In addition, Zn also enhances the wetting of Al on the Fe surface [9]. Peyre et al. [10] found that galvanized assemblies exhibited higher mechanical resistance than non-galvanized ones. However, the exact reason why intermetallic layers with added Zn are more shear-resistant (or more adherent) than pure Fe_2Al_5 is still under investigation. The improvement of the interface strength of η-phase-containing Zn does not have any direct evidence to be attributed to its better mechanical properties. According to Springer et al. [11], Zn has a significant effect on the metallurgical reaction rate of Fe-Al in a balanced condition, primarily the growth rate of the η phase. A more extensive and stable layer of the η phase may be the reason for improving mechanical performance. The presence of a thick Zn layer that vaporizes under laser irradiation significantly modifies the assemblies' microstructure near the interfacial zones and the failure modes, mainly located in the fusion zone.

The influence of Zn on the Fe-Al interface in Fe-Al laser cladding has not been extensively studied. Laser cladding of iron-based cladding layers on aluminum alloys involving a melting welding process can have chaotic characteristics at the interface. As described by Nguyen et al. [12], the morphology of the interface is caused by the molten material flow during the laser–powder interaction and non-repetitive formation of intermetallic layers. The degree of melting of the Al substrate is affected by both the high-temperature Fe melt and direct laser radiation. Furthermore, when high-temperature, high-density Fe powder comes into contact with the Al substrate melt pool, simultaneous metallurgical reactions and sinking occur, which results in a complex interface of the Al alloy surface laser-cladded with an Fe-based coating. As such, the role of Zn in this process requires further investigation.

DFT constitutes an integral tool in studying the mechanical and thermal properties of materials [13], and it can help reveal the effect of Zn atoms on the η phase. Currently, experimental and computational results have been able to determine the formation enthalpy and binding energy of Fe-Al IMCs [14,15]. A consistent conclusion based on the formation enthalpy of Fe-Al is that the η phase has a lower generation enthalpy and higher room temperature stability [16]. Zhang et al. [17] used the embedded-atom method (EAM) to calculate the mechanical properties of Fe-Al IMCs and found that the bulk moduli of the compounds increase with increasing Fe concentration. Li et al. [18] also computed the mechanical properties of Fe-Al IMCs and revealed that all the compounds, except for Fe_3Al and $FeAl_2$, are brittle materials. The joint's shear failure morphology displays a brittle characteristic with some localized ductile zones, which may primarily be associated with the brittle IMCs, including FeAl, Fe_2Al_5, $FeAl_3$, and Fe_4Al_{13}. However, there is a lack of research on the formation enthalpy of the η phase containing Zn, and computational results related to finite temperature are yet to be reported.

This paper aims to investigate the influence of Zn on the interface phase, microstructure, and shear resistance of the Fe-based laser cladding layer on an Al alloy surface. Through the DFT method, this study explicitly defines the mechanical performance, physical properties, and electronic distribution of Zn on the η phase interface. The findings of this study will provide both experimental and theoretical evidence to support the laser joining of an Fe-Al alloy.

2. Materials and Methods

2.1. Applied and Laser Alloying Processes of Zn Layer

Aluminum plates with a thickness of 4 mm were prepared, which featured an applied Zn powder layer and a laser alloying Zn layer. As shown in Figure 1a, secondary electron (SE) morphology demonstrates that Zn particles are spheroidal. The particle diameter (D/μm) distribution was characterized by a Mastersizer 3000E laser particle size analyzer (Malvern Panalytical, Malverncity, UK), where the particle diameter at 10% cumulative undersize is represented by a D_{10} of 3.350 μm, at 50% cumulative undersize is represented by a D_{50} of 5.406 μm, and at 90% cumulative undersize is represented by a D_{90} of 8.673 μm, as shown in Figure 1b. Zn powder was supplied by Beijing Zhongke Yannuo New Material Technology Co., Ltd., Beijing, China, and the purity was 99.9%. In contrast to the side-blowing shielding gas typically used in laser welding, the shielding gas blowing towards the substrate's surface in the laser cladding process can easily disperse. To prevent the blowout of Zn powder, which can be caused by high-flow direct protection gas, poly(vinyl alcohol) (PVA) was used as an adhesive to firmly attach the Zn powder to the substrate's surface. To prepare the adhesive, PVA powder was dispersed in anhydrous ethanol, and a 10% PVA aqueous solution was formed by pouring the PVA mixture into deionized water at 95 °C. After the PVA solution became clear, Zn powder was added and mixed into a suspension utilizing a magnetic stir bar. The amount of applied Zn powder ranged from 20 to 80 mg × cm^{-2}. The mixture was then poured onto the substrate's surface, and after drying, a high-quality Zn layer was obtained.

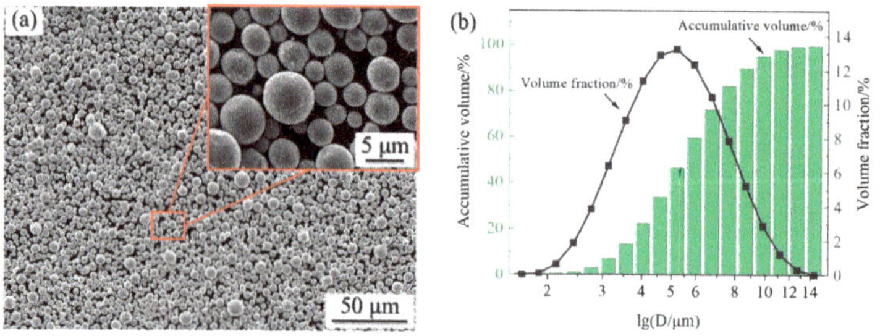

Figure 1. Characterization of Zn powders: (**a**) SE morphology and magnified view of Zn powders, and (**b**) particle diameter distribution.

Galvanized steel can be produced by either electroplating or hot plating. Similarly, two processes were proposed in this study: direct cladding with applied Zn powder layer (process 1) and alloying Zn layer (process 2). For process 2, prior to the cladding process, Zn powder was pre-scanned by a laser beam to prepare the Zn alloying layer on the Al substrate's surface. A 2500 W semiconductor laser with a 915 ± 10 nm wavelength was utilized, and its focal point had a diameter of 2 mm. Figure 2 illustrates the surface after laser scanning, revealing that the alloying Zn layer is less dense for a laser power lower than 1000 W (Figure 2a), whereas cracks appear when the laser power is 1250 W (Figure 2c). Consequently, a laser power of 1000 W was employed to produce the laser alloy Zn layer. Aluminum plates applied with 20 mg × cm^{-2} and 80 mg × cm^{-2} Zn powder were subjected to laser alloying at 1000 W, resulting in surface morphologies as shown in Figure 2b,d. The element composition of the cross-section was obtained through energy spectrum measurement of the cross-section, as shown in Table 1. The thickness of the alloying layer was approximately 10–20 μm, with sintered Zn clusters and Al-Zn eutectic being the primary component mixture, as shown in Figure 1e,f. As the thickness of the Zn layer increased, the radius of the Zn clusters also increased.

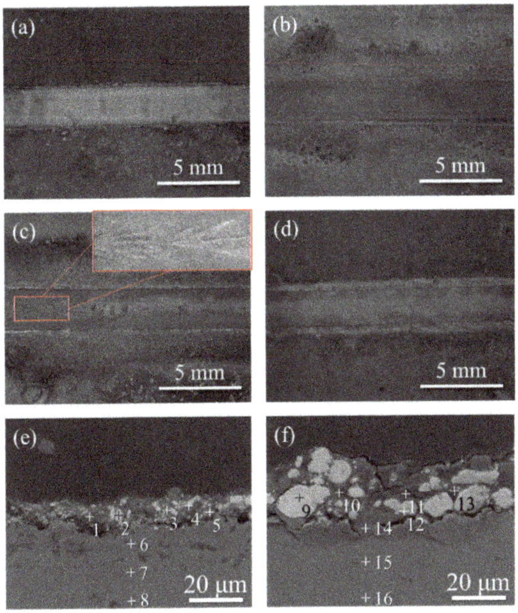

Figure 2. Surface morphology after laser alloying with varying amounts of applied Zn powder and laser powers: (**a**) 20 mg × cm^{-2} and 250 W, (**b**) 20 mg × cm^{-2} and 1000 W, (**c**) 20 mg × cm^{-2} and 1250 W with magnified view within the red boxed area revealing the presence of minute cracks, (**d**) 80 mg × cm^{-2} and 1000 W; cross-section morphologies of (**e**) 20 mg × cm^{-2} and (**f**) 80 mg × cm^{-2} applied Zn powder after laser alloying by laser power 1000 W.

Table 1. Element content of the test points, at.%.

	Point	1	2	3	4	5	6	7	8
20 mg × cm^{-2}	Al	95.67	6.20	55.97	42.23	24.71	96.54	99.26	99.57
	Zn	4.33	93.80	44.03	57.77	75.29	3.46	0.74	0.43
	Point	9	10	11	12	13	14	15	16
80 mg × cm^{-2}	Al	3.29	57.92	60.46	10.65	27.69	97.40	98.76	98.72
	Zn	96.71	42.08	39.54	89.35	72.31	2.60	1.24	1.28

2.2. Laser Cladding Method

The cladding material used in this study was 1Cr17Ni2B1Si1 martensitic stainless steel powder, which is illustrated in Figure 3a. Subsequently, the cladding experiment was carried out as depicted in the schematic diagram of Figure 3b, and the laser cladding equipment is displayed in Figure 3c. The same laser as the one mentioned in Section 2.1 was used. To ensure that the Fe-based powder was entirely melted, the laser's focus position was adjusted to adequately affect the powder flow. Additionally, the powder flow focus was placed 2 mm higher than the substrate, resulting in more thorough heating of the powder particles. The specific cladding parameters are available in Table 2.

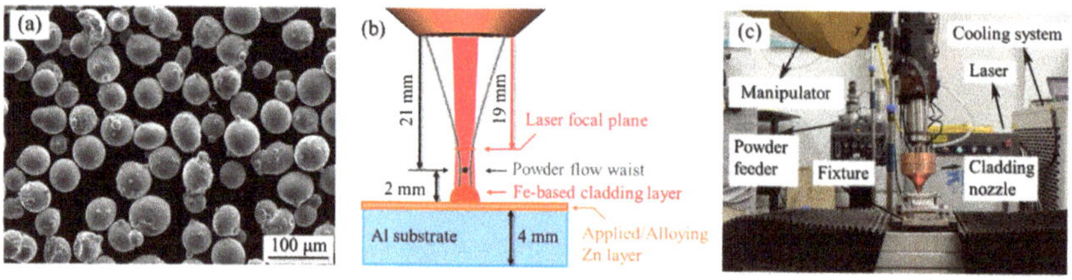

Figure 3. Laser cladding material and process: (**a**) SE morphology of 1Cr17Ni2B1Si1 martensitic stainless steel powder, (**b**) schematic of laser cladding process and (**c**) laser cladding equipment.

Table 2. Cladding parameters.

Parameter	Unit	Value
Laser power	W	2000
Cladding speed	$m \times min^{-1}$	1.0
Carrier gas pressure	MPa	0.1
Carrier gas flow rate	$L \times min^{-1}$	8.0
Powder feeding rate	$g \times min^{-1}$	28.8
Shielding gas pressure	MPa	0.2
Shielding gas flow rate	$m^3 \times h^{-1}$	1.0

2.3. Phase Characterization and Shear Test

The samples were polished using a diamond polishing compound after being polished to a 2000 grit size emery paper. Keller reagent was employed to corrode it until the interface between the cladding layer and the substrate became apparent. To investigate the interface characteristics, scanning electron microscopy (SEM; TESCAN MIRA3, TESCAN, Brno, Czech Republic) was used to examine the cross-sectional microstructures. The composition distribution at the interface was analyzed using the equipped backscattered electron (BSE, TESCAN, Brno, Czech Republic) and energy dispersive spectroscopy (EDS, TESCAN, Brno, Czech Republic) probes. In Figure 4, the interface shear resistance strength is determined by measuring the shear force of the cladding layer, which is then divided by the area of the section to calculate the shear stress. The phase of the fracture surface was tested using an X'Pert PRO X-ray diffractometer (Malvern Panalytical, Malverncity, UK).

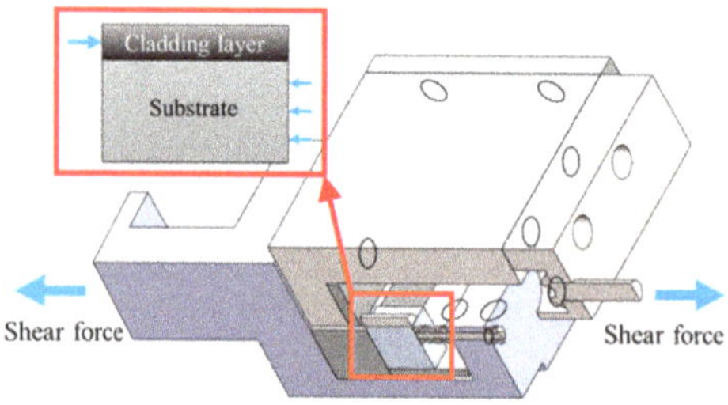

Figure 4. Shear fixture diagram and test process.

2.4. DFT Calculation

We used the Vienna Ab initio Simulation Package (VASP) to perform the first principle calculation of Fe_2Al_5 and $Fe_2Al_5Zn_x$ based on the density functional theory (DFT) [19–21]. The generalized gradient approximation (GGA) and projected affixed wave (PAW) methods proposed by Perdew, Burke and Ernzerhof (PBE) were used to evaluate the exchange correlation energy [22]. By using the method proposed by Monkhost-Pack, the energy integral was characterized as the first irreducible Brillouin zone [23]. For Fe_2Al_5 and $Fe_2Al_5Zn_x$ crystal cells, the grid size was $7 \times 8 \times 11$. The maximum energy cut-off value was used for plane wave expansion in reciprocating space, and the value was 500 eV. During the optimization process, the change of total energy converged to 10^{-5} eV.

Using the stress–strain method and generalized Hooker's law, we computed the elastic constants for Fe_2Al_5 and $Fe_2Al_5Zn_x$. We employed various strain patterns to determine the crystal structure and assessed the Cauchy stress tensor for each mode of strain. The elastic constants were then derived from the strain–stress relationships described in Equation (1):

$$\begin{pmatrix} \sigma_1 \\ \sigma_2 \\ \sigma_3 \\ \tau_4 \\ \tau_5 \\ \tau_6 \end{pmatrix} = \begin{pmatrix} C_{11} & C_{12} & C_{13} & C_{14} & C_{15} & C_{16} \\ C_{21} & C_{22} & C_{23} & C_{24} & C_{25} & C_{26} \\ C_{31} & C_{32} & C_{33} & C_{34} & C_{35} & C_{36} \\ C_{41} & C_{42} & C_{43} & C_{44} & C_{45} & C_{46} \\ C_{51} & C_{52} & C_{53} & C_{54} & C_{55} & C_{56} \\ C_{61} & C_{62} & C_{63} & C_{64} & C_{65} & C_{66} \end{pmatrix} \begin{pmatrix} \varepsilon_1 \\ \varepsilon_2 \\ \varepsilon_3 \\ \gamma_4 \\ \gamma_5 \\ \gamma_6 \end{pmatrix} \quad (1)$$

where σ_i is normal stress, τ_i is the shear stress, C_{ij} is the elastic constant, γ_i is the shear strain and ε_i is the corresponding normal strain. Anisotropic elastic properties were created by running the elastic anisotropy measures (ElAM) code [24].

3. Results and Discussion

3.1. Cross-Section of the Cladding Layer

Figure 5 illustrates the cross-section of the cladding layer, the thickness measurement results of the cladding layer, and its contact angle with the substrate. The reference line is the substrate surface, and the contact angle is the angle between the cladding layer's surface at the weld toe and the reference line. The thickness is the vertical length from the peak of the cladding layer to the reference line.

In process 1, the thickness of the cladding layer shows a tendency to decrease as the amount of applied Zn powder increases, as shown in Figure 5i. However, there may be a degree of randomness in the measurement outcomes, as evidenced by the slightly lower values of 564 μm and 592 μm. Conversely, in process 2, the thickness of the cladding layer increases, and there is no consistent change in thickness between the two processes. The heat absorbed during the melting of the Zn powder in process 1 causes the substrate temperature to decrease, which results in reduced efficiency in Fe-based cladding deposition and a subsequent decrease in the thickness of the cladding layer. In process 2, there is little correlation between the amount of applied Zn powder layer and the powder deposition efficiency. The thickness of the cladding layer is determined by its spreading when the contact angle of the Fe-based cladding layer is small and is restricted by surface tension [9]. This leads to an increase in the thickness of the cladding layer.

As the amount of applied Zn powder layer increases, the contact angle of the cladding layer decreases; process 1 shows a greater reduction than process 2, as shown in Figure 5j. This is because Zn is attached to the surface of the Al substrate in a solid state, and it absorbs a significant amount of heat from the Fe droplet, causing it to solidify before it spreads completely. An alloying Zn layer forms a surface agglomeration, making it easier to spread the liquid Fe-based droplet.

3.2. Interface Element Content and Phase Distribution

3.2.1. Interface of Process 1 (Applied Zn Powder Layer)

Figure 6 shows that with an increase in the amount of Zn powder, the local η phase thickens, with a maximum thickness of 29.3 μm and a minimum thickness of 1.6 μm. The η phase thickens with an increase in Zn content. Springer et al. [11] attributed this phenomenon to the improved growth kinetics of the η phase from Zn. Moreover, an increase in applied Zn powder causes an enhancement of the interface discontinuity. Figure 6c,d demonstrate a significant deposition of particles into the Al substrate.

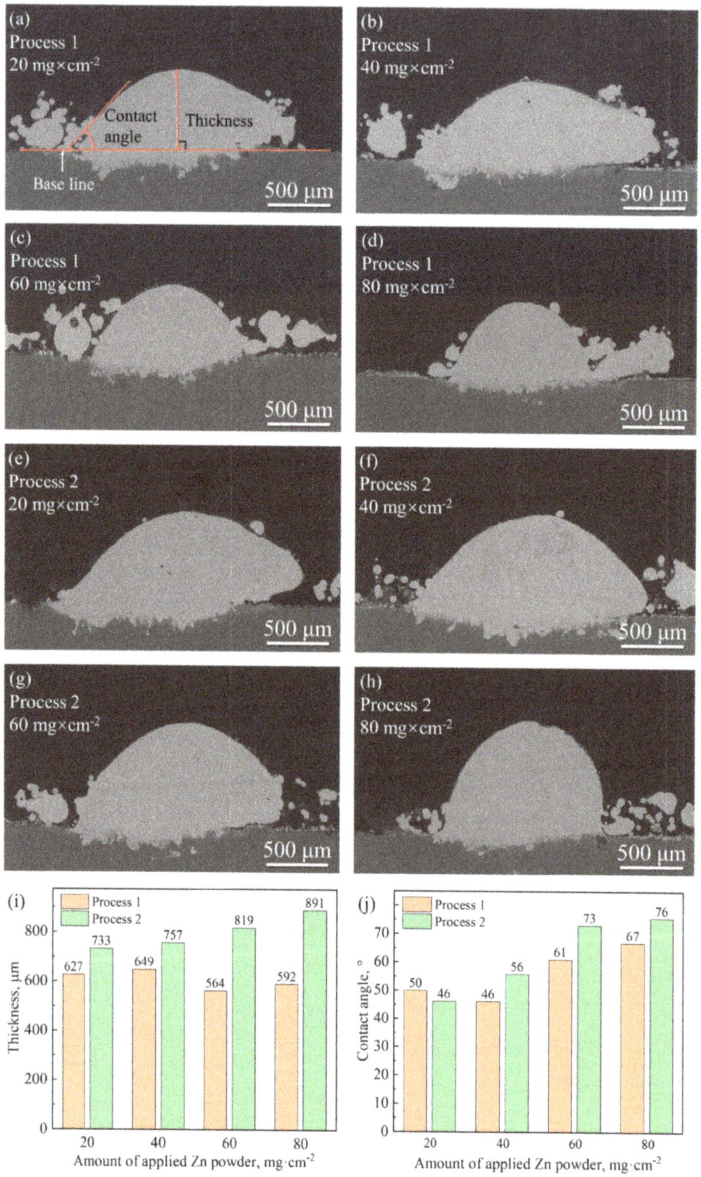

Figure 5. The (**a–h**) BSE morphologies, (**i**) thickness and (**j**) contact angle of the cross-section of the cladding layer. The red lines and arrows indicate the measurement methods for the thickness of the cladding layer and the contact angle.

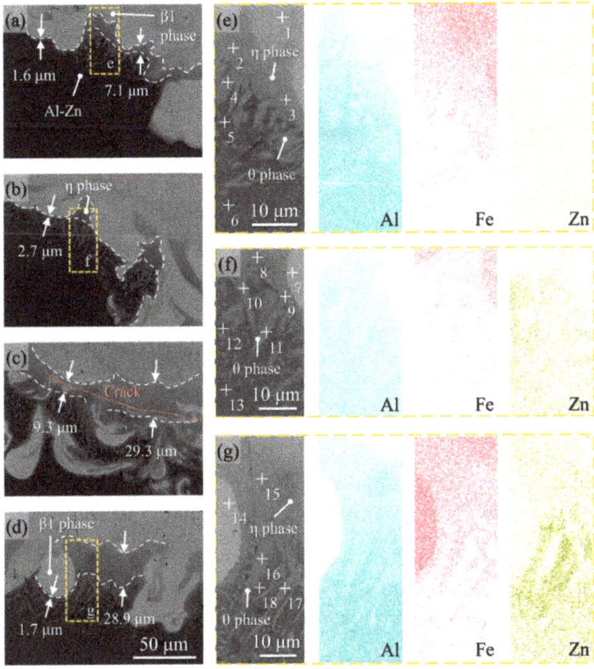

Figure 6. BSE morphologies of interface with applied Zn powder (**a**) 20 mg × cm^{-2}, (**b**) 40 mg × cm^{-2}, (**c**) 60 mg × cm^{-2}, (**d**) 80 mg × cm^{-2}; and magnified segments showing element distribution of specific areas from (**a**,**b**,**d**) corresponding to (**e**–**g**), respectively.

Figure 6e–g illustrate that a greater amount of applied Zn powder changes the θ phase on the near Al side from a pinkish needle to a lath. At the same time, there are distinct differences in the distribution of Zn at the interface. This difference is not observed in Figure 6e when the Zn content is low, but it is evident in Figure 6f,g as the Zn content increases. Firstly, compared to IMC phases, Zn significantly enriches the Al substrate, as demonstrated in Table 3 with a 40 mg × cm^{-2} concentration of Zn. The Zn content in the substrate is approximately 2.43%, which is higher than the concentrations observed in the η and θ phases (1.50% and 1.29%, respectively). Secondly, in different substrate regions, Zn manifests component segregation, as illustrated by Figure 6g.

Zn is concentrated in the Al substrate in the dendrite gap of the θ phase, indicating that Zn has a high affinity for the substrate. In conclusion, the Zn content in the θ phase is approximately 0.59% to 0.87%, and exceeds 2.77% in the η phase, as shown in Table 3.

3.2.2. Interface of Process 2 (Alloying Zn Layer)

As illustrated in Figure 7, an increase in Zn content resulted in the thickening of the η phase at the interface from approximately 9.6 μm to 43.6 μm and the coarsening of dendritic microstructure width adjacent to the aluminum interface from 1 μm to 5 μm. Energy spectrum testing at points 3 and 9 in Table 4 revealed that the dendritic microstructure's primary component is the θ phase, with 80% aluminum content. Furthermore, differences exist in the distribution of Zn at the interface, akin to the process seen in Figure 6. Table 4 shows that the substrate's Zn content ranges from 3.60% to 5.15%, higher than those found in the η and θ phases (2.56% and 1.48%, respectively). Consequently, Zn is significantly enriched in the aluminum substrate. According to Table 4, the Zn content in the θ phase is about 1.48% and 2.56% in the η phase.

Table 3. Element content at the test point of process 1, at.%.

Process	Position	Al	Cr	Fe	Zn	Phase
20 mg × cm^{-2}	1	16.30	13.32	70.39	0	β1-$(Fe\text{-}Cr)_3Al$
	2	70.80	3.58	25.18	0.45	η-$(Fe\text{-}Cr)_2Al_5Zn_x$
	3	69.13	7.78	22.49	0.61	η-$(Fe\text{-}Cr)_2Al_5Zn_x$
	4	81.89	1.19	16.26	0.65	θ-$(Fe\text{-}Cr)_4Al_{13}$
	5	98.55	0.07	0.66	0.72	Al-Zn
	6	97.74	0.35	1.55	0.36	Al-Zn
40 mg × cm^{-2}	7	2.46	16.10	80.41	1.03	M-Fe_4Cr
	8	71.63	4.53	22.34	1.50	η-$(Fe\text{-}Cr)_2Al_5Zn_x$
	9	71.06	4.89	22.93	1.12	η-$(Fe\text{-}Cr)_2Al_5Zn_x$
	10	78.95	1.40	18.37	1.29	θ-$(Fe\text{-}Cr)_4Al_{13}$
	11	79.69	1.95	17.32	1.05	θ-$(Fe\text{-}Cr)_4Al_{13}$
	12	93.50	0.11	0.65	5.74	Al-Zn
	13	95.72	0.55	1.31	2.43	Al-Zn
80 mg × cm^{-2}	14	10.16	19.38	70.16	0.29	M-Fe_4Cr
	15	63.78	4.85	28.60	2.77	η-$(Fe\text{-}Cr)_2Al_5Zn_x$
	16	77.74	2.02	19.66	0.59	θ-$(Fe\text{-}Cr)_4Al_{13}$
	17	80.35	1.78	17.00	0.87	θ-$(Fe\text{-}Cr)_4Al_{13}$
	18	92.44	1.06	1.41	5.09	Al-Zn

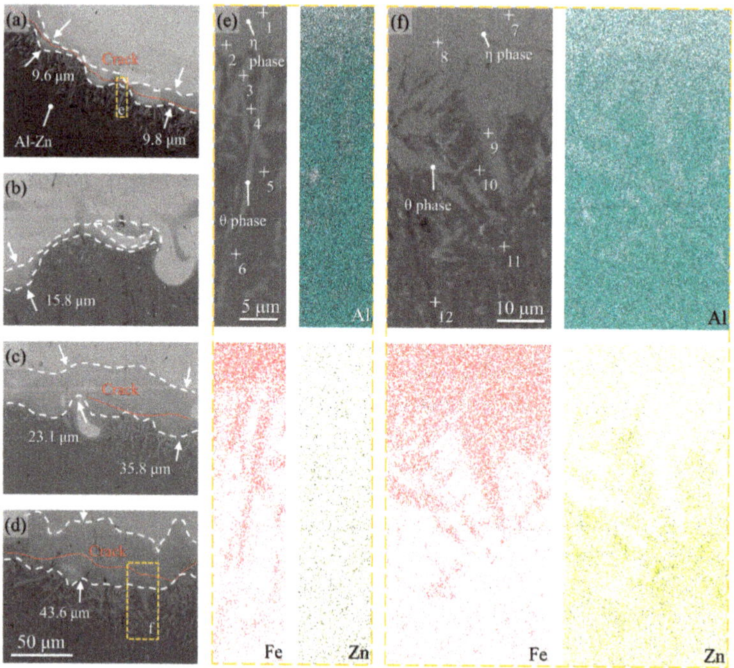

Figure 7. BSE morphologies of interface with applied Zn powder (**a**) 20 mg × cm^{-2}, (**b**) 40 mg × cm^{-2}, (**c**) 60 mg × cm^{-2}, (**d**) 80 mg × cm^{-2}; and magnified segments showing element distribution of specific areas from (**a**,**d**) corresponding to (**e**,**f**), respectively.

Table 4. Element content at the test point of process 2, at.%.

Process	Position	Al	Cr	Fe	Zn	Phase
40 mg × cm^{-2}	1	67.48	6.11	25.81	0.59	η-$(Fe\text{-}Cr)_2Al_5Zn_x$
	2	69.49	4.71	24.91	0.88	η-$(Fe\text{-}Cr)_2Al_5Zn_x$
	3	81.66	2.96	14.79	0.59	θ-$(Fe\text{-}Cr)_4Al_{13}$
	4	93.90	0.92	4.29	0.89	Al-Fe
	5	97.79	0.24	1.09	0.88	Al-Zn
	6	97.81	0.44	1.25	0.50	Al-Zn
80 mg × cm^{-2}	7	69.10	2.97	25.37	2.56	η-$(Fe_4Cr)_2Al_5Zn_x$
	8	70.03	6.09	20.92	2.96	η-$(Fe_4Cr)_2Al_5Zn_x$
	9	80.52	2.39	15.60	1.48	θ-$(Fe_4Cr)_4Al_{13}$
	10	94.38	0.38	1.31	3.93	Al-Zn
	11	94.51	0.14	0.20	5.15	Al-Zn
	12	94.97	0.21	1.22	3.60	Al-Zn

During both processes, a chaotic interface was formed when a larger Fe droplet entered the substrate. However, this droplet fails to diffuse and react fully, quickly solidifying and forming a bonding interface that has chaotic characteristics.

Process 1 and process 2 show a discrepancy in that Zn powders in process 1 absorb heat during cladding to shorten reaction time and reduce reaction temperature compared to process 2. This results in a thinner η phase layer and a shorter time for dendrite nucleation of the θ phase to grow, illustrated by the formation of a short rod-shaped θ phase in Figure 7e. The content of Zn in process 1 is slightly higher than in process 2 at the interface, indicating Zn diffusion during laser alloying. However, this may lead to a certain amount of burning loss, as evidenced in Table 1.

Figures 6c and 7a,c,d demonstrate the microcracks at the interface, particularly along the red line position. These cracks penetrate the η phase and are adjacent to the side of the aluminum substrate due to the higher aluminum content in the IMC interface adjacent to the aluminum side. Aluminum-rich intermetallics exhibit high brittleness. The composition assessment in Table 4 shows that the Al content of the IMC near the crack is about 70%, indicating an η phase. Discontinuous microcracks can also be observed at the η layer, which may lead to a reduction in the contact area of the cladding layer and a reduced interface bonding strength. However, subsequent tensile tests showed no significant correlation between microcracks and the shear strength of the cladding layer.

Zn is enriched in the substrate. There are two possible reasons for this: (1) The melting point of the Al substrate is lower than that of Fe, allowing the liquid to last longer after being acted upon by the molten Fe. This is conducive to Zn diffusion in the Al substrate. (2) The solid solubility of Zn in Al is high. The uneven distribution of Zn in the Al substrate is mainly due to the difference in temperature of the Fe droplets.

Although the θ phase and η phase undergo solid–liquid transformation at nearly the same time (their respective melting points, 1160 °C and 1166 °C, respectively, are similar [25]), the content of Zn in the two phases is distinct. This is because the [001] direction of the η phase encourages the diffusion of Zn atoms, leading to Zn atoms' rapid diffusion in the η phase even in the solid state at high temperatures [26].

3.2.3. Effect of Zn Content and Surface Energy Density on η Phase Thickness

Figure 8 illustrates the relationship between the thickness of the η phase layer and the amount of applied Zn powder in the two processes. Process 1 results in the thickest η phase layer at the bonding interface when the amount of Zn powder applied reaches 60 mg × cm^{-2}. On the other hand, process 2 exhibits a gradual thickening of the η phase layer with increasing amounts of Zn powder applied. Generally, Zn increases the thickness of the η phase layer significantly, whereas laser alloying the Zn layer can stabilize the thickness fluctuation of the η phase layer, as shown by the small error bar.

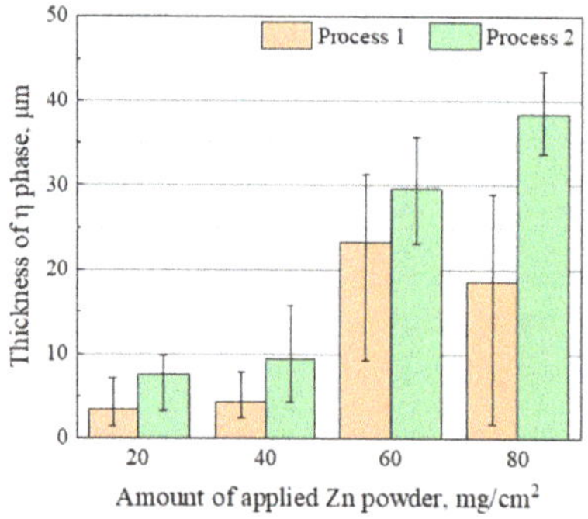

Figure 8. Comparison of the thickness of the η layer under two different processes at various amounts of applied Zn powder.

Springer et al. [11] suggests that under constant temperature conditions, the Zn layer enhances the reaction kinetics of the η phase. There may be no change in activation energy, but an interaction of Zn with the structural vacancies of the η phase likely causes its accelerated growth. Applying Zn coatings onto steel is advantageous for solid–liquid joining procedures, assisting bonding by enabling rapid dissolution and forming even and ordered reaction layers [27]. Laser cladding exhibits non-equilibrium metallurgical reactions. Based on the above results, there is a positive correlation between the growth thickness of the η phase and Zn at the interface. Ternary compounds of Zn, Fe, and Al have lower melting points. Thus, under the same temperature cycle, the lower the melting point, the longer the liquid duration, leading to continued Zn consumption within the reaction. Consequently, higher Zn content prolongs the reaction duration.

Various processes and material brands significantly influence the thickness of the η phase layer at the interface of a Fe-Al laser joint. Table 5 summarizes the Zn layer thickness (T_{Zn}), η phase thickness (T_{η}), velocity (v), laser power (P), laser spot diameter (d), and material brand of different Fe-Al laser joining technologies.

Suder et al. [34] proposed a definition of specific point energy to analyze the effects of welding parameters and processes. However, the formula's applicability is restricted to small beam diameters, requiring a laser beam with uniform intensity distribution and constant interaction time across the laser spot. When investigating the influence of altered laser power density, the impact of spot diameter on power density is negatively correlated. On the basis of the linear energy density described by Yang et al. [5], we divided it by the equivalent spot diameter of the laser spot to produce the area energy density. By analyzing the relationship between surface energy density and η phase thickness, the relationship is determined using the surface energy density expressed as Equation (2):

$$\rho = \frac{P}{vd} \tag{2}$$

where ρ is surface energy density, P is laser power, v is welding velocity, and d is laser spot diameter.

Figure 9 summarizes and compares the impacts of surface energy density and the Zn coating on the thickness of the η phase at the interface. The thickness of the η phase is influenced by both the presence of Zn at the interface and the surface energy density. A higher surface energy density clearly increases the thickness of the η phase at the interface. Under the same surface energy density, a Zn layer attached to the surface before processing significantly boosts the thickness of the η phase at the interface, exemplified by the results of Ogura et al. [30], who imposed a 7.1 μm Zn layer. However, Al-1.88Si wire is used for wire filling connection where Si inhibits the growth of the η phase. The results of Rathod et al. [28] adopted the rolling method, resulting in the formation of a thicker η layer despite the absence of Zn addition. Notably, Ozaki et al. [33] adopted a similar method to

Rathod et al. [28]. Still, the η layer formed under the same surface energy density is thinner when compared to Rathod. This may be due to the different distance between the laser center and the contact point of the roller (17 mm [28] vs. 25 mm [33]).

Table 5. Zn layer thickness (T_{Zn}), thickness of η phase (T_η), velocity (v), laser power (P), laser spot diameter (d), materials brand of some Fe-Al laser joining technologies.

Initial Surface	T_{Zn}, μm	T_η, μm	v, mm×s^{-1}	P, W	d, mm	Materials Brand	Ref.
Without Zn coating	0	4 5 8	20 16.7 11.7	3000	1	6016/DC 04	[10]
		10 2	25 40	1500	2.5–3.5	5052/low carbon steel	[28]
		5 3 1	10 10 10	2200	0.85	6061/S235-JR	[29]
Zn coated	10	23 8 4	10 20 23.3	3000	1	6016/DC 04	[10]
	7.1	30 3	5 3.33	1500	1.5	A5052/SPCC-GA	[30]
	20	40 5	10 23.3	3000	~2	6016/DC 04	[31]
	10	5 9 12	11.7 11.7 11.7	1000 1300 1600	2 × 0.8	7075/DP590	[32]
	20	3.5 1.2	8.3 10	1500	~1.5	GI/A6000	[33]

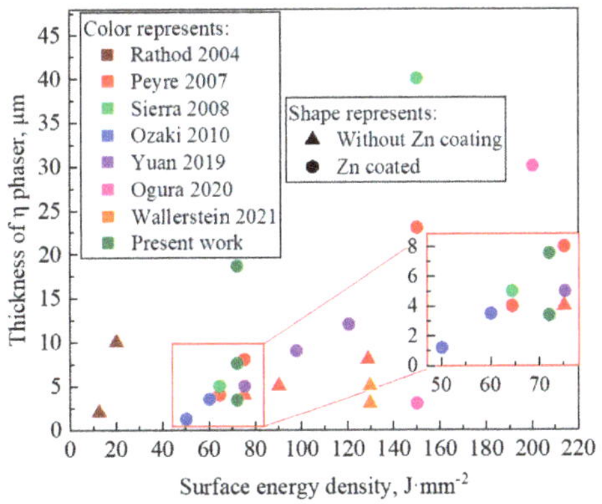

Figure 9. Impact of surface energy density and Zn coating on η phase thickness at interface [10,28–33].

3.3. Shear Resistance

3.3.1. Shear Strength and Fracture Analysis

The information shown in Figure 10 indicates that the amount of applied Zn powder has a noteworthy impact on the shear strength of the cladding layer. Initially, the shear strength of the cladding layer increases with the rise in the amount of Zn powder up to a particular point, then it begins to decrease.

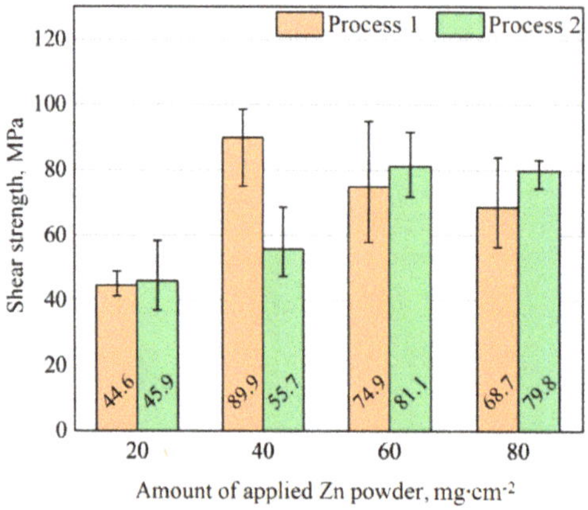

Figure 10. Shear strength of the interface.

Process 1 resulted in the highest shear strength of the cladding layer at 40 mg × cm^{-2} of applied Zn powder, which was 89.9 MPa. On the other hand, the greatest average shear strength of the cladding layer for process 2 was produced by the implementation of 60 mg × cm^{-2} of Zn powder, which resulted in 81.19 MPa. However, the cladding layer's maximum shear strength slightly decreases when the amount of applied Zn powder increases to 80 mg × cm^{-2}, but the standard deviation decreases, indicating more stable shear resistance. Therefore, the recommended applied Zn powder quantities can vary depending on the process used, and it should be meticulously adjusted to gain the desired shear strength and stability.

Xia et al. [35] observed that cross dual laser beams generated satisfactory weld formation and the most homogeneity of τ_5, thus resulting in joints with improved tensile strength compared to single laser beams and in-line dual laser beams. Ozaki et al. [33] discovered that specimen failure happens at the base metal, which was made through laser roll welding with a 150 MPa roll pressure.

Figure 11 shows the XRD results and macroscopic morphology of the shear fracture. As shown in Figure 11a, the shear fracture zone consists mainly of the Al phase, the $Fe_2Al_5Zn_x$ phase, the Fe_4Al_{13} phase, and a small quantity of the Zn phase. This composition is consistent with the EDS analysis results mentioned above. The composition of the shear fracture surface is mainly composed of Al-riched IMCs and the Al phase, indicating that the fracture occurred at the IMCs layer, and due to the peeling effect, part of the aluminum substrate is exposed.

The shear process creates shear stress, which exposes the Al substrate in the right fracture zone. Figure 11b,c show that for process 1, when the amount of Zn powder is 20 mg × cm^{-2}, there is a point peeling area, which mainly comes from the separation of the droplet part connected with the cladding layer and the substrate together with the cladding layer. When the amount of Zn increases to 80 mg × cm^{-2}, it is mainly valley tearing.

Figure 11d,e show that for process 2, when the amount of Zn powder is 20 mg × cm^{-2}, there is not only a point peeling area, but also a valley tearing at the right part, and with the increase in the amount of applied Zn powder, the valley tear is more obvious, indicating that the shear strength of the interface is closer to the shear strength of the substrate. In addition, the flatter and more uniform bonding interface also reduces the area of the spot peeling area.

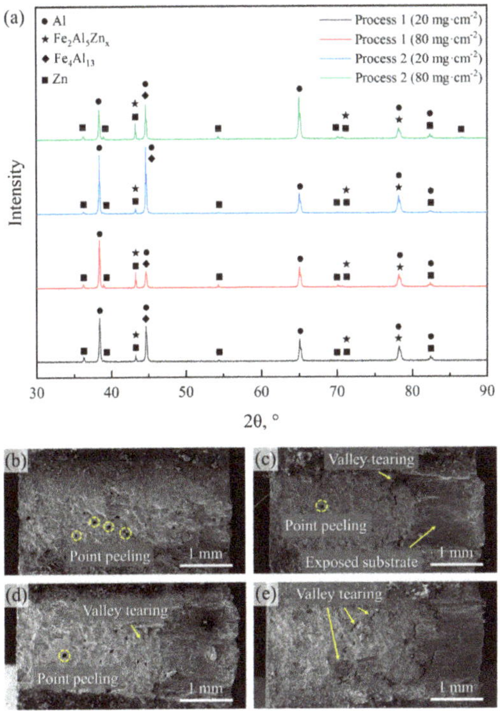

Figure 11. Shear fracture analysis of samples: (**a**) XRD results and macroscopic morphology of (**b**) process 1 with 20 mg × cm^{-2} applied Zn powder layer; (**c**) process 1 with 80 mg × cm^{-2} applied Zn powder layer; (**d**) process 2 with 20 mg × cm^{-2} applied Zn powder layer; (**e**) process 2 with 80 mg × cm^{-2} applied Zn powder layer.

3.3.2. Shear Section of Process 1 (Applied Zn Powder Layer)

Figure 12 shows the BSE and SE morphologies of the fracture section morphology with 20 mg × cm^{-2} and 40 mg × cm^{-2} applied Zn powder. The average shear strength is 44.6 MPa for 20 mg × cm^{-2} and 89.9 MPa for 40 mg × cm^{-2} applied Zn powder amounts, respectively. Table 6 shows the element content at the corresponding position in Figure 12.

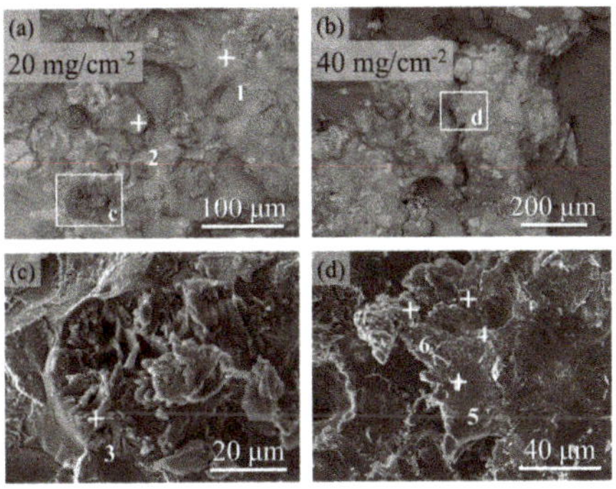

Figure 12. Fracture section morphologies of process 1: (**a**) BSE morphology with 20 mg × cm^{-2} applied Zn powder; (**b**) BSE morphology with 40 mg × cm^{-2} applied Zn powder; (**c**) SE morphology of coarse dendrites θ phase; (**d**) SE morphology of tear area.

Table 6. Element content at the corresponding position of process 1, at.%.

Process	0	Al	Cr	Fe	Zn	Phase
20 mg × cm^{-2}	1	61.75	7.31	30.11	0.84	η-$(Fe\text{-}Cr)_2Al_5Zn_{0.1}$
	2	77.13	3.4	19.27	0.19	θ-$(Fe\text{-}Cr)_4Al_{13}$
	3	84.17	3.48	11.94	0.40	θ-$(Fe\text{-}Cr)_4Al_{13}$
40 mg × cm^{-2}	4	43.94	8.79	45.88	0.39	β2-(Fe-Cr)Al
	5	66.01	5.97	25.57	2.45	η-$(Fe\text{-}Cr)_2Al_5Zn_{0.2}$
	6	78.58	2.38	19.04	0	θ-$(Fe\text{-}Cr)_4Al_{13}$

According to the composition of the sample with an applied Zn powder of 20 mg × cm^{-2}, the η and θ phases mainly exist in the section. As shown in Figure 12c, the θ phase with 84.17% Al content is at point 3, which remains on the surface of the substrate in the form of coarse dendrites. Due to the small amount of Zn, it has little effect on the formation of the Zn-rich η phase, and there is a large θ phase area.

Figure 12b shows that the sample section with an applied Zn powder of 40 mg × cm^{-2} has a long tear area. As shown in Figure 12d, the crack initiation site (point 6) of the tear area is θ phase, and the rest (point 4 and point 5) are mainly β2 and η phases, and the fracture surface presented brittle fracture. The interface strength is improved due to the higher Zn content at the interface.

With the further increase in the applied Zn powder amount reaching up to 60 mg × cm^{-2} and 80 mg × cm^{-2}, η and θ exist at the same time, mainly due to the better mechanical properties of the η phase, and its better strength. However, the discontinuity of the interface is improved and microcracks appear, as shown in Figure 6, which results in the average shear strength of the interface being lower than that of 40 mg × cm^{-2} applied Zn powder amount, but it is still higher than that of the 20 mg × cm^{-2} applied Zn powder amount.

3.3.3. Shear Section of Process 2 (Alloying Zn Layer)

Figure 13 shows the BSE and SE morphologies of the fracture section morphology with an applied Zn powder amount of 20 mg × cm^{-2} and 80 mg × cm^{-2}, yielding an average shear strength of 45.7 MPa and 79.8 MPa, respectively. Table 7 shows the element content at the corresponding position in Figure 13.

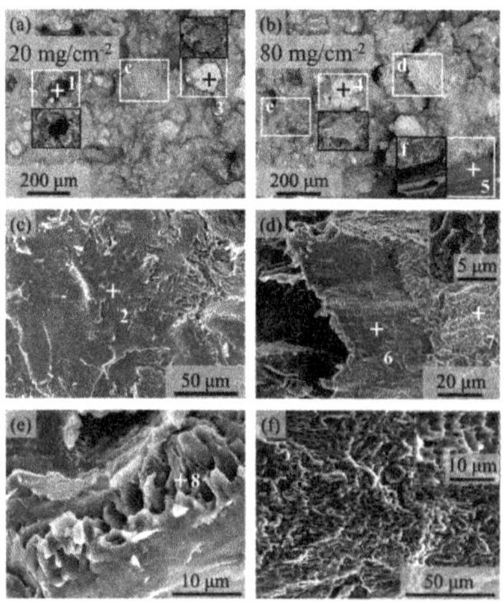

Figure 13. Fracture section morphologies of process 2: (**a**) BSE morphology with 20 mg × cm^{-2} applied Zn powder; (**b**) BSE morphology with 80 mg × cm^{-2} applied Zn powder; (**c**) SE morphology of brittle fracture; (**d**) SE morphology of tear area; (**e**) SE morphology of coarse dendrite θ phase; (**f**) SE morphology of torn Al substrate area.

Table 7. Element content at the corresponding position of process 2, at.%.

Process	Point	Al	Cr	Fe	Zn	Phase
	1	97.25	0.05	2.62	0.98	Al-Fe-Zn
20 mg × cm^{-2}	2	61.67	6.17	31.48	0.68	η-$(Fe\text{-}Cr)_2Al_5$
	3	12.05	16.17	71.03	0.75	α-(Fe-Cr)
	4	5.97	18.58	75.37	0.09	α-(Fe-Cr)
	5	97.61	0.08	0.74	1.58	Al-Fe-Zn
80 mg × cm^{-2}	6	55.77	6.70	32.70	4.83	η-$(Fe\text{-}Cr)_2Al_5Zn_x$
	7	0.81	20.07	78.53	0.59	α-(Fe-Cr)
	8	77.09	2.85	18.97	1.08	θ-$(Fe\text{-}Cr)_4Al_{13}$

In Figure 13a, there is a point peeling section in area 1, and the droplet part connected with the cladding layer is separated from the substrate along with the cladding layer. According to the composition scanning in Table 7, it is found that the point peeling section is mainly composed of the Al substrate. As shown in Figure 13f, the completely torn area exposes the Al substrate, showing dense dimple morphology. In area 5 shown in Figure 13b, friction marks formed on the surface of the Al substrate due to ploughing are found.

Figure 13c displays the region on the fracture surface where the η phase is situated with an applied Zn powder of 20 mg × cm^{-2}. The fracture surface is relatively flat and presents a typical brittle fracture morphology. As shown in Figure 13d, the fracture surface with 80 mg × cm^{-2} applied Zn powder has a horizontal tear area, such as the valley area in Figure 11 above. The content of Zn in the upper part of the initial crack zone of the tear zone is 4.83% higher than that in other areas (area 6). The Al substrate also cracked under the pulling action of the Zn-rich η phase and the alpha phase during the shear test. Meanwhile, the right side of this area is the α phase, and its fracture surface is ductile. Figure 13e shows a short rod-shaped fracture structure, and its main component is θ-$(Fe\text{-}Cr)_4Al_{13}$ with high Zn content, which is derived from the residue on the substrate after the fracture of the coarse dendrite θ phase shown in Figure 7.

Under different amounts of Zn powder preform, the shear fracture surface contains two types of fracture areas with brittle and ductile fracture characteristics. The brittle fracture zone presents a fluvial pattern, as shown in the area near point 2 in Figure 13, which is mainly an exposed η phase (area 2 and area 6). The ductile fracture area is characterized by dimples. The main components of the ductile fracture zone are alpha-Fe4Cr and Al substrate areas rich in Fe and Cr, such as the areas near points 3, 4, 5, and 7. The area near point 8, which is dominated by the θ phase, presents irregular fracture morphology, and its essence is a brittle fracture.

3.4. DFT Calculation of Fe_2Al_5 and $Fe_2Al_5Zn_x$

3.4.1. Calculation Model

Figure 14 shows the crystal structure of η-Fe_2Al_5, which belongs to the orthorhombic crystal system. The space group is Cmcm (No. 63) and the lattice constant a = 7.6559 (8), b = 6.4154 (6), and c =4.2184 (4) Å. The crystal structure of Fe_2Al_5 is homologous with that of $MnAl_6$ and Co_2Al_5. When optimizing the structure of the Fe_2Al_5 crystal, Burkhardt et al. [26] pointed out that the model of the Al atom at position 4 (b) (occupied by the Al2 atom) cannot explain the continuous overlapping small peaks of electron density near y = 0.5 on the (100) crystal plane. This phenomenon can be reasonably described by adding at least one Al atom (Al3) at the position of 8 (f). Among them, the occupation factor of the Al2 atom is 0.36, and that of the Al3 atom is 0.23, as shown in Figure 14a.

Due to the interaction between Fe atoms and Al1 atoms, the Al2 and Al3 atoms in the [001] direction form a disordered distribution. As shown in Figure 14b, after the Al2 and Al3 atoms are ignored, an octagonal channel structure (or pentagonal antiprism shape) is formed in the [001] direction. The Al2 atom and Al3 atom are in the same cell and generally do not appear at the same time [26]. The reason for the emergence of channels of disordered or incommensurately spaced atoms can be explained by the combination of preferred electron counts tied to stoichiometry and continuous paths of CP quadrupoles [36].

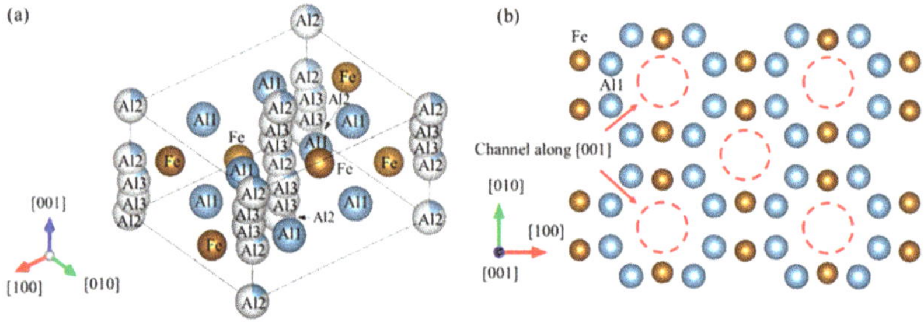

Figure 14. Crystal structure of η-Fe_2Al_5: (**a**) Atom position of Fe, Al1, Al2 and Al3, and (**b**) channel in [001] direction.

The adopted η-Fe_2Al_5 crystal structure model is shown in Figure 15a, and the aluminum atom at the Al3 position is omitted from its crystal structure. Figure 15b shows the crystal structure model of η-$Fe_2Al_5Zn_x$. In the crystal structure model, the content of Zn is close to 0.4%, and Zn atoms are in the [001] direction, replacing 1/4 of the Al2 atom. Chen et al. [37] pointed out that the vacancy in the [001] direction is an important way for Zn atoms to stay and diffuse.

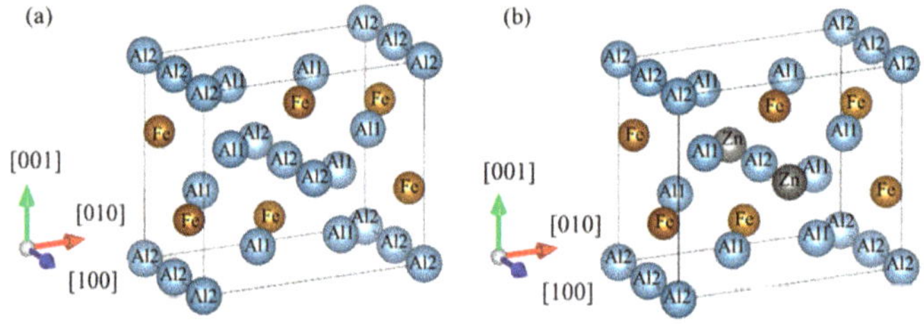

Figure 15. Calculation model of (**a**) Fe_2Al_5 and (**b**) $Fe_2Al_5Zn_x$.

As shown in Table 8, the lattice constant and formation enthalpy of Fe_2Al_5 and $Fe_2Al_5Zn_x$ after structural optimization are calculated ΔH_f. The enthalpy of the formation of both crystals is negative, indicating that the two crystal structures are thermodynamically stable. The calculated enthalpy of the formation of Fe_2Al_5 is about -16.99 kJ × mol^{-1}, which is close to that calculated by the EAM method. $Fe_2Al_5Zn_x$ shows a lower enthalpy of formation, mainly because Zn has a higher formation energy than Al. The lattice constant of $Fe_2Al_5Zn_x$ is close to that of Fe_2Al_5, but the lattice constant α increases slightly. This may be caused by the introduction of Zn atoms which destroyed the original crystal structure symmetry.

Table 8. Calculated lattice constant and formation enthalpy of Fe_2Al_5 and $Fe_2Al_5Zn_x$.

Phase	Method	a (Å)	b (Å)	c (Å)	Angle	ΔH_f (kJ × mol^{-1})
Fe_2Al_5	This work EAM	7.476 7.622	6.214 6.323	4.741 4.178	α = β = γ = 90°	−16.99 −14.09 [17]
$Fe_2Al_5Zn_x$	This work	7.477	6.236	4.700	α = 90.125° β = γ = 90°	−17.62

3.4.2. Elastic Properties

The anisotropic elastic properties of Fe_2Al_5 and $Fe_2Al_5Zn_x$ are calculated. Table 9 shows the elastic constants of Fe_2Al_5 and $Fe_2Al_5Zn_x$. Fe_2Al_5 has 9 elastic constants because of its orthorhombic structure, and $Fe_2Al_5Zn_x$ has 13 elastic constants because of its monoclinic structure. Figure 16 shows the Young's modulus E and shear modulus G. The Young's modulus has a maximum value in the ⟨100⟩ crystallographic family, the shear modulus has an extreme value in the [100] and [001] crystallographic family, and a maximum value in the ⟨111⟩ crystallographic family, which is consistent with the symmetry of the orthorhombic system. The shear modulus of Fe_2Al_5 and $Fe_2Al_5Zn_x$ are basically the same, but the Young's modulus is different in the direction of [010] and [001]. After Zn is added, the Young's modulus in the direction of [010] and [001] decreases.

Table 9. Elastic constants of Fe_2Al_5 and $Fe_2Al_5Zn_x$.

Fe_2Al_5	C_{11}	C_{12}	C_{13}	C_{22}	C_{23}	C_{33}	C_{44}	C_{55}	C_{66}
	240.91	66.08	53.07	283.08	63.37	272.50	70.26	52.30	86.33
$Fe_2Al_5Zn_x$	C_{11}	C_{12}	C_{13}	C_{22}	C_{23}	C_{33}	C_{44}	C_{55}	C_{66}
	233.56	64.31	55.16	277.00	66.83	265.66	75.92	54.22	86.07
	C_{15}	C_{25}	C_{35}	C_{46}					
	4.15	−1.03	2.07	1.38					

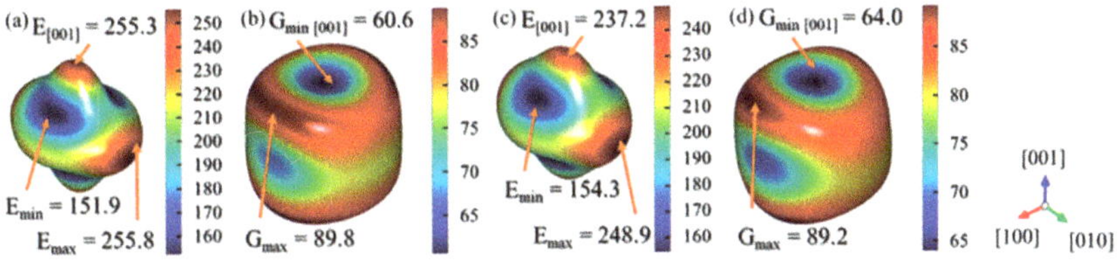

Figure 16. Calculated anisotropic elastic moduli of Fe_2Al_5 and $Fe_2Al_5Zn_x$: (**a**) Young's modulus and (**b**) shear modulus of Fe_2Al_5; (**c**) Young's modulus and (**d**) shear modulus of $Fe_2Al_5Zn_x$. (Unit: GPa).

3.4.3. Differential Charge Density

After Zn is added, the Young's modulus of the η phase decreases in the b and c directions, which can be explained by the charge distribution. Figure 17 shows the differential charge density distribution of η-Fe_2Al_5 and η-$Fe_2Al_5Zn_x$ on (100) and $(1\bar{3}0)$ crystal planes. As described by Burkhardt et al. [26], the peak of the charge distribution appears at the Fe atom. According to Figure 17a,c, the differential charge density also shows that the charge is mainly concentrated near the Fe atom, which improves the bond strength of the Fe-Al binary IMC and makes the IMC have properties similar to ceramics. Also affected by the Al atom in the [001], the high charge density region near the Fe atom is mainly arranged along the [001], and the charge distribution is non-uniform. This difference will also cause the mechanical properties of IMC to show anisotropy. The large atomic spacing and small charge density difference in the direction of a are the reasons why the Young's modulus in the direction of a is smaller than that in the direction of [010] and [001], as shown in Figure 16 above. Comparing Figure 17a,b, the charge density of Fe atoms on the (100) crystal plane decreases after the presence of Zn atoms in the channel along the [001] direction, indicating that Zn atoms also have strong binding ability on electrons. Similar phenomena can also be found on the $(1\bar{3}0)$ crystal plane by comparing Figure 17c,d.

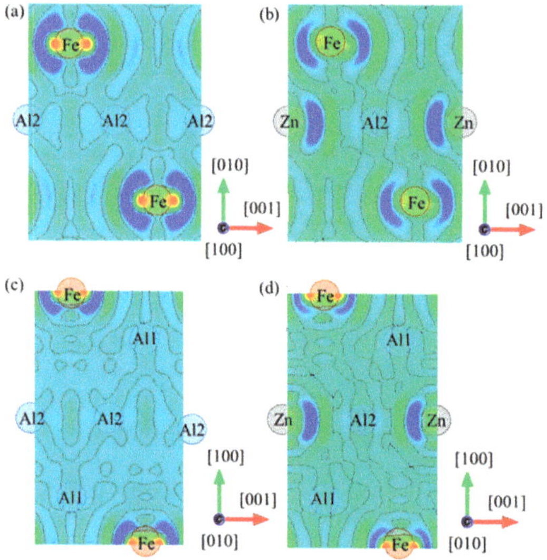

Figure 17. Differential charge density on crystal planes, (**a**) (100) plane of Fe_2Al_5, (**b**) (100) plane of $Fe_2Al_5Zn_x$, (**c**) $(1\bar{3}0)$ plane of Fe_2Al_5 and (**d**) $(1\bar{3}0)$ plane of $Fe_2Al_5Zn_x$.

4. Conclusions

In summary, this study explored the impact of the Zn content on the Al-Fe laser cladding interface. The findings indicate that adding Zn powder to the surface of the Al substrate enables the production of a continuous Fe-based cladding layer by laser cladding. However, exceeding 60 mg $\times$ cm^{-2} of applied Zn powder can limit the spread of the Fe-based cladding layer and increase the contact angle of the Fe-based cladding layer. The alloying Zn layer is useful for spreading the Fe-based cladding layer on the Al substrate.

Moreover, the results demonstrate that the Zn content at the interface is positively correlated with the amount of applied Zn powder layer. Pre-alloying increases the thickness of the η phase at the interface, with a maximum thickness of 43.6 μm. Zn is mainly enriched in the substrate, followed by the IMC layer. The Zn content is higher in the η phase compared to the θ phase in the [001] due to the vacancy of the η phase. Additionally, we found that Zn can improve the shear strength of the interface and increase the content of Zn-Al eutectic in the applied Zn layer through laser alloying, enhancing the shear strength stability. While the shear fracture is mainly a brittle fracture dominated by the η phase, there is also a localized ductile fracture between the cladding layer and the substrate. After laser alloying, the highest shear strength is obtained when the amount of applied Zn powder is 60 mg $\times$ cm^{-2}, with an average of 89.9 MPa.

Finally, based on DFT calculations, the mechanical properties of Fe_2Al_5 and $Fe_2Al_5Zn_x$ are found to be quite similar, with the Young's modulus and shear modulus in [010] and [001] being only slightly reduced. The study also found that the effect of Zn on the strength of the η phase is limited and that the introduction of Zn can improve the thickness and stability of the η phase at the Al-Fe interface. Overall, these findings provide important insights into the effect of Zn content at the Al-Fe laser cladding interface and can be useful in developing new high-performance materials for various industrial applications.

Author Contributions: Investigation and writing—original draft, K.W.; Methodology, supervision and Funding acquisition, X.Y.; Project administration, supervision and funding acquisition, C.L.; Resources and Funding acquisition, K.D. All authors have read and agreed to the published version of the manuscript.

Funding: This research was funded by National Key R&D Program of China (grant number 2018YFB2002000) and National Key R&D Program of China (grant number 2022YFB4601803).

Institutional Review Board Statement: Not applicable.

Informed Consent Statement: Not applicable.

Data Availability Statement: All data that support the findings of this study are included within the article.

Conflicts of Interest: The authors declare no conflict of interest.

References

1. Lampa, C.; Smirnov, I. High speed laser cladding of an iron based alloy developed for hard chrome replacement. *J. Laser Appl.* **2019**, *31*, 022511. [CrossRef]
2. Sadeghian, A.; Iqbal, N. A review on dissimilar laser welding of steel-copper, steel-aluminum, aluminum-copper, and steel-nickel for electric vehicle battery manufacturing. *Opt. Laser Technol.* **2021**, *146*, 107595. [CrossRef]
3. Edrisy, A.; Perry, T.; Alpas, A.T. Investigation of scuffing damage in aluminum engines with thermal spray coatings. *Wear* **2005**, *259*, 1056–1062. [CrossRef]
4. Wang, K.Y.; Xie, C.; Du, K.P.; Li, C.; Yin, X. A novel rotatable inner-surface laser cladding method. In Proceedings of the 10th International Symposium on Advanced Optical Manufacturing and Testing Technologies: Advanced and Extreme Micro-Nano Manufacturing Technologies, Chengdu, China, 13 December 2021. [CrossRef]
5. Shang, S.-L.; Sun, H.; Pan, B.; Wang, Y.; Krajewski, A.M.; Banu, M.; Li, J.; Liu, Z.-K. Forming mechanism of equilibrium and non-equilibrium metallurgical phases in dissimilar aluminum/steel (Al–Fe) joints. *Sci. Rep.* **2021**, *11*, 24251. [CrossRef] [PubMed]
6. Šlapáková, M.; Křivská, B.; Fekete, K.; Králík, R.; Grydin, O.; Stolbchenko, M.; Schaper, M. The influence of surface on direction of diffusion in Al-Fe clad material. *Mater. Charact.* **2022**, *190*, 112005. [CrossRef]
7. Zhou, D.W.; Liu, J.S.; Lu, Y.Z.; Xu, S. Effect of adding powder on joint properties of laser penetration welding for dual phase steel and aluminum alloy. *Opt. Laser Technol.* **2017**, *94*, 171–179. [CrossRef]
8. Yan, F.; Zhang, K.; Yang, B.; Chen, Z.; Zhu, Z.; Wang, C. Interface characteristics and reaction mechanism of steel/Al welds produced by magnetic field assisted laser welding-brazing. *Opt. Laser Technol.* **2021**, *138*, 106843. [CrossRef]
9. Cao, R.; Chang, J.H.; Huang, Q.; Zhang, X.; Yan, Y.; Chen, J. Behaviors and effects of Zn coating on welding-brazing process of Al-Steel and Mg-steel dissimilar metals. *J. Manuf. Process.* **2018**, *31*, 674–688. [CrossRef]
10. Peyre, P.; Sierra, G.; Deschaux-Beaume, F.; Stuart, D.; Fras, G. Generation of aluminium–steel joints with laser-induced reactive wetting. *Mater. Sci. Eng. A* **2007**, *444*, 327–338. [CrossRef]
11. Springer, H.; Szczepaniak, A.; Raabe, D. On the role of zinc on the formation and growth of intermetallic phases during interdiffusion between steel and aluminium alloys. *Acta Mater.* **2015**, *96*, 203–211. [CrossRef]
12. Nguyen, D.-S.; Park, H.-S.; Lee, C.-M. Applying Selective Laser Melting to Join Al and Fe: An Investigation of Dissimilar Materials. *Appl. Sci.* **2019**, *9*, 3031. [CrossRef]
13. Zhang, J.L.; Liu, Y.C.; Hu, C.H.; Jiang, Y.; Addad, A.; Ji, G.; Song, C.; Zhai, Q. The effect of Cr content on intragranular κ-carbide precipitation in Fe-Mn-Al-(Cr)-C low-density steels: A multiscale investigation. *Mater. Charact.* **2022**, *186*, 111801. [CrossRef]
14. Desai, P.D. Thermodynamic Properties of Selected Binary Aluminum Alloy Systems. *J. Phys. Chem. Ref. Data* **1987**, *16*, 109–124. [CrossRef]
15. Baker, H. *ASM Handbook, Volume 3: Alloy Phase Diagrams*, 10th ed.; ASM International: Phoenix, AZ, USA, 1993; pp. 293–294.
16. Liu, Y.; Chong, X.; Jiang, Y.; Zhou, R.; Feng, J. Mechanical properties and electronic structures of Fe-Al intermetallic. *Phys. B* **2017**, *506*, 1–11. [CrossRef]
17. Zhang, C.; Huang, S.; Shen, J.; Chen, N.-X. Structural and mechanical properties of Fe–Al compounds: An atomistic study by EAM simulation. *Intermetallics* **2014**, *52*, 86–91. [CrossRef]
18. Li, Y.; Liu, Y.; Yang, J. First principle calculations and mechanical properties of the intermetallic compounds in a laser welded steel/aluminum joint. *Opt. Laser Technol.* **2020**, *122*, 105875. [CrossRef]
19. Kresse, G.; Furthmüller, J. Efficient iterative schemes for *ab initio* total-energy calculations using a plane-wave basis set. *Phys. Rev. B* **1996**, *54*, 11169–11186. [CrossRef]
20. Kresse, G.; Furthmüller, J. Efficiency of ab-initio total energy calculations for metals and semiconductors using a plane-wave basis set. *Comput. Mater. Sci.* **1996**, *6*, 15–50. [CrossRef]
21. Perdew, J.P. Erratum: Density-functional approximation for the correlation energy of the inhomogeneous electron gas. *Phys. Rev. B* **1986**, *34*, 7406. [CrossRef]
22. Perdew, J.P.; Burke, K.; Ernzerhof, M. Generalized gradient approximation made simple. *Phys. Rev. Lett.* **1996**, *77*, 3865–3868. [CrossRef]
23. Chadi, D.J. Special points for Brillouin-zone integrations. *Phys. Rev. B* **1977**, *16*, 1746–1747. [CrossRef]
24. Marmier, A.; Lethbridge, Z.A.; Walton, R.I.; Smith, C.W.; Parker, S.C.; Evans, K.E. ElAM: A computer program for the analysis and representation of anisotropic elastic properties. *Comput. Phys. Commun.* **2010**, *181*, 2102–2115. [CrossRef]

25. Li, X.; Scherf, A.; Heilmaier, M.; Stein, F. The Al-Rich part of the Fe-Al phase diagram. *J. Phase Equilib. Diff.* **2016**, *37*, 162–173. [CrossRef]
26. Burkhardt, U.; Grin, Y.; Ellner, M.; Peters, K. Structure refinement of the iron–aluminium phase with the approximate composition Fe_2Al_5. *Acta Crystallogr. Sect. B* **1994**, *50*, 313–316. [CrossRef]
27. Szczepaniak, A.; Fan, J.; Kostka, A.; Raabe, D. On the Correlation Between Thermal Cycle and Formation of Intermetallic Phases at the Interface of Laser-Welded Aluminum-Steel Overlap Joints. *Adv. Eng. Mater.* **2012**, *14*, 464–472. [CrossRef]
28. Rathod, M.; Kutsuna, M. Joining of aluminum Alloy 5052 and low-carbon steel by laser roll welding. *Weld. J.* **2004**, *83*, 16S–26S.
29. Wallerstein, D.; Lusquiños, F.; Comesaña, R.; del Val, J.; Riveiro, A.; Badaoui, A.; Pou, J. Dissimilar unbeveled butt joints of AA6061 to S235 structural steel by means of standard single beam fiber laser welding-brazing. *J. Mater. Process. Technol.* **2020**, *291*, 116994. [CrossRef]
30. Ogura, T.; Wakazono, R.; Yamashita, S.; Saida, K. Dissimilar laser brazing of aluminum alloy and galvannealed steel and defect control using interlayer. *Weld. World* **2020**, *64*, 697–706. [CrossRef]
31. Sierra, G.; Peyre, P.; Beaume, F.D.; Stuart, D.; Fras, G. Galvanised steel to aluminium joining by laser and GTAW processes. *Mater. Charact.* **2008**, *59*, 1705–1715. [CrossRef]
32. Yuan, R.; Deng, S.; Cui, H.; Chen, Y.; Lu, F. Interface characterization and mechanical properties of dual beam laser welding-brazing Al/steel dissimilar metals. *J. Manuf. Process.* **2019**, *40*, 37–45. [CrossRef]
33. Ozaki, H.; Kutsuna, M.; Nakagawa, S.; Miyamoto, K. Laser roll welding of dissimilar metal joint of zinc coated steel to aluminum alloy. *J. Laser Appl.* **2010**, *22*, 1–6. [CrossRef]
34. Suder, W.J.; Williams, S.W. Investigation of the effects of basic laser material interaction parameters in laser welding. *J. Laser Appl.* **2012**, *24*, 032009. [CrossRef]
35. Xia, H.; Tao, W.; Li, L.; Tan, C.; Zhang, K.; Ma, N. Effect of laser beam models on laser welding–brazing Al to steel. *Opt. Laser Technol.* **2020**, *122*, 105845. [CrossRef]
36. Vinokur, A.I.; Hilleke, K.P.; Fredrickson, D.C. Principles of weakly ordered domains in intermetallics: The cooperative effects of atomic packing and electronics in Fe_2Al_5. *Acta Crystallogr. A* **2019**, *75*, 297–306. [CrossRef]
37. Chen, M.J.; Huang, J.K.; He, C.C.; Shi, Y.; Fan, D. Thermodynamic analysis of the formation of Fe-Al-Zn intermetallic compounds in Al/galvanized steel interface. *Acta Metall. Sinica* **2016**, *52*, 113–119. [CrossRef]

coatings

Review

Corrosion of Laser Cladding High-Entropy Alloy Coatings: A Review

Yusheng Liu [1], Dingding Xiang [1,2,3,*], Kaiming Wang [2,4] and Tianbiao Yu [1,*]

1 School of Mechanical Engineering and Automation, Northeastern University, Shenyang 110057, China
2 State Key Laboratory of Tribology in Advanced Equipment, Tsinghua University, Beijing 100084, China
3 Key Laboratory for Precision and Non-Traditional Machining Technology of Ministry of Education, Dalian University of Technology, Dalian 116081, China
4 College of Automotive and Mechanical Engineering, Changsha University of Science and Technology, Changsha 410114, China
* Correspondence: xiangdd@mail.neu.edu.cn (D.X.); tbyu@me.neu.edu.cn (T.Y.)

Abstract: Material corrosion is a common phenomenon. Severe corrosion not only causes material failure, but may also lead to unexpected catastrophic accidents. Therefore, reducing the loss caused by corrosion has become a problem faced by countries around the world. As a surface modification technology, laser cladding (LC) can be used to prepare coatings that can achieve metallurgical bonding with the substrate. High-entropy alloys (HEAs) are a new material with superior anti-corrosion ability. Therefore, HEA coatings prepared by LC have become a research hotspot to improve the anti-corrosive ability of material surfaces. In this work, the effects of LC process parameters, post-processing, and the HEA material system on the anti-corrosion ability of HEA coatings and their mechanisms are reviewed. Among them, the LC process parameters influence the anti-corrosion ability by affecting the macroscopic quality, dilution rate, and uniformity of the coatings. The post-processing enhances the anti-corrosion ability of the coatings by improving the internal defects and refining the grain structure. The anti-corrosion ability of the coatings can be improved by appropriately adding transition metal elements such as Ni, Cr, Co, and rare earth elements such as Ce and Y. However, the lattice distortion, diversification of phase composition, and uneven distribution caused by excess elements will weaken the corrosion protection of the coatings. We reviewed the impact of corrosion medium on the anti-corrosion ability of coatings, in which the temperature and pH value of the corrosion medium affect the quality of the passive film on the surface of the coatings, thereby affecting the anti-corrosion ability of the coatings. Finally, to provide references for future research, the development trend of preparing HEA coatings by LC technology is prospected.

Keywords: laser cladding; high-entropy alloys; corrosion resistance; coating preparation

1. Introduction

With the progress of society, more and more countries have realized the importance of marine economy. However, equipment is prone to corrosion in seawater environments. Severe corrosion will not only lead to material failure and economic loss, but may even endanger human life. Therefore, corrosion protection is related to safety, economy, people's livelihood, ecology, and conservation. As early as the last century, the total direct cost of corrosion in the United States was estimated to be USD 276 billion per year [1]. It is thus urgent to improve the anti-corrosion ability of materials. Surface technology is the use of chemical and physical technologies to modify the surface structure or composition of materials. It can prolong the service life of components and enhance their surface properties, while reducing corrosion costs. As a new star in the material field, high-entropy alloys (HEAs) are also called multi-principal alloys. Compared with traditional alloys, HEAs have four core characteristics [2]: a thermodynamically high entropy effect, a structural lattice distortion effect, a kinetics slow diffusion effect, and a performance "cocktail" effect. Therefore, HEAs have exceptional properties such as good strength [3], wearability [4], anti-corrosion [5], and resistance to oxidation at high temperatures [6]. HEAs are frequently employed in aerospace, nuclear fusion, and other industries [7]. Scholars typically favor HEAs as surface modification materials. Physical vapor deposition, chemical vapor deposition, laser surface alloying, and laser cladding are some of the current techniques used to prepare HEA coatings. High energy density, adjustable dilution rate, quick processing speed, good metallurgical bonding with the matrix, high solidification rate, and a variety of cladding materials are all advantages of LC over previous preparation techniques [8,9]. LC is frequently employed in functional coating preparation [10] and part repair [11].

Generally, the uniformity of element distribution of materials has a significant influence on the anti-corrosion ability of coatings, and LC technology has the advantage that the coating material can be uniformly clad on the surface of the substrate, and the segregation of elements in the coatings can be easily avoided. The choice of coating material is also one of the factors affecting the anti-corrosion of the coatings, and one of the HEAs' characteristics is that several kinds of elements can be selected as the main element of the alloy. Due to the high entropy effect of HEAs, their phase composition is relatively simple, and it is not easy to form intermetallic compounds [12,13], and due to its "cocktail" effect, elements with exceptional anti-corrosion ability can be selected to match each other and improve the anti-corrosion ability of the alloys. Using LC technology to prepare HEA coatings can organically combine both advantages and significantly enhance the anti-corrosion ability of the coatings.

At present, the research and application of laser cladding high-entropy alloy coatings (LC-HEACs) in the field of corrosion protection are still in the early stage. In this paper, taking LC-HEACs as the research object, the research status of LC-HEACs in the field of corrosion protection is reviewed and the current research insufficiency and future development trends are pointed out, providing a certain reference for the future development of LC-HEACs in the field of corrosion protection.

2. Influence of LC Process Parameters

Regardless of the production process, it is crucial to optimize the process parameters to achieve the best performance [14,15]. In the LC process, high-energy laser beam radiation is used to melt cladding materials and deposit them on the substrate. To obtain coatings with no macroscopic defects, uniform composition distribution, and exceptional mechanical properties, it is necessary to select appropriate LC process parameters. At present, the effects of laser power, scanning speed, and laser energy density on the anti-corrosion of HEA coatings have been reported.

In general, a smaller molten pool is easily formed by a smaller laser power, which will lead to surface coatings that are melted but substrate that is not melted, so the metallurgical bonding between the substrate and the coatings cannot be achieved. At the same time, too small of a laser power will also lead to many pores in the cladding layer and a large number of cracks between the coatings and the substrate, as shown in Figure 1a. However, the molten pool will be too big if the laser power is too high. In this way, as shown in Figure 1b, the dilution rate will increase, and a large number of elements in the matrix will enter the coatings, decreasing the coatings' properties as a result.

Qiu et al. [16] successfully prepared Al_2-Co-Cr-Cu-Fe-Ni-Ti HEA coatings with different laser powers by LC technology. The anti-corrosion ability of HEA coatings is the best when the laser power is 2500 W in the corrosive medium of 1 mol/L NaCl solution. When the laser power is low (2000 W), the HEAs are not melted, the cladding layer is not well combined with the substrate, and the corrosion resistance of the HEA coatings decreases. However, when the laser power is higher (3000 W), the matrix is seriously diluted, and a large amount of Fe in the matrix enters the coatings, which weakens the anti-corrosion ability of the coatings.

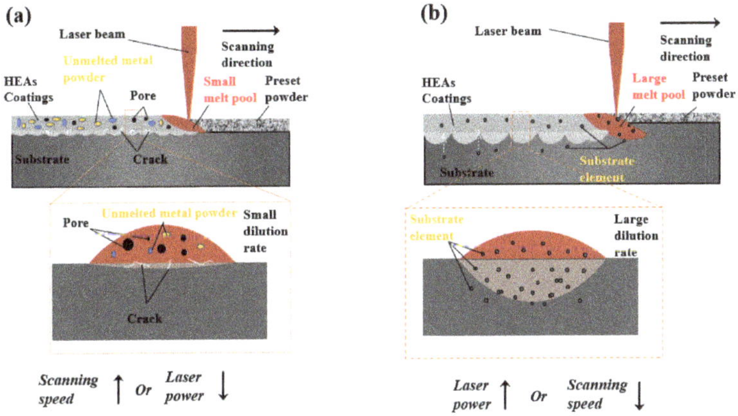

Figure 1. Influence mechanism of laser power and scanning speed on the preparation of HEA coatings: (**a**) laser power is too small or scanning speed is too fast; (**b**) laser power is too large or scanning speed is too slow.

The scanning speed has a similar effect to laser power. If the scanning speed is too fast, the coating materials cannot be completely melted, and the effect of cladding cannot be achieved. Meanwhile, with a faster scanning speed, the coatings solidify rapidly, resulting in a large number of structural defects such as pores and cracks in the coatings. However, the molten pool will exist for too long if the scanning speed is too slow, which will lead to the loss of alloy elements due to over-sintering of the coating materials. The increase in heat input is accompanied by the increase in dilution rate and a large number of elements in the matrix will enter the coatings, which will weaken the properties of the coatings.

Qiu et al. [17] prepared Al-Cr-Fe-Cu-Co HEA coatings with different scanning speeds. XRD results demonstrated that the HEA coatings are composed of BCC and FCC phases. Meanwhile, in solutions of 0.5 mol/L H_2SO_4 and 1 mol/L NaCl, electrochemical experiments were conducted. The results show that with the rise in scanning speed, the corrosion potential of the coatings will increase and then decrease, and the corrosion current will decrease and then increase. The reason is that with the increase in scanning speed and when it is appropriate, the time for the laser beam to irradiate the coatings becomes shorter, and the molten pool cools rapidly and produces fine dendrites, which is conducive to grain refinement. This will be beneficial to the improvement of corrosion performance. However, if the scanning speed is too quick, the molten pool is too small, the convection in the molten pool increases, the surface of the coatings is rough, and the anti-corrosion ability is weakened.

Laser energy density (E) is not a direct parameter in LC process parameters, and can be obtained by the formula $E = P/(DV)$. In this formula, P represents laser power, V represents scanning speed, and D represents laser spot size. He et al. [18] prepared a series of Fe-Cr-Ni-Mn-Al HEA coatings by changing the value of E (50~70 J/mm^2). The results show that the macro appearance of the coatings changes with the increase in E, and the anti-corrosion ability of the coatings increases first and then decreases. The reason is that the appropriate E value (65 J/mm^2) can effectively improve the structural defects and optimize the grain structure in the coatings, thereby enhancing the anti-corrosion ability of the coatings. In addition to laser power, scanning speed, and laser energy density, the LC process also includes important process parameters such as shielding gas rate, defocus distance, powder feeding rate, laser spot size, and overlap ratio. The selection of these process parameters also affects the structure and properties of the HEA coatings. However, there are few reports on the relevant research in this field at present. Therefore, in future study, it is necessary to fully reveal the influence of LC process parameters on the HEA coatings.

3. Influence of Post-Processing Technology

By selecting appropriate process parameters, the defects of the HEA coatings can be effectively reduced, and the anti-corrosion ability of the HEA coatings can be improved. Recently, numerous scholars have also prepared HEA coatings with better properties by combining post-processing technology.

Ultrasonic surface mechanical rolling treatment (USMRT) can form gradient nanostructures in the surface layer, refine grains, and generate compressive residual stress. Cui et al. [19] prepared Fe-Co-Ni-Cr-Mn HEA coatings by LC technology, and applied USMRT to the coatings. The process of LC-HEACs and USMRT was described in Figure 2. After USMT, the grain of the coatings is refined, and the internal stress that inhibits the growth of the coatings' oxide film is effectively reduced by generating compressive residual stress, which greatly enhances the stability of the oxide film and improves the anti-corrosion ability of the coatings.

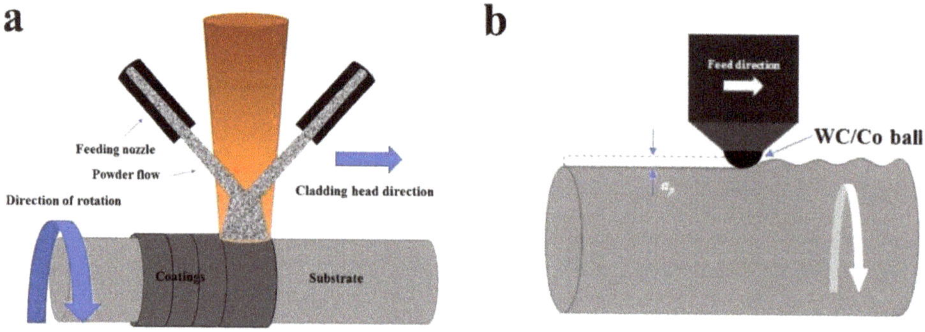

Figure 2. (**a**) Schematic of LC-HEACs and (**b**) ultrasonic surface mechanical rolling treatment. Reprinted with permission from [19] 2020 Elsevier.

Ultrasonic impact treatment (UIT) is also a post-processing technology that forms nanocrystals and a work-hardened layer through high strain rate compressive plastic deformation. Figure 3 shows a schematic diagram of the comparison between LC-HEACs and UIT layers. Due to the high strain rate produced by the UIT, the matrix's plastic deformation causes the grain borders and gaps to close and the size of the precipitates between the grain boundaries to shrink, effectively inhibiting intergranular corrosion of the coatings [20]. Li et al. [21] performed UIT on the cladding layer after preparing the $Al_{0.5}$-Co-Cr-Fe-Mn-Ni HEA coatings. The experimental results show that the coatings after UIT have exceptional anti-corrosion ability. The reason is that the precipitation at the grain boundary of the cladding layer after UIT will be broken into fine precipitates, and at the same time, due to the high strain rate produced by the UIT, the cladding layer's plastic deformation causes the narrowing of the gaps, the grain boundaries to close, and the further reduction in the size of the precipitates, thereby improving the anti-corrosion ability of the coatings.

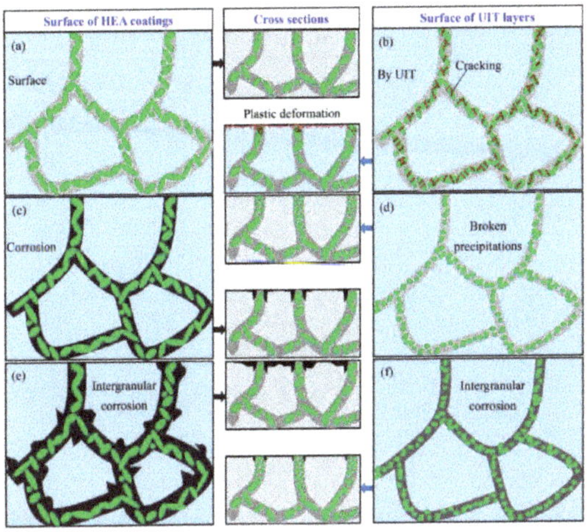

Figure 3. Diagrammatic comparison of the UIT layers and HEA coatings. (**a**) Surface of HEA coatings; (**b**) surface of UIT layers; (**c**) HEA coating is corroded; (**d**) fractured precipitations and narrower grain boundaries; (**e**) substantial intergranular corrosion of HEA coatings; (**f**) mild intergranular corrosion of UIT layers. Reprinted with permission from [21] 2021 Elsevier.

4. Influence of Alloying Elements

As described in Section 2, after selecting appropriate LC process parameters, HEA coatings with a defect-free macrostructure, uniform microstructure, and exceptional anti-corrosion ability can be prepared. The requirements for anti-corrosion of the coatings cannot be fully met only by optimizing LC process parameters, so a large number of scholars improve the anti-corrosion of the coatings by adjusting the types of HEAs in LC-HEACs. Initially, scholars usually studied the corrosion resistance of LC-HEACs with equal or approximate equal molar ratio, such as the Fe-Co-Ni-Cr system composed of transition elements [22] and the Ti-Nb-Zr-Mo system composed of high melting point elements [23], and found that the coatings have adequate anti-corrosion ability. The influence of different material systems on the anti-corrosion ability of the coatings is shown in Table 1. With the development of LC-HEACs, more LC-HEACs with unequal molar ratios have been discovered and reported, and many scholars have begun to research the influence of different elements on the anti-corrosion ability of coatings. This section summarizes the influence of different elements on the anti-corrosion ability of LC-HEACs, which is intended to provide reference for future research.

4.1. Chromium

One of the best components for improving anti-corrosion ability is Cr, which can create a thick layer of Cr_2O_3 oxide on the surface of coatings. Jiang et al. [40] investigated the influence of different Cr additions on the anti-corrosion ability of Al-Co-Cr_x-Fe-Ni (x = 0.5, 0.75, 1.0, 1.5, 2.0) LC-HEACs. The findings of the experiments demonstrate that as the amount of Cr added increases, coatings' anti-corrosion ability initially rises and then falls. Adding the appropriate amount of Cr element is conducive to the formation of continuous and uniform passivation film of the coatings, effectively protecting the coatings from corrosion. However, the excess Cr element leads to the dispersion of the oxide layer, which reduces the anti-corrosion ability of the coatings.

Table 1. Effects of different material systems on the corrosion resistance of LC-HEACs.

HEAs	Substrate	Corrosive Medium Solution	Electrochemical Parameters of the Substrate		Electrochemical Parameters of HEA Coatings		Ref.
-	-	-	E_{corr} (V)	I_{corr} (A/cm^2)	E_{corr} (V)	I_{corr} (A/cm^2)	-
$Cu_{0.5}$NiAlCoCrFeSi	AZ91D	3.5 wt % NaCl	−1.46	6.20×10^{-4}	−0.76	3.3×10^{-4}	[24]
$Cu_{0.9}$NiAlCoCrFe	AZ91D	3.5 wt % NaCl	−1.46	6.20×10^{-4}	−1.40	2.56×10^{-8}	[25]
$Al_{0.5}$CoCrCuFeNi	AZ91D	3.5 wt % NaCl	−1.46	6.20×10^{-4}	−0.998	1.60×10^{-4}	[26]
AlCrFeNiCuCo	Aluminum alloy	1 mol/L H_2SO_4	−0.596	8.056×10^{-5}	−0.493	5.253×10^{-5}	[27]
AlCoCrFeNiTi	AISI1045 steel	1 mol/L NaCl	−0.793	2.98×10^{-6}	−0.470	7.09×10^{-7}	[28]
$Al_2CoCrFe_{2.7}$MoNi	Pure iron	3.5 wt % NaCl	N/A	N/A	0.586	2.762×10^{-9}	[29]
AlCrFeNiCu+5 at % Nb	316 SS	0.5 M H_2SO_4	N/A	N/A	−0.020	9.58×10^{-5}	[30]
AlCoCrFeNiCu	Q235 steel	3.5 wt % NaCl	−0.825	2.482×10^{-5}	−0.061	2.56×10^{-7}	[31]
FeCrNiMnAl	17-4PH stainless steel	3.5 wt % NaCl	−0.860	1.082×10^{-4}	−0.688	1.156×10^{-5}	[18]
CoCrFeNiTi	40CrNiMoA alloy steel	1 mol/L NaCl	−0.70	2.12×10^{-6}	−0.37	1.62×10^{-6}	[32]
AlCoCrFeNiSi	AISI 304 stainless steel	3.5% NaCl	−0.482	7×10^{-6}	−0.326	6.48×10^{-7}	[33]
$Ti_{0.8}$CrFeCoNiCu	Aluminum alloy	3.5% NaCl	−1.296	1.93×10^{-5}	−0.432	5.12×10^{-7}	[34]
$Al_2CoCrCuFeNiTi_2$	Q235 steel	0.5 mol/L HCl	−0.41	2.1×10^{-6}	−0.35	1.7×10^{-7}	[35]
$Al_2CrFeCo_2CuNiTi$	Q235 steel	0.5 mol/L HNO_3	−0.35	2.4×10^{-4}	−0.05	2.9×10^{-5}	[5]
Co-Cr-Fe-Ni-$Mo_{0.2}$	Ti6Al4V	3.5 wt % NaCl	−0.594	9.9×10^{-8}	−0.329	9.87×10^{-8}	[36]
AlCoCrFeNiCu	N/A	3.5 wt % NaCl	N/A	N/A	−0.7019	7.21×10^{-8}	[37]
AlCoCrFeNiTi	N/A	3.5 wt % NaCl	N/A	N/A	−0.4654	7.63×10^{-8}	[37]
CrFeNiNbTi	40CrNiMoA alloy steel	3.5 wt % NaCl	−0.702	2.12×10^{-6}	−0.555	3.58×10^{-6}	[38]
$CoCrFeMnTi_{0.2}$	15CrMn Steel	1 mol/L NaCl	−0.6705	4.889×10^{-6}	−0.4720	3.408×10^{-6}	[39]

4.2. Nickel

Ni element is also one of the most effective elements for improving the anti-corrosion ability of alloys. Usually, Ni element and Cr element cooperate with each other, which can significantly improve the anti-corrosion ability of the coatings. In 1 mol/L NaOH solution and 3.5 wt % NaCl solution, Qiu et al. [41] investigated the corrosion characteristics of Al_2-Cr-Fe-Co-Cu-Ti-Ni_x (x = 0, 0.5, 1, 1.5, 2.0) LC-HEACs. The anti-corrosion ability of the coatings increased initially and then dropped as Ni content increased. Despite the anti-corrosion ability of the Ni element, when the Ni content is high, the alloy suffers from significant lattice distortion, which has an impact on the microstructure and reduces the coatings' anti-corrosion ability.

4.3. Molybdenum

As one of the most effective elements to improve corrosion resistance, Mo is usually combined with Ni and Cr elements [42,43], which are beneficial to improving the anti-corrosion ability of coatings. Fu et al. [44] investigated the influence of different Mo additions on Co-Cr_2-Fe-Ni-Mo_x (x = 0, 0.1, 0.2, 0.3, 0.4) LC-HEACs. The findings demonstrate that adding the right amount of Mo can increase the coatings' anti-corrosion ability, but adding too much Mo encourages the development of inter-dendritic structures and the σ-CrMo phase, which speeds up galvanic corrosion and lowers the coatings' anti-corrosion ability. Meanwhile, to further investigate the actual anti-corrosion ability of coatings, a 6-month immersion test was carried out on HEA coating samples and 2605 N duplex stainless steel (stainless steel specially developed for chlorine-containing solutions) in an acidic solution of pH = 2. The surface of 2605 N duplex stainless steel is covered in numerous pits, and after the Mo element was added, the HEACs were able to efficiently control the size of the pits.

4.4. Aluminum

The density of Al is light, and the Al element can form a dense oxide film to improve the corrosion resistance of the material. As a result, many researchers have chosen the Al element as the central component of HEAs [45,46]. Sun et al. [47] used LC technology to prepare Co-Cr-Fe-Ni-Al_x-$Mn_{(1-x)}$ (x = 0, 0.2, 0.4, 0.6, 0.8, 1.0) HEA coatings with different ratios of Al and Mn elements. The phase structure of the HEA coatings changes with the addition of the Al element from a single FCC phase to a dual phase made up of FCC and BCC, and lastly into a single BCC phase. With an increase in Al concentration, the anti-corrosion ability of the HEA coatings initially declined suddenly and then slowly increased. The coatings only have one FCC phase when no additional Al element is present, and microstructures with uniform phase structures often have superior anti-corrosion ability. The coatings contain FCC and BCC dual phases when a tiny amount of Al element (0.2 at %) is applied. Corrosion micro-cells are created as a result of the equilibrium potential difference between the FCC phase and the BCC phase, which lowers anti-corrosion ability. The possibility of the Al_2O_3 passivation film forming on the surface of the HEA coatings grows as the addition of Al element increases, which leads to a gradual improvement in the coatings' anti-corrosion ability. When the addition amount of Al element is 1.0 at %, the HEA coatings have a single BCC phase composition. At this time, due to the uniform and single microstructure and the existence of the Al_2O_3 oxide film, the anti-corrosion ability of the coatings is the best.

4.5. Titanium

Ti element has strong passivation ability and exceptional corrosion resistance. Therefore, it has been widely studied by scholars. Table 2 summarizes the influence of different Ti additions on the anti-corrosion ability of LC-HEACs.

Table 2. Effects of different Ti additions on the corrosion resistance of LC-HEACs.

HEAs	Substrate	Corrosive Medium Solution	Electrochemical Parameters of the Substrate		Electrochemical Parameters of HEA Coatings		Ref.
-	-	-	E_{corr} (V)	I_{corr} (A/cm^2)	E_{corr} (V)	I_{corr} (A/cm^2)	-
$Al_2CrFeNiCoCuTi_x$ x = 0, 0.5, 1.0, 1.5, 2.0	Q235 steel	0.5 mol/L HNO_3	−0.35	2.4×10^{-1}	−0.18 for Ti_0 −0.30 for $Ti_{0.5}$ −0.33 for $Ti_{1.0}$ −0.30 for $Ti_{1.5}$ −0.15 for $Ti_{2.0}$	3.8×10^{-2} for Ti_0 2.2×10^{-2} for $Ti_{0.5}$ 7.3×10^{-3} for $Ti_{1.0}$ 4.4×10^{-3} for $Ti_{1.5}$ 2.7×10^{-3} for $Ti_{2.0}$	[48]
$Al_2CoCrCuFeNiTi_x$ x = 0, 0.5, 1.0, 1.5	Q235 steel	0.5 mol/L H_2SO_4	−0.38	3.6×10^{-5}	−0.32 for Ti_0 −0.40 for $Ti_{0.5}$ −0.52 for $Ti_{1.0}$ −0.37 for $Ti_{1.5}$	4.2×10^{-6} for Ti_0 3.1×10^{-7} for $Ti_{0.5}$ 2.3×10^{-8} for $Ti_{1.0}$ 1.1×10^{-7} for $Ti_{1.5}$	[49]
$Al_2CoCrCuFeNiTi_x$ x = 0, 0.5, 1.0, 1.5	Q235 steel	3.5% NaCl	−0.57	7.1×10^{-7}	−0.58 for Ti_0 −0.45 for $Ti_{0.5}$ −0.22 for $Ti_{1.0}$ −0.63 for $Ti_{1.5}$	6.3×10^{-7} for Ti_0 6.7×10^{-7} for $Ti_{0.5}$ 1.3×10^{-8} for $Ti_{1.0}$ 5.4×10^{-7} for $Ti_{1.5}$	[49]
$CoCrFeNiTi_x$ x = 0, 0.2, 0.4, 0.6, 0.8	45 steel	3.5 wt % NaCl	N/A	N/A	−0.848 for Ti_0 −0.792 for $Ti_{0.2}$ −0.650 for $Ti_{0.4}$ −0.509 for $Ti_{0.6}$ −0.598 for $Ti_{0.8}$	3.66×10^{-6} for Ti_0 1.99×10^{-6} for $Ti_{0.2}$ 8.13×10^{-7} for $Ti_{0.4}$ 4.22×10^{-7} for $Ti_{0.6}$ 5.76×10^{-7} for $Ti_{0.8}$	[50]
$CoCrFeNiTi_x$ x = 0.1, 0.3,0.5, 0.7	Q235 steel	3.5 wt % NaCl	N/A	N/A	−1.18 for $Ti_{0.1}$ −1.16 for $Ti_{0.3}$ −1.13 for $Ti_{0.5}$ −1.11 for $Ti_{0.7}$	5.64×10^{-6} for $Ti_{0.1}$ 4.23×10^{-6} for $Ti_{0.3}$ 3.81×10^{-6} for $Ti_{0.5}$ 1.04×10^{-6} for $Ti_{0.7}$	[51]
$CoCr_{2.5}FeNi_2Ti_x$ x = 0, 0.5, 1.0	Q235 steel	Simulated saturated salty water mud solution	−0.902	4.531×10^{-14}	−0.714 for Ti_0 −0.578 for $Ti_{0.5}$ −0.228 for $Ti_{1.0}$	3.648×10^{-15} for Ti_0 4.892×10^{-15} for $Ti_{0.5}$ 1.386×10^{-15} for $Ti_{1.0}$	[52]

It can be seen from Table 2 that in different corrosive media and HEA material systems, the appropriate addition of Ti element can improve the anti-corrosion ability of HEA coatings. Liu et al. [53] reported the influence of Ti element content on the anti-corrosion ability of Al-Co-Cr-Fe-Ni-Ti_x (x = 0, 0.2, 0.4, 0.6, 0.8, 1.0) LC-HEACs. The experimental results demonstrated that in the HEA coatings, the distribution of each element is uniform and the phase structure composition is relatively single, which inhibits the formation of corrosion micro-cells, and the coatings exhibit exceptional anti-corrosion ability.

4.6. Copper

The Cu element helps FCC solid solution develop and it aggregates readily in the intergranular space to produce a Cu-rich phase, which gives HEAs their favorable characteristics.

Li et al. [54] reported the influence of Cu addition on the microstructure and anti-corrosion ability of $Al_{0.8}$-Cr-Fe-Co-Ni-Cu_x LC-HEACs. The experimental results demonstrated that with the increase in Cu addition, the phase structure of the coatings changes from two BCC phases to two BCC phases and one FCC phase. The anti-corrosion ability of the coatings decreases with the increase in Cu addition. The reason is that the additional Cu element added causes an increase in the Cu-rich phase, which in turn causes a rise in the intergranular corrosion area, which causes the contact area between the corrosion solution and the coating surface to continuously grow. As a result, the electrode reaction's coating corrosion rate is accelerated, and the coatings' corrosion resistance is decreased.

4.7. Tantalum

Zhang et al. [55] reported the influence of Ta amount on the microstructure and anti-corrosion ability of Ni-Ti-Cr-Nb-Ta_x (x = 0.1, 0.3, 0.5, 1) LC-HEACs. The results show that the formation of BCC phase and the precipitation of Laves phase (Cr_2Nb) are promoted with the increase in Ta element content. In 3.5 wt % NaCl solution, the HEA coatings exhibit exceptional anti-corrosion ability, especially the Ni-Ti-Cr-Nb-$Ta_{0.5}$ and Ni-Ti-Cr-Nb-Ta coatings, which have a wider passivation zone and show better pitting corrosion resistance.

4.8. Carbon

To report the influence of C on the anti-corrosion of LC-HEACs, Shang et al. [56] prepared $Al_{6.25}$-$Mo_{6.25}$-$Ti_{29.2}$-$Zr_{29.2}$-$Nb_{29.2}$ + xC at % (x = 0, 0.4, 0.8) LC-HEACs on the surface of pure titanium alloy. The distribution of the elements in the HEA coatings becomes more uniform with the proper addition of C, which significantly increases the coatings' anti-corrosion ability. However, when the amount of C is too high, the HEACs are made of BCC solid solution, MC carbide, and HCP-α phase, which speeds up corrosion by causing the development of micro-cells and decreases the anti-corrosion ability of the HEA coatings.

4.9. Ceramic Reinforcement

Many scholars have found that the anti-corrosion ability of LC-HEACs can be effectively improved by introducing reinforcing phases. Table 3 summarizes the effect of enhancing the anti-corrosion ability of relative LC-HEACs. Guo et al. [57] prepared in situ TiN ceramic particles which improved Co-Cr_2-Fe-Ni-Ti_x (x = 0, 0.5, 1) LC-HEACs. According to the experimental findings, the HEA coatings' phase structure is composed primarily of FCC + TiN with a trace amount of Laves phase. The appropriate addition of Ti element, which can dissolve in the FCC phase to improve the anti-corrosion ability of the HEA coatings, results in the higher positive corrosion potential and lower corrosion current density of the HEA coatings. However, an excessive amount of Ti will result in TiN particles and Laves phase, which will create corrosion micro-cells and heighten the susceptibility to corrode. Liu et al. [58] prepared Fe-Co-Ni-Cr-Mn-Ti_x (x = 0, 0.5, 1.0, 1.5) HEA coatings by LC technology under the protection of nitrogen. Electrochemical experiments show that the coating corrosion potential is high and its corrosion current is low when the addition amount of Ti is 0. With the increase in Ti addition, TiN ceramic particles and Laves phase were formed in situ in the HEA coatings. Due to the different electrode potentials, micro-cells will appear between both the TiN particles and the Laves phase and γ-phase, resulting in galvanic corrosion and weakening the anti-corrosion ability of the HEA coatings.

Table 3. Effect of enhanced relative LC-HEACs' corrosion resistance.

Substrate	HEAs	Enhancement Phase	Corrosive Medium	Electrochemical Parameters of the Substrate		Electrochemical Parameters of HEA Coatings		Ref.
-	-	-	-	E_{corr} (V)	I_{corr} (A/cm^2)	E_{corr} (V)	I_{corr} (A/cm^2)	-
45 steel	FeCoCrNiCu +Cf	Cf	5% H_2SO_4	−0.437	2.317×10^{-2}	−0.087	2.371×10^{-4}	[59]
304SS	CoCrNiMo +1 at %B_4C	B_4C	3.5 wt % NaCl	−0.289	2.21×10^{-7}	−0.255	1.86×10^{-7}	[60]
304SS	CoCrNiMo +2 at %B_4C	B_4C	3.5 wt % NaCl	−0.289	2.21×10^{-7}	−0.292	1.82×10^{-7}	[60]
304SS	CoCrNiMo +3 at %B_4C	B_4C	3.5 wt % NaCl	−0.289	2.21×10^{-7}	−0.265	2.22×10^{-7}	[60]
304SS	CoCrNiMo +4.2 at %B_4C	B_4C	3.5 wt % NaCl	−0.289	2.21×10^{-7}	−0.199	2.84×10^{-7}	[60]
AZ91D	$Ti_{50}Zr_{25}Al_{15}Cu_{10}$ +2 wt %LaB_6	LaB_6	3.5 wt % NaCl	−1.37	5.31×10^{-4}	−0.42	1.48×10^{-6}	[61]
AZ91D	$Ti_{50}Zr_{25}Al_{15}Cu_{10}$ +4 wt %LaB_6	LaB_6	3.5 wt % NaCl	−1.37	5.31×10^{-4}	−0.34	8.52×10^{-7}	[61]
AZ91D	$Ti_{50}Zr_{25}Al_{15}Cu_{10}$ +8 wt %LaB_6	LaB_6	3.5 wt % NaCl	−1.37	5.31×10^{-4}	−0.65	1.14×10^{-5}	[61]
AISI1045 steel	$Al_{0.5}CoCrFeNiTi_{0.5}$ +2 at %SiC	SiC	3.5 wt % NaCl	N/A	N/A	−0.787	2.116×10^{-8}	[62]
AISI1045 steel	$Al_{0.8}CoCrFeNiTi_{0.2}$ +2 at %SiC	SiC	3.5 wt % NaCl	N/A	N/A	−0.573	1.005×10^{-8}	[62]
304SS	$FeCoNiCrMnTi_{0.5}$	In situ TiN	3.5% NaCl	−0.217	5.45×10^{-9}	−0.209	1.15×10^{-10}	[58]
304SS	FeCoNiCrMnTi	In situ TiN	3.5% NaCl	−0.217	5.45×10^{-9}	−0.247	3.18×10^{-10}	[58]
304SS	$FeCoNiCrMnTi_{1.5}$	In situ TiN	3.5% NaCl	−0.217	5.45×10^{-9}	−0.412	3.965×10^{-8}	[58]
904L stainless steel	$CoCr_2FeNiTi_{0.5}$	In situ TiN	Simulated phosphoric acid reactor solution	−0.213	3.3×10^{-7}	−0.307	3.3×10^{-10}	[57]
904L stainless steel	$CoCr_2FeNiTi$	In situ TiN	Simulated phosphoric acid reactor solution	−0.213	3.3×10^{-7}	−0.231	2.7×10^{-8}	[57]

4.10. Rare Earth Elements

As beneficial chemical and physical modifiers, rare earth elements and their oxides are referred to as the "vitamins of metals". Cui et al. [63] successfully prepared Fe-Co-Ni-Cr-Mo LC-HEACs with an appropriate amount of CeO_2 on the surface of titanium alloys. Electrochemical studies in a solution of 3.5% NaCl revealed that the addition of CeO_2 increased the corrosion protection of the HEA coatings. The electrochemical corrosion mechanism of HEA coatings and those augmented with CeO_2 in 3.5% NaCl solution is shown in Figure 4. A higher density and more stable passivation film is formed as a result of the addition of CeO_2 powder, which also refines the coating's grain structure, adds more nucleation points for the passivation film to form, and improves the passivation film's uniformity and formation rate. This increases the coatings' resistance to corrosion.

Due to Ce element's exceptional qualities, Y element, another rare earth element, has also captured the interest of academics. The corrosion protection of Mg-Mo-Nb-Fe-Ti_2-Y_x (x = 0, 0.4, 0.8, 1.2) LC-HEACs was examined by Gu et al. [64] in relation to the effects of various Y element levels. The results demonstrated that the added Y element improves the corrosion resistance of the HEA coatings, as evidenced by the higher corrosion potential and lower corrosion current density compared to the HEA coatings without the addition of Y element. When Y element is added, structural flaws such microcracks in HEA coatings are somewhat reduced, and Cl^- ion entry into the coatings is blocked, which significantly increases HEA coatings' corrosion protection. The amount of Y supplied does not, however, positively correlate with the anti-corrosion ability of HEA coatings. The alloying of Ti and Mo elements, increased lattice distortion, and decreased corrosion protection of HEA coatings are all effects of adding too much Y.

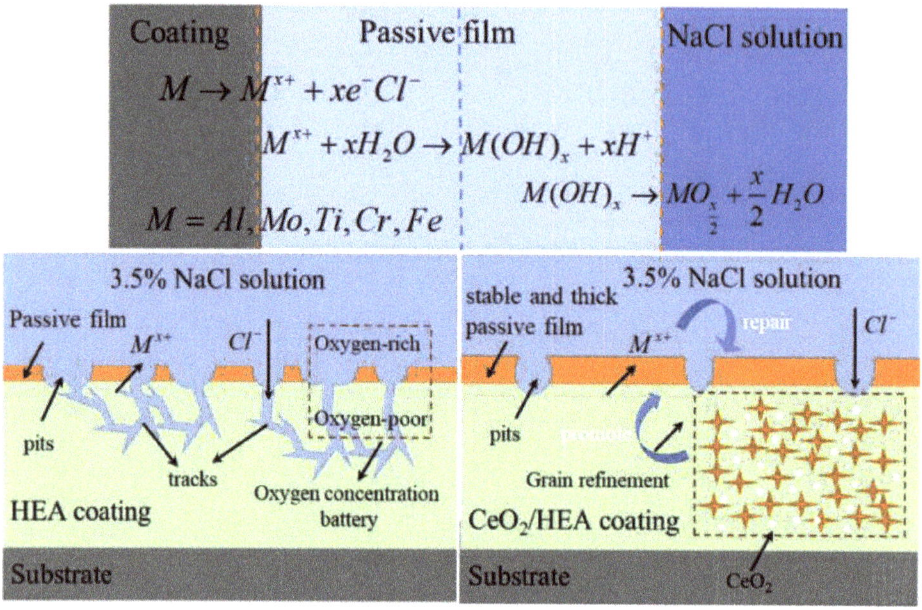

Figure 4. Schematic diagram of corrosion in 3.5% NaCl solution. Reprinted with permission from [63] 2021 Elsevier.

5. Influence of Corrosive Environment

In general, the coating material system, microstructure, and external environment coming into contact with the coatings are the elements that influence the corrosion of metal coatings. Among these, the impact of the coating's microstructure and material system on its anti-corrosion ability has been thoroughly discussed. The corrosion environment, including temperature, humidity, and pH, greatly affects the corrosion performance of the coating. According to current reports, the influence of acid, alkaline, salt solution, room temperature environment, and non-room temperature environment on the performance of HEAs has been reported by scholars. The effects of salt environment (pH = 7) and acid environment (pH < 7) at different temperatures on the corrosion resistance of HEA coatings are shown in Tables 4 and 5, respectively.

The pH value of the corrosive liquid has a considerable influence on the corrosion resistance of the coatings. Qiu et al. [65,66] discussed the effects of acidic environment (pH < 7), alkaline environment (pH > 7), and salt environment (pH = 7) on the corrosion properties of Al_2-Cr-Fe-Co_x-Cu-Ni-Ti HEA coatings. It was found that the corrosion potential and corrosion current of the same HEA coatings were greatly affected under the corrosion environment with different pH values. At the same time, the report also found that there were some differences in corrosion resistance of different HEA material systems under the same corrosion solution, which also confirmed the content of the Section 4 of this paper. Therefore, the pH value of the corrosion solution and the material composition of HEAs significantly affect the corrosion performance of the coatings. Hao et al. [67] explained why it is difficult to form a passive film on the surface of Co-Cr-Fe-Mn-Ni HEAs in a chloride containing salt environment, and proposed an effective method to solve this problem. The alloys are given a nitriding treatment to cause N to react with H^+ ions in the corrosion medium, with H^+ ions preventing Mn and Co elements from creating the passive film, thus improving the alloy's capacity to produce the passive film and boosting its anti-corrosion ability.

The temperature of the corrosion medium also affects the corrosion resistance of the alloy. Wang et al. [68] reported the influence of different corrosion solutions (3.5 wt % NaCl and 0.5 M H_2SO_4) and the external environment at different temperatures on the corrosion performance of Cr-Fe-Co-Ni HEA coatings. The results show that the corrosion potential and corrosion current of the coating worsen and the corrosion resistance of the coating is weakened to a certain extent with the increase in temperature under the same corrosion solution. At the same time, the report also found that the influence of temperature on the corrosion performance of the coating was different in different corrosion solutions. Xue et al. [69] reported the corrosion behavior of Al_2-Cr_5-Cu_5-Fe_{53}-Ni_{35} HEAs at different temperatures and chloride solution concentrations. The report pointed out that at low temperatures (<25 °C), the coating has exceptional corrosion resistance due to the existence of the oxide film. However, with the increase in temperature, the passive film of the coating is damaged, resulting in accelerated corrosion of the coating. Similarly, higher chloride concentration solutions will considerably weaken the corrosion resistance of the coating

At present, more and more engineering structural materials need to be used in complex working conditions. Therefore, in addition to the corrosion performance of LC-HEACs in standard acidic, alkali, and salt environments, the corrosion resistance of LC-HEACs in simulated actual working conditions is more valuable. The 20 G steel used in the boiler water-cooling environment is often used in a high-temperature and H_2S-containing environment, which will lead to serious corrosion of the 20 G steel and even a tube burst accident. To solve this problem, Liu et al. [70] prepared Ni-Cr-Mo-W-Si_x LC-HEACs with different Si additions on the surface of 20 G steel, and corrosion experiments were carried out on the HEA coatings in a simulated high-temperature (500–600 °C) H_2S environment. The experimental findings demonstrate the superior anti-corrosion ability of HEA coatings, and particularly when the Si concentration is 1.0 wt %, a protective oxide film forms on the coating's surface to further increase the anti-corrosion ability. However, if Si is added in excess, it can result in large cracks along grain boundaries and poor anti-corrosion ability. Low-carbon steel is widely used in engineering due to its low price. However, the poor corrosion resistance of low-carbon steel seriously affects its service life. Andri Isak Thorhallsson et al. [71] successfully prepared Co-Cr-F-eNi-$Mo_{0.85}$ and $Al_{0.5}$-Co-Cr-Fe-Ni HEA coatings on the surface of low-carbon steel, and tested the anti-corrosion ability of the coatings under three actual geothermal environments. The results show that the corrosion environment has a significant impact on the anti-corrosion ability of the coatings, especially when the HEA coatings have manufacturing defects, as it will further aggravate the corrosion of the coatings. X70 pipeline steel has suitable strength and toughness, as well as weldability, and is widely used in the petroleum industry. However, in the long-term service process, it is easily corroded, resulting in oil leakage, economic losses, and even endangering the personal safety of workers. Therefore, Wan et al. [72] successfully prepared $Al_{0.4}$-Co-$Cu_{0.6}$-Ni-$Si_{0.2}$-$Ti_{0.25}$ LC-HEACs on the X70 pipeline steel substrate. The anti-corrosion ability of HEA coatings in challenging sulfur-containing conditions was investigated in order to prolong the useful life of X70 pipeline steel. The findings reveal that HEA coatings have higher impedance and impedance modulus than X70 steel, which indicates that HEA coatings have significantly superior anti-corrosion ability than X70 steel. HEA coatings also have a more accurate corrosion potential and lower current density. The thorough results demonstrate that HEA coatings have much greater resistance to corrosion than X70 steel.

Table 4. The effects of 3.5 wt % NaCl (pH = 7) at different temperatures on the corrosion resistance of HEA coatings.

HEAs	Temperature (°C)	Electrochemical Parameters of HEA Coatings		Corrosion Mechanism	Ref.
-	-	E_{corr} (V)	I_{corr} (A/cm^2)	-	-
Cr-Fe-Co-Ni	20	−0.166	7.15×10^{-8}	N/A	[68]
Cr-Fe-Co-Ni	40	−0.190	1.33×10^{-7}	N/A	[68]
Cr-Fe-Co-Ni	60	−0.133	2.99×10^{-7}	N/A	[68]
Cr-Fe-Co-Ni-$Mo_{0.2}$	N/A	−0.192	1.84×10^{-9}	N/A	[73]
Fe-Co-Cr-Ni-Al	N/A	−0.485	1.29×10^{-5}	N/A	[74]
Fe-Co-$Cr_{1.5}$-Ni-Al	N/A	−0.297	2.56×10^{-6}	N/A	[74]
Fe-Co-$Cr_{2.0}$-Ni-Al	N/A	−0.423	6.983×10^{-6}	N/A	[74]
$Al_{0.5}$-Co-Cr-Fe-Ni-$Ti_{0.5}$	27	−0.610	9.91×10^{-9}	Pitting corrosion	[62]
$Al_{0.5}$-Co-Cr-Fe-Ni-$Ti_{0.5}$+2 at % Si	27	−0.787	2.11×10^{-8}	Pitting corrosion	[62]
$Al_{0.8}$-Co-Cr-Fe-Ni-$Ti_{0.2}$	27	−0.509	1.37×10^{-8}	Pitting corrosion	[62]
$Al_{0.8}$-Co-Cr-Fe-Ni-$Ti_{0.2}$+2 at % Si	27	−0.573	1.00×10^{-8}	Pitting corrosion	[62]
Fe-Ni-Co-Cr	22~28	−0.188	8.77×10^{-9}	Intergranular corrosion	[43]
Fe-Ni-Co-Cr-$Mo_{0.15}$	22~28	−0.161	7.63×10^{-9}	Intergranular corrosion	[43]
Fe-Ni-Co-Cr-$Mo_{0.2}$	22~28	−0.142	3.68×10^{-9}	Intergranular corrosion	[43]
Fe-Ni-Co-Cr-$Mo_{0.25}$	22~28	−0.192	5.37×10^{-9}	Intergranular corrosion	[43]
$Al_{0.3}$-$Cu_{0.3}$-Co-Cr-Fe-Ni	20~24	−0.294~ −0.282	$2.42\text{~}3.04 \times 10^{-7}$	Pitting and galvanic corrosion	[75]
$Al_{0.3}$-$Cu_{0.4}$-Co-Cr-Fe-Ni	20~24	−0.271~ −0.251	$2.92\text{~}3.88 \times 10^{-7}$	Pitting and galvanic corrosion	[75]
$Al_{0.3}$-$Cu_{0.5}$-Co-Cr-Fe-Ni	20~24	−0.263~ −0.245	$6.90\text{~}7.96 \times 10^{-7}$	Pitting and galvanic corrosion	[75]
$Al_{0.3}$-$Cu_{0.6}$-Co-Cr-Fe-Ni	20~24	−0.255~ −0.241	$9.53\text{~}9.99 \times 10^{-7}$	Pitting and galvanic corrosion	[75]

Table 5. The effects of acidic environment (pH < 7) at different temperatures on the corrosion resistance of HEA coatings.

HEAs	Medium	Temperature (°C)	Electrochemical Parameters of HEA Coatings		Ref.
-	-	-	E_{corr} (V)	I_{corr} (A/cm^2)	-
Cr-Fe-Co-Ni	0.5 M H_2SO_4	20	−0.737	7.07×10^{-5}	[68]
Cr-Fe-Co-Ni	0.5 M H_2SO_4	40	−0.767	4.97×10^{-4}	[68]
Cr-Fe-Co-Ni	0.5 M H_2SO_4	60	−0.759	1.26×10^{-3}	[68]
Cr-Fe-Co-Ni	0.5 M H_2SO_4	80	−0.758	3.32×10^{-3}	[68]
Cr-Fe-Co-Ni-$Mo_{0.2}$	1 M H_2SO_4	N/A	−0.406	2.28×10^{-8}	[73]
Cr-Mn-Fe-Co-Ni	0.1 M H_2SO_4	25	−0.509	6.21×10^{-4}	[76]

6. Corrosion Resistance Enhancement Mechanism

The factors affecting metal corrosion include metal material itself and external environment. The metal material itself includes the composition of the metal and the surface finish of the metal material, while the external environment includes the temperature, humidity, and pH value of the contact environment [77–79]. When the external environment cannot be changed, the alloy can be formed by improving the metal structure and selecting elements with exceptional anti-corrosion ability. In fact, LC-HEACs have a suitable coating structure and a flexible selection of coating materials, thus significantly improving the anti-corrosion ability of the substrate surface. The following three reasons (as shown in Figure 5) are primarily responsible for the superior corrosion resistance of LC-HEACs:

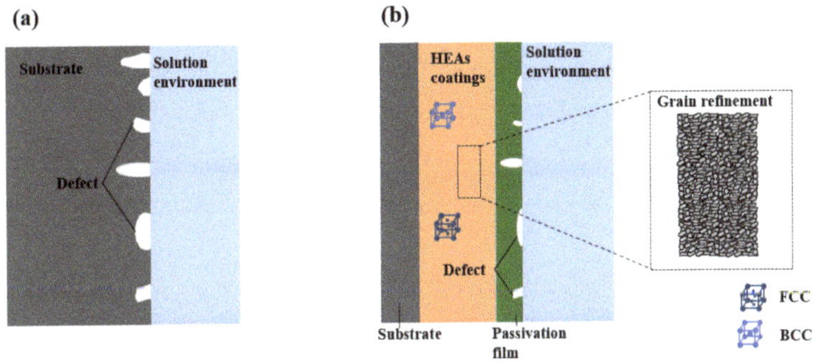

Figure 5. Schematic representation of corrosion resistance mechanism: (**a**) bare substrate; (**b**) LC-HEACs.

(1) From the perspective of LC technology: LC technology can produce coatings with exceptional microstructures. Generally, coatings with fine microstructures perform well in anti-corrosion, because this not only reduces the impurity concentration at the grain boundary, but also reduces the component segregation during rapid cooling (Figure 5b).

(2) From the perspective of HEAs: HEAs are composed of a variety of main elements, some of which can react with oxygen in the air, such as Cr, Al, and Mo, to produce a dense Cr_2O_3, Al_2O_3, and MoO_2 passivation film on the surface of the HEA coatings (Figure 5b), so that under the protection of passive film, it can effectively prevent corrosion ions from entering and corroding the coatings, which greatly inhibits the damage of the corrosion ions to the coatings and the substrate.

(3) From the perspective of LC-HEACs: LC-HEACs frequently form simple FCC or BCC phase under the circumstances of high mixing entropy impact of various primary components and quick solidification of LC [66,80], which further lessens the development of corroded galvanic cells and reduces the coating's propensity for corrosion.

7. Conclusions

In this work, the influence rule and mechanism of anti-corrosion of LC-HEACs are summarized from four aspects: laser cladding process parameters, post-processing technology, high-entropy alloy material system, and service environment. The corrosion behavior of LC-HEACs was studied in the following ways: (1) electrochemical tests, (2) characterizing and analyzing passive film, and (3) analysis of corrosion morphology and corrosion mechanism. At present, with the research of many scholars, the influence laws and mechanisms of different elements on the anti-corrosion of LC-HEACs have been gradually clarified, and HEA material systems with exceptional corrosion resistance have been discovered. At the same time, various workable approaches to enhance the anti-corrosion ability of LC-HEACs are also proposed. However, it is important to prepare LC-HEACs with greater anti-corrosion ability in order to increase the application range of LC-HEACs in the field of anti-corrosion. The following aspects may be the focus of future investigations.

(1) For different complex engineering practical application environments, design corresponding LC-HEACs as needed, and carry out the corrosion resistance test in the simulated actual service environment to study the anti-corrosion ability of the coatings in the corrosive environment.

(2) Develop a new LC-HEACs material system, e.g., a lightweight LC-HEACs material system composed of light elements such as Al, Mg, or Sc, which is more in line with the trend of lightweight design and manufacturing. To reduce the preparation cost of LC-HEACs, focus on the research of affordable elements to realize the industrialization of LC-HEACs. At the same time, when developing a corrosion-resistant LC-HEACs, it is necessary to take into account the wearability and hardness of the HEA coatings.

(3) At present, there are a few studies on the influence of LC process parameters such as laser power, scanning speed, and laser spot size on the anti-corrosion ability of

HEA coatings. Moreover, the lack of a theoretical model for the primary selection of process parameters leads to unnecessary waste of materials in the process of seeking the best process parameters. Therefore, a general theoretical model for selecting appropriate process parameters should be established in future studies.

(4) Many types of equipment are used in harsh environments. The surface of the equipment is not affected by corrosion alone, but rather the coupling effect of wear and corrosion. Therefore, the exacerbation of corrosion due to wear should be considered in future studies.

Funding: The project was supported by Project Funded by China Postdoctoral Science Foundation (2022M710633), the Liaoning Province Natural Science Foundation (2022-YGJC-79, 2022-BS-078), the Tribology Science Fund of State Key Laboratory of Tribology (SKLTKF21B03), the Science Fund of Key Laboratory for Precision and Non-Traditional Machining Technology of Ministry of Education (B202101) and the Fundamental Research Funds for the Central Universities (N2203017).

Institutional Review Board Statement: Not applicable.

Informed Consent Statement: Not applicable.

Data Availability Statement: Not applicable.

Acknowledgments: The authors thank Northeastern University Zhao Ji for his useful comments and suggestions on this manuscript.

Conflicts of Interest: The authors declare that they have no conflict of interest.

References

1. Koch, G.; Brongers, M.; Thompson, N.; Virmani, Y.; Payer, J. *Corrosion Cost and Preventive Strategies in the United States*; Federal Highway Administration: McLean, VA, USA, 2002.
2. Fu, Y.; Li, J.; Luo, H.; Du, C.; Li, X. Recent advances on environmental corrosion behavior and mechanism of high-entropy alloys. *J. Mater. Sci. Technol.* **2021**, *80*, 217–233. [CrossRef]
3. Ren, J.; Zhang, Y.; Zhao, D.; Chen, Y.; Guan, S.; Liu, Y.; Liu, L.; Peng, S.; Kong, F.; Poplawsky, J.; et al. Strong yet ductile nanolamellar high-entropy alloys by additive manufacturing. *Nature* **2022**, *608*, 62–68. [CrossRef] [PubMed]
4. Li, Y.; Wang, K.; Fu, H.; Guo, X.; Lin, J. Microstructure and wear resistance of in-situ TiC reinforced AlCoCrFeNi-based coatings by laser cladding. *Appl. Surf. Sci.* **2022**, *585*, 152703. [CrossRef]
5. Qiu, X. Microstructure and corrosion properties of $Al_2CrFeCo_xCuNiTi$ high entropy alloys prepared by additive manufacturing. *J. Alloys Compd.* **2021**, *887*, 161422. [CrossRef]
6. Li, S.; Yamaguchi, T. High-temperature oxidation behavior of laser-cladded refractory $NiSi_{0.5}CrCoMoNb_{0.75}$ high-entropy coating. *J. Mater. Res. Technol.* **2022**, *17*, 1616–1627. [CrossRef]
7. Zhang, Y.; Zuo, T.; Tang, Z.; Gao, M.; Dahmen, K.; Liaw, P.; Lu, Z. Microstructures and properties of high-entropy alloys. *Prog. Mater. Sci.* **2014**, *61*, 1–93. [CrossRef]
8. Zhu, L.; Xue, P.; Lan, Q.; Meng, G.; Ren, Y.; Yang, Z.; Xu, P.; Liu, Z. Recent research and development status of laser cladding: A review. *Opt. Laser Technol.* **2021**, *138*, 106915. [CrossRef]
9. Chen, L.; Zhao, Y.; Song, B.; Yu, T.; Liu, Z. Modeling and simulation of 3D geometry prediction and dynamic solidification behavior of Fe-based coatings by laser cladding. *Opt. Laser Technol.* **2021**, *139*, 107009. [CrossRef]
10. Zhao, Y.; Zhang, T.; Chen, L.; Yu, T.; Sun, J.; Guan, C. Microstructure and mechanical properties of Ti-C-TiN-reinforced Ni204-based laser-cladding composite coating. *Ceram. Int.* **2021**, *47*, 5918–5928. [CrossRef]
11. Chen, L.; Zhao, Y.; Chen, X.; Yu, T.; Xu, P. Repair of spline shaft by laser-cladding coarse TiC reinforced Ni-based coating: Process, microstructure and properties. *Ceram. Int.* **2021**, *47*, 30113–30128. [CrossRef]
12. Cui, Y.; Shen, J.; Hu, S.; Geng, K. Oxidation and wear mechanisms of $FeCoCrNiMnAl_x$ cladding layers at high-temperature condition. *Coatings* **2020**, *10*, 1136. [CrossRef]
13. Li, X.; Feng, Y.; Liu, B.; Yi, D.; Yang, X.; Zhang, W.; Chen, G.; Liu, Y.; Bai, P. Influence of NbC particles on microstructure and mechanical properties of AlCoCrFeNi high-entropy alloy coatings prepared by laser cladding. *J. Alloys Compd.* **2019**, *788*, 485–494. [CrossRef]
14. Chen, Z.; Lu, Y.; Luo, F.; Zhang, S.; Wei, P.; Yan, S.; Wang, Y. Effect of laser scanning speed on the microstructure and mechanical properties of laser-powder-bed-fused K418 nickel-based alloy. *Materials* **2022**, *15*, 3045. [CrossRef]
15. Yu, H.; Meng, X.; Wang, Z.; Chen, C. Influence of scanning speed on the microstructure and wear resistance of laser alloying coatings on Ti-6Al-4V substrate. *Materials* **2022**, *15*, 5819. [CrossRef]
16. Qiu, X.; Liu, C. Effect of laser processing parameters on quality of $Al_2CoCrCuFeNiTi$ high-entropy alloys coating. *Mater. Sci. Eng. Powder Metall.* **2015**, *20*, 59–64.

17. Qiu, X.; Zhang, Y.; He, L.; Liu, C. Microstructure and corrosion resistance of AlCrFeCuCo high entropy alloy. *J. Alloys Compd.* **2013**, *549*, 195–199. [CrossRef]
18. He, R.; Wu, M.; Cui, C.; Miao, X.; Gong, Y. Effects of laser energy density on microstructure and corrosion resistance of FeCrNiMnAl high entropy alloy coating. *Opt. Laser Technol.* **2022**, *152*, 108188.
19. Cui, Z.; Qin, Z.; Dong, P.; Mi, Y.; Gong, D.; Li, W. Microstructure and corrosion properties of FeCoNiCrMn high entropy alloy coatings prepared by high speed laser cladding and ultrasonic surface mechanical rolling treatment. *Mater. Lett.* **2020**, *259*, 126769. [CrossRef]
20. Zhang, S.; Han, B.; Li, M.; Hu, C.; Zhang, Q.; Liu, X.; Wang, Y. Investigation on microstructure and properties of laser cladded AlCoCrCuFeNi high entropy alloy coating by ultrasonic impact treatment. *Intermetallics* **2021**, *128*, 107017. [CrossRef]
21. Li, M.; Zhang, Q.; Han, B.; Song, L.; Li, J.; Zhang, S. Effects of ultrasonic impact treatment on structures and properties of laser cladding $Al_{0.5}CoCrFeMnNi$ high entropy alloy coatings. *Mater. Chem. Phys.* **2021**, *258*, 123850. [CrossRef]
22. Yue, T.M.; Zhang, H. Laser cladding of $FeCoNiCrAlCu_xSi_{0.5}$ high entropy alloys on AZ31 Mg alloy substrates. *Mater. Res. Innov.* **2014**, *18* (Suppl. 2), S624–S628. [CrossRef]
23. Huang, Y.; Wang, Z.; Xu, Z.; Zang, X.; Chen, X. Microstructure and properties of TiNbZrMo high entropy alloy coating. *Mater. Lett.* **2021**, *285*, 129004. [CrossRef]
24. Yue, T.M.; Huang, K.J. Laser cladding of $Cu_{0.5}NiAlCoCrFeSi$ high entropy alloy on AZ91D magnesium substrates for improving wear and corrosion resistance. *World J. Eng.* **2012**, *9*, 119–124. [CrossRef]
25. Huang, K.J.; Lin, X.; Wang, Y.Y.; Xie, C.S.; Yue, T.M. Microstructure and corrosion resistance of $Cu_{0.9}NiAlCoCrFe$ high entropy alloy coating on AZ91D magnesium alloys by laser cladding. *Mater. Res. Innov.* **2014**, *18* (Suppl. 2), S1008–S1011. [CrossRef]
26. Huang, K.; Chen, L.; Lin, X.; Huang, H.; Tang, S.; Du, F. Wear and corrosion resistance of $Al_{0.5}CoCrCuFeNi$ high-entropy alloy coating deposited on AZ91D magnesium alloy by laser cladding. *Entropy* **2018**, *20*, 915. [CrossRef]
27. Shi, Y.; Ni, C.; Liu, J.; Huang, G. Microstructure and properties of laser clad high-entropy alloy coating on aluminium. *Mater. Sci. Technol.* **2018**, *34*, 1239–1245.
28. Liu, H.; Liu, J.; Chen, P.; Yang, H.; Hao, J.; Tian, X. Microstructure and properties of AlCoCrFeNiTi high-entropy alloy coating on AISI1045 steel fabricated by laser cladding. *J. Mater. Eng. Perform.* **2019**, *28*, 1544–1552. [CrossRef]
29. Sha, M.; Jia, C.; Qiao, J.; Feng, W.; Ai, X.; Jing, Y.; Shen, M.; Li, S. Microstructure and properties of high-entropy $Al_xCoCrFe_{2.7}MoNi$ alloy coatings prepared by laser cladding. *Metals* **2019**, *9*, 1243. [CrossRef]
30. Pityana, S.; Malatji, N.; Popoola, A.; Lengopeng, T.; Pityana, S. Effect of Nb addition on the microstructural, mechanical and electrochemical characteristics of AlCrFeNiCu high-entropy alloy. *Int. J. Miner. Metall. Mater.* **2020**, *27*, 1332–1340.
31. Li, L.; Ye, H.; Liu, Y. Process optimization and corrosion resistance of laser cladding AlCoCrFeNiCu high-entropy alloy. *Surf. Technol.* **2022**, *51*, 388–396.
32. Liu, H.; Gao, W.; Liu, J.; Du, X.; Li, X.; Yang, H. Microstructure and properties of CoCrFeNiTi high-entropy alloy coating fabricated by laser cladding. *J. Mater. Eng. Perform.* **2020**, *29*, 7170–7178. [CrossRef]
33. Zhang, G.; Liu, H.; Tian, X.; Chen, P.; Yang, H.; Hao, J. Microstructure and properties of AlCoCrFeNiSi high-entropy alloy coating on AISI 304 stainless steel by laser cladding. *J. Mater. Eng. Perform.* **2020**, *29*, 278–288. [CrossRef]
34. Li, Y.; Shi, Y. Phase assemblage and properties of laser cladded $Ti_xCrFeCoNiCu$ high-entropy alloy coating on aluminum. *Mater. Res. Express* **2020**, *7*, 36519. [CrossRef]
35. Qiu, x. Structure and electrochemical properties of laser cladding $Al_2CoCrCuFeNiTi_x$ high-entropy alloy coatings. *Met. Mater. Int.* **2019**, *26*, 998–1003. [CrossRef]
36. Deng, C.; Wang, C.; Chai, L.; Wang, T.; Luo, J. Mechanical and chemical properties of $CoCrFeNiMo_{0.2}$ high entropy alloy coating fabricated on Ti6Al4V by laser cladding. *Intermetallics* **2022**, *144*, 107504. [CrossRef]
37. Dada, M.; Popoola, P.; Mathe, N.; Pityana, S.; Adeosun, S.; Aramide, O. The comparative study of the microstructural and corrosion behaviour of laser-deposited high entropy alloys. *J. Alloys Compd.* **2021**, *866*, 158777. [CrossRef]
38. Zhang, T.; Liu, H.; Hao, J.; Chen, P.; Yang, H. Evaluation of microhardness, tribological properties, and corrosion resistance of CrFeNiNbTi high-entropy alloy coating deposited by laser cladding. *J. Mater. Eng. Perform.* **2021**, *30*, 9245–9255. [CrossRef]
39. Liu, H.; Li, X.; Liu, J.; Gao, W.; Du, X.; Hao, J. Microstructural evolution and properties of dual-layer $CoCrFeMnTi_{0.2}$ high-entropy alloy coating fabricated by laser cladding. *Opt. Laser Technol.* **2021**, *134*, 106646. [CrossRef]
40. Jing, Y.; Li, J.; Juan, Y.; Lu, Z.; Jia, W. Evolution in microstructure and corrosion behavior of $AlCoCr_xFeNi$ high-entropy alloy coatings fabricated by laser cladding. *J. Alloys Compd.* **2019**, *775*, 1–14. [CrossRef]
41. Qiu, X.; Liu, C. Microstructure and properties of $Al_2CrFeCoCuTiNi_x$ high-entropy alloys prepared by laser cladding. *J. Alloys Compd.* **2013**, *553*, 216–220. [CrossRef]
42. Gu, Z.; Peng, W.; Guo, W.; Zhang, Y.; Hou, J.; He, Q.; Zhao, K.; Xi, S. Design and characterization on microstructure evolution and properties of laser-cladding $Ni_{1.5}CrFeTi_2B_{0.5}Mo_x$ high-entropy alloy coatings. *Surf. Coat. Technol.* **2021**, *408*, 126793.
43. Wu, H.; Zhang, S.; Wang, Z.; Zhan, C.; Chen, H.; Chen, J. New studies on wear and corrosion behavior of laser cladding $FeNiCoCrMo_x$ high entropy alloy coating: The role of Mo. *Int. J. Refract. Met. Hard Mater.* **2022**, *102*, 105721. [CrossRef]
44. Fu, Y.; Huang, C.; Du, C.; Li, J.; Dai, C.; Luo, H.; Liu, Z.; Li, X. Evolution in microstructure, wear, corrosion, and tribocorrosion behavior of Mo-containing high-entropy alloy coatings fabricated by laser cladding. *Corros. Sci.* **2021**, *191*, 109727. [CrossRef]
45. Xu, Y.; Li, Z.; Liu, J.; Chen, Y.; Zhang, F.; Wu, L.; Hao, J.; Liu, L. Microstructure evolution and properties of laser cladding $CoCrFeNiTiAl_x$ high-entropy alloy coatings. *Coatings* **2020**, *10*, 373. [CrossRef]

46. Li, Y.; Shi, Y. Microhardness, wear resistance, and corrosion resistance of $Al_xCrFeCoNiCu$ high-entropy alloy coatings on aluminum by laser cladding. *Opt. Laser Technol.* **2021**, *134*, 106632.
47. Sun, S.; Liu, H.; Hao, J.; Yang, H. Microstructural evolution and corrosion behavior of $CoCrFeNiAl_xMn_{(1-x)}$ dual-phase high-entropy alloy coatings prepared by laser cladding. *J. Alloys Compd.* **2021**, *886*, 161251. [CrossRef]
48. Qiu, X.; Zhang, Y.; Liu, C. Effect of Ti content on structure and properties of $Al_2CrFeNiCoCuTi_x$ high-entropy alloy coatings. *J. Alloys Compd.* **2014**, *585*, 282–286. [CrossRef]
49. Qiu, X. Microstructure, hardness and corrosion resistance of $Al_2CoCrCuFeNiTi_x$ high-entropy alloy coatings prepared by rapid solidification. *J. Alloys Compd.* **2018**, *735*, 359–364. [CrossRef]
50. Zuo, R.; Sun, R.; Niu, W. Microstructure and properties of $CoCrFeNiTi_x$ high entropy alloy coated by laser cladding. *Surf. Technol.* **2022**, *51*, 363–370.
51. Wang, X.; Liu, Q.; Huang, Y.; Xie, L.; Xu, Q.; Zhao, T. Effect of Ti content on the microstructure and corrosion resistance of $CoCrFeNiTi_x$ high entropy alloys prepared by laser cladding. *Materials* **2020**, *13*, 2209. [CrossRef]
52. Gu, Z.; Xi, S.; Sun, C. Microstructure and properties of laser cladding and $CoCr_{2.5}FeNi_2Ti_x$ high-entropy alloy composite coatings. *J. Alloys Compd.* **2020**, *819*, 152986. [CrossRef]
53. Liu, J.; Liu, H.; Chen, P.; Hao, J. Microstructural characterization and corrosion behaviour of $AlCoCrFeNiTi_x$ high-entropy alloy coatings fabricated by laser cladding. *Surf. Coat. Technol.* **2019**, *361*, 63–74. [CrossRef]
54. Li, Y.; Shi, Y.; Olugbade, E. Microstructure, mechanical, and corrosion resistance properties of $Al_{0.8}CrFeCoNiCu_x$ high-entropy alloy coatings on aluminum by laser cladding. *Mater. Res. Express* **2020**, *7*, 26504. [CrossRef]
55. Zhang, X.; Cui, X.; Jin, G.; Ding, Q.; Zhang, D.; Wen, X.; Jiang, L.; Wan, S.; Tian, H. Microstructure evolution and properties of $NiTiCrNbTa_x$ refractory high-entropy alloy coatings with variable Ta content. *J. Alloys Compd.* **2022**, *891*, 161756. [CrossRef]
56. Shang, X.; Bo, S.; Guo, Y.; Liu, Q. ZrC reinforced refractory-high-entropy-alloy coatings: Compositional design, synthesis, interstitials, and microstructure evolution effects on wear, corrosion and oxidation behaviors. *Appl. Surf. Sci.* **2021**, *564*, 150466. [CrossRef]
57. Guo, Y.; Shang, X.; Liu, Q. Microstructure and properties of in-situ TiN reinforced laser cladding $CoCr_2FeNiTi_x$ high-entropy alloy composite coatings. *Surf. Coat. Technol.* **2018**, *344*, 353–358. [CrossRef]
58. Liu, S.; Zhang, M.; Zhao, G.; Wang, X.; Wang, J. Microstructure and properties of ceramic particle reinforced FeCoNiCrMnTi high entropy alloy laser cladding coating. *Intermetallics* **2022**, *140*, 107402. [CrossRef]
59. Zhao, M.; Wu, T.; Liu, D.; Huang, Y.; Zhao, L.; Tang, Y.; Shen, M.; Hu, Y.; Zhang, J.; Li, J.; et al. Effect of carbon fiber on microstructure evolution and surface properties of FeCoCrNiCu high-entropy alloy coatings. *Mater. Corros.* **2020**, *71*, 430–439. [CrossRef]
60. Jiang, D.; Cui, H.; Chen, H.; Zhao, X.; Ma, G.; Song, X. Wear and corrosion properties of B_4C-added CoCrNiMo high-entropy alloy coatings with in-situ coherent ceramic. *Mater. Des.* **2021**, *210*, 110068. [CrossRef]
61. Jiang, L.; Cui, X.; Jin, G.; Tian, Z.; Wen, X.; Tian, H.; Wang, S. Design and characterization of in-situ TiB reinforced $TiB/Ti_{50}Zr_{25}Al_{15}Cu_{10}$ non-equiatomic medium-entropy alloy composite coating on magnesium alloy by laser cladding. *Opt. Laser Technol.* **2022**, *156*, 108494. [CrossRef]
62. Xu, Y.; Wang, G.; Song, Q.; Lu, X.; Li, Z.; Zhao, Q.; Chen, Y. Microstructure, mechanical properties, and corrosion resistance of SiC reinforced $Al_xCoCrFeNiTi_{1-x}$ high-entropy alloy coatings prepared by laser cladding. *Surf. Coat. Technol.* **2022**, *437*, 128349. [CrossRef]
63. Cui, C.; Wu, M.; Miao, X.; Zhao, Z.; Gong, Y. Microstructure and corrosion behavior of CeO_2/FeCoNiCrMo high-entropy alloy coating prepared by laser cladding. *J. Alloys Compd.* **2022**, *890*, 161826. [CrossRef]
64. Gu, Z.; Mao, P.; Gou, Y.; Chao, Y.; Xi, S. Microstructure and properties of $MgMoNbFeTi_2Y_x$ high entropy alloy coatings by laser cladding. *Surf. Coat. Technol.* **2020**, *402*, 126303. [CrossRef]
65. Qiu, X.; Wu, M.; Liu, C.; Zhang, Y.; Huang, C. Corrosion performance of $Al_2CrFeCo_xCuNiTi$ high-entropy alloy coatings in acid liquids. *J. Alloys Compd.* **2017**, *708*, 353–357. [CrossRef]
66. Qiu, X. Corrosion behavior of $Al_2CrFeCo_xCuNiTi$ high-entropy alloy coating in alkaline solution and salt solution. *Results Phys.* **2019**, *12*, 1737–1741. [CrossRef]
67. Feng, H.; Li, H.; Dai, J.; Han, Y.; Qu, J.; Jian, Z.; Zhan, Y.; Zhan, T. Why CoCrFeMnNi HEA could not passivate in chloride solution?—A novel strategy to significantly improve corrosion resistance of CoCrFeMnNi HEA by N-alloying. *Corros. Sci.* **2022**, *204*, 110396. [CrossRef]
68. Wang, C.; Yu, Y.; Shao, M.; Zhang, H. Effect of temperature on corrosion behavior of laser-remelted CrFeCoNi coating. *Metals* **2022**, *12*, 970. [CrossRef]
69. Xue, L.; Ding, Y.; Pradeep, K.; Case, R.; Castaneda, H.; Paredes, M. Development of a non-equimolar AlCrCuFeNi high-entropy alloy and its corrosive response to marine environment under different temperatures and chloride concentrations. *J. Alloys Compd.* **2022**, *928*, 167112. [CrossRef]
70. Liu, C.; Liu, Z.; Gao, Y.; Zheng, C.; Wang, X. Investigation on the corrosion behavior of Ni-Cr-Mo-W-xSi laser cladding coating in H_2S corrosion environment. *Appl. Surf. Sci.* **2022**, *578*, 152061. [CrossRef]
71. Thorhallsson, A.; Fanicchia, F.; Davison, E.; Paul, S.; Davidsdottir, S.; Olafsson, D. Erosion and corrosion resistance performance of laser metal deposited high-entropy alloy coatings at hellisheidi geothermal site. *Materials* **2021**, *14*, 3071. [CrossRef]

72. Wan, H.; Song, D.; Shi, X.; Cai, Y.; Li, T.; Chen, C. Corrosion behavior of $Al_{0.4}CoCu_{0.6}NiSi_{0.2}Ti_{0.25}$ high-entropy alloy coating via 3D printing laser cladding in a sulphur environment. *J. Mater. Sci. Technol.* **2021**, *60*, 197–205. [CrossRef]
73. Wang, Q.; Amar, A.; Jiang, C.; Luan, H.; Zhao, S.; Zhang, H.; Le, G.; Liu, X.; Wang, X.; Yang, X.; et al. $CoCrFeNiMo_{0.2}$ high entropy alloy by laser melting deposition: Prospective material for low temperature and corrosion resistant applications. *Intermetallics* **2020**, *119*, 106727. [CrossRef]
74. Ben, Q.; Zhang, Y.; Sun, L.; Wang, L.; Wang, Y.; Zhan, X. Wear and corrosion resistance of $FeCoCr_xNiAl$ high-entropy alloy coatings fabricated by laser cladding on Q345 welded joint. *Metals* **2022**, *12*, 1428. [CrossRef]
75. Chen, X.; Qian, H.; Lou, Y.; Yang, B.; Cui, T.; Zhang, D. Effects of Cu-content and passivation treatment on the corrosion resistance of $Al_{0.3}Cu_xCoCrFeNi$ high-entropy alloys. *J. Alloys Compd.* **2022**, *920*, 165956. [CrossRef]
76. Zhang, Z.; Yuan, T.; Li, R. Striped non-uniform corrosion behavior of non-equiatomic FeMnCoCr high-entropy alloy prepared by laser melting deposition in 0.1 M H_2SO_4 solution. *Materials* **2020**, *13*, 5554. [CrossRef]
77. Croll, S. Surface roughness profile and its effect on coating adhesion and corrosion protection: A review. *Prog. Org. Coat.* **2020**, *148*, 105847. [CrossRef]
78. Umoren, S.; Solomon, M. Protective polymeric films for industrial substrates: A critical review on past and recent applications with conducting polymers and polymer composites/nanocomposites. *Prog. Mater. Sci.* **2019**, *104*, 380–450. [CrossRef]
79. Aslam, R.; Mobin, M.; Zehra, S.; Aslam, J. A comprehensive review of corrosion inhibitors employed to mitigate stainless steel corrosion in different environments. *J. Mol. Liq.* **2022**, *364*, 119992. [CrossRef]
80. Zhang, P.; Xu, Z.; Yao, Z.; Liu, Y.; Lin, S.; He, M.; Lu, S.; Wu, X. A high-corrosion-resistant high-entropy alloys (HEAs) coatings with single BCC solid solution structure by laser remelting. *Mater. Lett.* **2022**, *324*, 132728. [CrossRef]

www.ingramcontent.com/pod-product-compliance
Lightning Source LLC
Chambersburg PA
CBHW061848210926
51850CB00006B/1004
9783725862122

MDPI AG
Grosspeteranlage 5
4052 Basel
Switzerland
Tel.: +41 61 683 77 34

Coatings Editorial Office
E-mail: coatings@mdpi.com
www.mdpi.com/journal/coatings